Mechanical Properties and Performance of Engineering Ceramics and Composites VII

Mechanical Properties and Performance of Engineering Ceramics and Composites VII

A Collection of Papers Presented at the 36th International Conference on Advanced Ceramics and Composites
January 22-27, 2012
Daytona Beach, Florida

Edited by
Dileep Singh
Jonathan Salem

Volume Editors
Michael Halbig
Sanjay Mathur

A John Wiley & Sons, Inc., Publication

Published by John Wiley & Sons, Inc., Hoboken, New Jersey.
Published simultaneously in Canada.

For general information on our other products and services or for technical support, please contact our
Customer Care Department within the United States at (800) 762-2974, outside the United States at
(317) 572-3993 or fax (317) 572-4002.

Wiley also publishes its books in a variety of electronic formats. Some content that appears in print may
not be available in electronic formats. For more information about Wiley products, visit our web site at
www.wiley.com.

Library of Congress Cataloging-in-Publication Data is available.

ISBN: 978-1-118-20588-4
ISSN: 0196-6219

10 9 8 7 6 5 4 3 2 1

Contents

WEAR, CHIPPING, AND FATIGUE OF CERAMICS AND COMPOSITES

MICROSTRUCTURE AND MECHANICAL PROPERTIES OF
MONOLITHIC AND COMPOSITE SYSTEMS

Preface

This volume is a compilation of papers presented in the Mechanical Behavior and Performance of Ceramics & Composites symposium during the 36th International Conference & Exposition on Advanced Ceramics and Composites (ICACC) held January 22–27, 2012, in Daytona Beach, Florida.

This long-standing symposium received presentations on a wide variety of topics providing the opportunity for researchers in different areas of related fields to interact. This volume emphasizes some practical aspects of real-world engineering applications of ceramics such as oxidation, fatigue, wear, nondestructive evaluation, and mechanical behavior as associated with systems ranging from diamond reinforced silicon carbide to rare earth pyrosilicates. Symposium topics included:

- Composites: Fibers, Matrices, Interfaces and Applications
- Fracture Mechanics, Modeling, and Mechanical Testing
- Nondestructive Evaluation
- Processing-Microstructure-Properties Correlations
- Tribological Properties of Ceramics and Composites

Significant time and effort is required to organize a symposium and publish a proceeding volume. We would like to extend our sincere thanks and appreciation to the symposium organizers, invited speakers, session chairs, presenters, manuscript reviewers, and conference attendees for their enthusiastic participation and contributions. Finally, credit also goes to the dedicated, tireless and courteous staff at The American Ceramic Society for making this symposium a huge success.

DILEEP SINGH
Argonne National Laboratory

JONATHAN SALEM
NASA Glenn Research Center

Introduction

This issue of the Ceramic Engineering and Science Proceedings (CESP) is one of nine issues that has been published based on content presented during the 36th International Conference on Advanced Ceramics and Composites (ICACC), held January 22–27, 2012 in Daytona Beach, Florida. ICACC is the most prominent international meeting in the area of advanced structural, functional, and nanoscopic ceramics, composites, and other emerging ceramic materials and technologies. This prestigious conference has been organized by The American Ceramic Society's (ACerS) Engineering Ceramics Division (ECD) since 1977.

The 36th ICACC hosted more than 1,000 attendees from 38 countries and had over 780 presentations. The topics ranged from ceramic nanomaterials to structural reliability of ceramic components which demonstrated the linkage between materials science developments at the atomic level and macro level structural applications. Papers addressed material, model, and component development and investigated the interrelations between the processing, properties, and microstructure of ceramic materials.

The conference was organized into the following symposia and focused sessions:

Symposium 1	Mechanical Behavior and Performance of Ceramics and Composites
Symposium 2	Advanced Ceramic Coatings for Structural, Environmental, and Functional Applications
Symposium 3	9th International Symposium on Solid Oxide Fuel Cells (SOFC): Materials, Science, and Technology
Symposium 4	Armor Ceramics
Symposium 5	Next Generation Bioceramics

Symposium 6	International Symposium on Ceramics for Electric Energy Generation, Storage, and Distribution
Symposium 7	6th International Symposium on Nanostructured Materials and Nanocomposites: Development and Applications
Symposium 8	6th International Symposium on Advanced Processing & Manufacturing Technologies (APMT) for Structural & Multifunctional Materials and Systems
Symposium 9	Porous Ceramics: Novel Developments and Applications
Symposium 10	Thermal Management Materials and Technologies
Symposium 11	Nanomaterials for Sensing Applications: From Fundamentals to Device Integration
Symposium 12	Materials for Extreme Environments: Ultrahigh Temperature Ceramics (UHTCs) and Nanolaminated Ternary Carbides and Nitrides (MAX Phases)
Symposium 13	Advanced Ceramics and Composites for Nuclear Applications
Symposium 14	Advanced Materials and Technologies for Rechargeable Batteries
Focused Session 1	Geopolymers, Inorganic Polymers, Hybrid Organic-Inorganic Polymer Materials
Focused Session 2	Computational Design, Modeling, Simulation and Characterization of Ceramics and Composites
Focused Session 3	Next Generation Technologies for Innovative Surface Coatings
Focused Session 4	Advanced (Ceramic) Materials and Processing for Photonics and Energy
Special Session	European Union – USA Engineering Ceramics Summit
Special Session	Global Young Investigators Forum

The proceedings papers from this conference will appear in nine issues of the 2012 Ceramic Engineering & Science Proceedings (CESP); Volume 33, Issues 2-10, 2012 as listed below.

- Mechanical Properties and Performance of Engineering Ceramics and Composites VII, CESP Volume 33, Issue 2 (includes papers from Symposium 1)
- Advanced Ceramic Coatings and Materials for Extreme Environments II, CESP Volume 33, Issue 3 (includes papers from Symposia 2 and 12 and Focused Session 3)
- Advances in Solid Oxide Fuel Cells VIII, CESP Volume 33, Issue 4 (includes papers from Symposium 3)
- Advances in Ceramic Armor VIII, CESP Volume 33, Issue 5 (includes papers from Symposium 4)

- Advances in Bioceramics and Porous Ceramics V, CESP Volume 33, Issue 6 (includes papers from Symposia 5 and 9)
- Nanostructured Materials and Nanotechnology VI, CESP Volume 33, Issue 7 (includes papers from Symposium 7)
- Advanced Processing and Manufacturing Technologies for Structural and Multifunctional Materials VI, CESP Volume 33, Issue 8 (includes papers from Symposium 8)
- Ceramic Materials for Energy Applications II, CESP Volume 33, Issue 9 (includes papers from Symposia 6, 13, and 14)
- Developments in Strategic Materials and Computational Design III, CESP Volume 33, Issue 10 (includes papers from Symposium 10 and from Focused Sessions 1, 2, and 4)

The organization of the Daytona Beach meeting and the publication of these proceedings were possible thanks to the professional staff of ACerS and the tireless dedication of many ECD members. We would especially like to express our sincere thanks to the symposia organizers, session chairs, presenters and conference attendees, for their efforts and enthusiastic participation in the vibrant and cutting-edge conference.

ACerS and the ECD invite you to attend the 37th International Conference on Advanced Ceramics and Composites (http://www.ceramics.org/daytona2013) January 27 to February 1, 2013 in Daytona Beach, Florida.

MICHAEL HALBIG AND SANJAY MATHUR
Volume Editors
July 2012

Nondestructive Evaluation of Ceramics Systems

DAMAGE SENSITIVITY AND ACOUSTIC EMISSION OF SIC/SIC COMPOSITE DURING TENSILE TEST AND STATIC FATIGUE AT INTERMEDIATE TEMPERATURE AFTER IMPACT DAMAGE

Matthieu Picard, Emmanuel Maillet, Pascal Reynaud, Nathalie Godin, Mohamed R'Mili, Gilbert Fantozzi, Jacques Lamon
INSA-Lyon, MATEIS CNRS UMR 5510, 7 Avenue Jean Capelle, F-69 621 Villeurbanne, France.

ABSTRACT

This paper discusses the tensile resistance of an impact-damaged SiC/SiC based ceramic composite. As-received and impact-damaged specimens were subjected to static fatigue tests at 650°C and 450°C and to monotonous tensile tests at room temperature. Damage induced during the tensile and fatigue tests was characterized by the evolution of linear density of acoustic emission events (AE) during unloading-reloading cycles. Results of tensile tests at room temperature show the material impact insensitivity and the analysis of AE signals has demonstrated efficiency for proper investigation of damage evolution in the impacted specimens. During fatigue at 450°C, 650°C and under 80 MPa, both impact-damaged and as-fabricated samples survived 1000 hours. Residual strengths indicated insensitivity to impact damage. At 650°C, damage evolution seemed to slow down from matrix self-healing.

INTRODUCTION

Due to low weight/mechanical properties ratios and high temperature strength, ceramic matrix composites (CMCs) are very attractive candidates for civil aircrafts applications. For these applications, resistance to foreign object damage (FOD) is a key issue to insure structural reliability in service. In the literature, a few authors have investigated the FOD response of 2D woven CMC[1-5]. They have shown that low energy impact was equivalent to quasi-static indentation and that a conical damage zone was created.

Meanwhile, investigations by Ogi et al.[6] and Herb et al.[7] on 3D woven CMC have shown that tri-dimensional fibre architectures prevent the material from delamination so that the damaged cone remains limited and well delineated. After indentation (i.e. ballistic[1-6] or quasi-static impact tests[7]), residual strengths were measured using tensile or flexural tests at room temperature[1-4,6,7].

Few works on the effect of impact damage on composite lifetime under fatigue at elevated temperature have been reported. Recently, Verrilli et al.[5] have performed cyclic fatigue tests on 2D cross ply SiC/SiC composite at 1316°C after impact tests at 1200°C. They have observed that lifetime decreases tremendously with increasing impact energy. After high energy impact damage the average lifetime was 40 times smaller than that obtained after low energy impact tests. However, impact damage evolution during fatigue has not been studied in real time.

This paper investigates the evolution of impact damage during fatigue at high temperature on a 3D SiC/SiC composite using acoustic emission signals and the sensitivity to impact damage. Acoustic emission data were analyzed using homemade software that determines the spatial distribution of AE events during the tests. As-fabricated and impact-damaged specimens were tested at room and at high temperatures (i.e. 450°C and 650°C).

3

EXPERIMENTAL PROCEDURE

Material and specimen preparation
    The SiC/SiC composite investigated (manufactured by Snecma Propulsion Solide - SAFRAN Group (Bordeaux, France)) was made of an interlock preform of plies of 0/90 yarns woven in a 8 HSW pattern[8] with a self-healing [Si-B-C] matrix. The fibres were coated by a PyC layer deposited via chemical vapour infiltration[8]. This 3D fibrous preform (Guipex® preform) improves the through thickness properties[9]. Yarns contain 500 SiC Nicalon fibres. The fibre volume fraction was between 35% and 40% and the porosity volume fraction was around 12%. A barrier coating protected sample surface against oxidation.
    Rectangular test specimens were machined out of 1.8 mm thick panels: specimens' dimensions were 24 mm in width and 200 mm in length. Impact damage was generated by quasi-static indentation by Herb et al.[7]. A hemispherical steel punch with a 9 mm diameter was used. The specimens were clamped as shown on Figure 1.

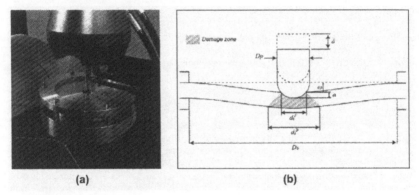

(a)   (b)

Figure 1. (a) Quasi-static indentation indentation with D_s=18 and D_p= 9 mm. (b) Schematic view of specimen loading (from Herb et al.[7]).

    Figure 2 shows the cone crater that had been created. Breakage of fibre bundles was observed on back side (Figure 2a), and the impact side (Figure 2b). On the impact side, the sample displayed a neat circular mark. Images of the impact and back sides were post-treated using the Image J software, in order to measure cone area on each side of samples. It was found to be about 33 mm^2 on the front side which is equivalent to a hole with a diameter of 6.5 mm. On the back side, the damaged area was about 85 mm^2 which is equivalent to a hole with a 10.5 mm diameter. The cone crater was sharply delineated, which can be attributed to the 3D fibrous architecture of material and the resulting absence of delamination.

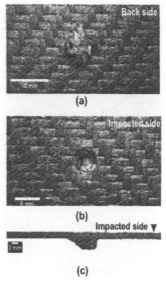

(a)

(b)

Impacted side ▼

(c)

Figure 2. Optical photographs of the SiC/SiC specimens after quasi static indentation: (a) back side (d_i^B = 10.5 mm) (b) impacted side (d_i^F = 6.5 mm) (c) thickness view.

Post-indentation mechanical testing

Static fatigue and tensile experiments were performed using a 25kN uniaxial pneumatic tensile loading machine with one direction of fibres parallel to the loading direction. Specimen elongation was measured using an extensometer (gauge length = 32 mm).

Static fatigue tests were carried out at 450°C and 650°C under air, with uniform temperature in the gauge length (cold grips). Specimens were heated up to the test temperature at a rate of 20°C/min. The load was applied after 2 hours, when the gauge length was expected to be at the test temperature. During the static fatigue tests, specimens were first loaded at a constant rate of 1200 N/min up to the test load corresponding to 80 MPa, close to the elastic limit of composite.

Tensile tests under monotonous loading were performed at room temperature, on impacted specimens (post-impact strength) and on specimens after 1000 hours of static fatigue (residual strength). For all the tests, damage was characterized using periodical unloading-loading cycles (every 12 hours during the static fatigue tests) and also acoustic emission signals. For comparison purposes, a few tests were performed on as-fabricated specimens.

Acoustic emission monitoring

AE was monitored using a MISTRAS 2001 data acquisition system (Euro Physical Acoustics). Two MICRO-80 sensors were positioned directly on the specimen, inside the grips, using vacuum grease with a medium viscosity as a coupling agent. Acquisition parameters were set as follows: pre-amplification 40 dB, threshold 36 dB, peak definition time 50 µs, hit definition time 100 µs, hit lockout time 1000 µs[10,11]. AE signal parameters (amplitude, energy, duration, counts, location...) as well as time, load and strain were measured in real time by the data acquisition system.

As depicted by Figure 3, the zone of interest for the AE events was 80 mm long and located on each side of the mid-plane. Since two sensors only were used, the longitudinal positions of AE origins were determined (see Figure 3a).

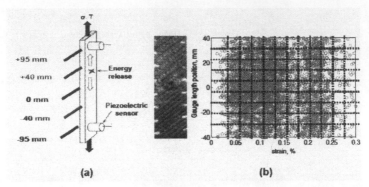

(a)   (b)

Figure 3. (a) Specimen and Acoustic Emission setting (b) Locations of AE sources in a tensile specimen versus strain. Also shown is the grid defined in order to determine the linear density of counts of acoustic events along specimen axis during tensile tests or static fatigue tests.

AE sources locations were derived using AE wave velocity, that was measured using a pencil lead break procedure: 9100 m/s for both the as-fabricated and the impact-damaged specimens. Morscher et al.[12] showed that in ceramic matrix composites wave velocity decreases with increasing stress-induced damage. Wave velocity value was corrected using the attenuation coefficient

$$\gamma = (E/E_0)^{0.5} \qquad\qquad (1)$$

E is the secant elastic modulus determined from unloading-reloading hysteresis loops and E_0 is the initial elastic modulus.
AE data were post-treated using dedicated software, which provides linear density along specimen axis of acoustic events during the tests[13]. This analysis of acoustic emission is non-trivial. It is useful to identify the most active zones during the tests.

RESULTS AND DISCUSSION

Effect of impact-damage on tensile behaviour
During the tensile tests, the impacted specimens failed from the mid-plane whereas the as-fabricated samples failed from the upper or lower parts of the gauge length. Figure 4a shows typical tensile curves. For both the as-fabricated and the impact-damaged specimens, the stresses were determined from the net-section of samples:

$$\sigma = F/\, b\, (w\text{-}a) \qquad\qquad (2)$$

where F is the force on specimen, w is specimen width, a is the average diameter of crater, $b = 1.1$ mm is specimen effective thickness (sealcoat was not taken into account since it did not contribute to the mechanical resistance). As shown by Figure 4a, for all the samples, the tensile response was non-linear, beyond the proportional limit, as a result of stress-induced matrix cracking. The two impacted specimens exhibited higher stresses (or smaller deformations) and a slightly smaller ultimate strength compared to as-fabricated specimens (Figure 4a): ultimate strength was 80% of the reference, whereas strain-to-failure was reduced by 50% (Table 1).

Table 1. Failure stresses and strains measured on the impact-damaged and as-fabricated specimens at room temperature.

Sample	Failure strain (%)	Failure stress * (MPa)
Not-impacted	0.64	275
Impacted-1	0.30	245
Impacted-2	0.28	231

To evaluate the impact damage sensitivity strength data were plotted with respect to average cone diameter size according the classical equation for notch sensitivity, which indicates that the stress at failure at hole tip is equal to the strength of specimen without hole:

$$\sigma_R(a) = \sigma_R(a = 0)\, (1\text{-}a/w) \qquad\qquad (3)$$

where a is average diameter of cone (figure 1b), σ_R is the failure stress given by F/wb.

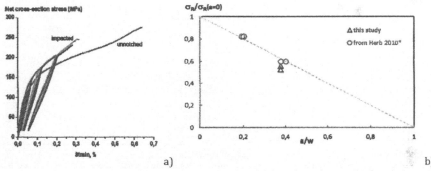

a) b)

Figure 4. (a) Tensile behaviour of impact-damaged and as-fabricated specimens. (b) Post-indentation relative strength of the specimens versus the relative diameter of damage cone (comparison with results from Herb et al.[7]).

Figure 4a shows that equation (3) fits the strength data which indicates that the material is insensitive to impact damage. It means that there is no stress concentration induced by the presence of the impact cone. The σ_R strength dependence on a results instead from reduction in specimen section.

Figures 5a and 5b show the evolution of the linear density of acoustic events during monotonous loading. It can be noticed that acoustic emission was very significant in the zone of impact (*i.e.* between the dotted lines) for both samples at the beginning of loading (*i.e.* for strain level values between 0 and 0.10%). Failure did not occur under these strains but much later under larger deformation. Instead, under these larger strains, the AE activity in the impact area slowed down whereas it increased progressively in other parts of specimen, but looked quite homogeneous as load increased, indicating diffuse stress-induced damage. These results are consistent with impact damage insensitivity previously indicated. They indicate that the impacted specimens experience damage exactly like the as-received specimens under tensile load: stress driven diffuse matrix cracking and, in a second step, fibre failures.

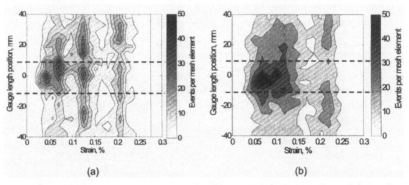

(a)   (b)

Figure 5. Evolution of linear densities of acoustic events along specimen axis during tensile tests at room temperature for the two impacted specimens (a) Impacted sample1. (b) Impacted sample2. The dotted lines bound the damage cone area.

Effect of impact-damage on static-fatigue

Very little deformation and damage were observed during fatigue on the as fabricated specimens. Besides, these specimens did not fail even after 1000 hours. Thus, the fatigue tests were interrupted and the tensile residual strength was measured at room temperature. The results are discussed below. As the static fatigue behaviour of SiC/SiC composites has been well documented in the literature[10,11] this part of the paper will focus on the impact-damaged specimens. Figures 6a and 6b compare the evolution of damage parameter ($D = 1 - E/E_0$) deduced from the strain measurements and the cumulative count of acoustic events. They highlight two interesting features.

First, the cumulative count of AE events mimics the evolution of damage under constant load. Second, damage and acoustic emission were less intense and quite constant at 650°C. This result demonstrates the effect of the self-healing matrix which protects the composite against oxidation at 650°C. Figure 7 shows also the location of AE events reflecting the occurrence of diffuse damage at 450°C as well as at 650°C. It is obvious on Figure 7 that acoustic emission was less intense at 650°C. The location of events is homogeneous all along the test, which agrees with the above-mentioned feature of stress-induced damage which is essentially diffuse, and also with the feature of notch insensitivity, since it does not appear a concentration of events in the vicinity of impact cone. Fig. 7 also shows significant slowing down of acoustic emission after 400 hours of static fatigue. This AE behaviour was observed at 450°C and 650°C. On Figures 7a and 7b, concentration of acoustic sources was not detected during the first 400 hours. To check this result, the linear density of AE events was plotted on Figures 8 and 9.

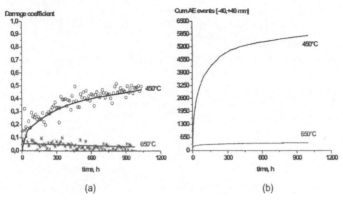

(a)   (b)

Figure 6. Plots of (a) evolution of damage parameter D = 1 − E/E$_0$ (b) cumulative count of acoustic events during static fatigue tests at 450°C and 650°C and under a constant stress (80 MPa) on impact-damaged specimens.

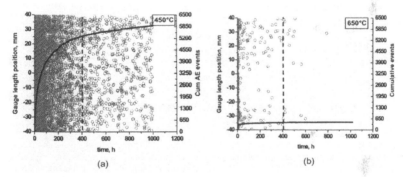

(a)   (b)

Figure 7. Location of the AE signals and the cumulative count of AE events in impact-damaged specimens during static fatigue tests under 80MPa, and at 450°C (a) and 650°C (b).

Figures 8 and 9 confirm that there was no acoustic activity concentration and that AE activity was less intense at 650°C. Nevertheless, in both cases, the impacted zone was more active at the beginning of tests, and then it was less active than the other parts of specimens after 100h. These results suggest that fatigue behaviour was not influenced by impact damage.

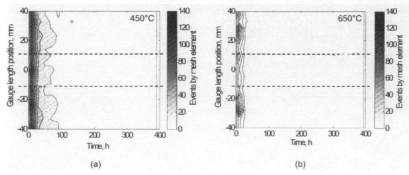

(a)　　　　　　　　　　　　(b)

Figure 8. Linear density along specimen axis of AE events between 0 and 400 hours during static fatigue (a) at 450°C and (b) at 650°C. The dotted lines delineate the impact-damaged zone.

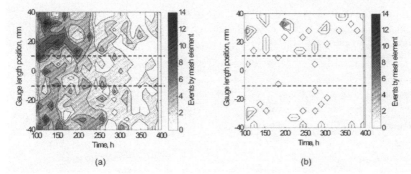

(a)　　　　　　　　　　　　(b)

Figure 9. Linear density along specimen axis of AE events between 100 and 400 hours of static fatigue (a) at 450°C and (b) at 650°C. The dotted lines delineate the impact-damaged zone.

The impacted specimens did not break during the 1000 hours of static-fatigue whatever the test temperature. Residual strength was also measured using tensile tests at room temperature. Figure 10 compares initial and residual strengths for both as-fabricated specimens and impact-damaged specimens.

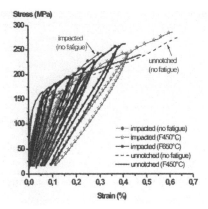

Figure 10. Comparison of the initial and residual tensile behaviour at room temperature of as-fabricated and impact-damaged specimens.

Though the tensile behaviour exhibits some variation (probably a material feature), the stress-strain curves highlight three important features. First, for both the as-fabricated and the impact-damaged specimens, the residual strengths were not significantly affected by fatigue tests at high temperature. Second, though material protection by matrix healing was not effective at 450°C, residual strength of the impacted specimens after 1000 hours of static fatigue was unchanged whatever the test temperature. Third, the residual failure stresses and strains for the impact-damaged specimens were close to those obtained on the as-fabricated samples after fatigue (Table 2). These features again indicate that under current experimental conditions the SiC/SiC composite fatigue behaviour was impact damage insensitive.

Table 2. Residual strengths and strains measured on specimens with or without impact damage after 1000 hours at 650°C and 450°C, under 80 MPa.

Sample	Failure strain (%)	Failure stress * (MPa)
Not-impacted (F450°C)	0.48	240
Impacted (F450°C)	0.61	288
Impacted (F650°C)	0.40	263

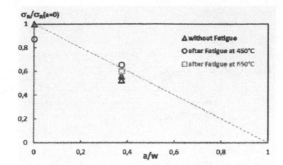

Figure 11. Relative residual strength of specimens versus the relative diameter of damage cone. Also given are the reference data of figure 4(b)

In order to assess the damage insensitivity of residual strengths, the data from tensile tests on specimens after fatigue were plotted on damage sensitivity diagram (figure 11). Figure 11 shows that the data pertain to the damage insensitivity line corresponding to equation (3) which agrees perfectly with the above results.

CONCLUSION
This paper investigated the effect of impact damage on SiC/SiC composite tensile behaviour at room temperature and static fatigue resistance at two intermediate temperatures (450°C and 650°C). Homemade software for post-treatment of AE signals from specimens gauge length during tests provided maps of linear density of acoustic events. These maps imaged damage during the tests. They showed that emission was intense near the impact cone at beginning of tests only. Specimens failed under larger strains. Their subsequent response was comparable to the non-linear mechanical behaviour of as-fabricated specimens. This observation illustrates the insensitivity to damage of this material, which is consistent with the notch insensitivity of long fibers reinforced composites[14].
Whatever the test temperature, no acoustic emission concentration was detected in the vicinity of impact cones during static fatigue at 80MPa. Under this stress the impacted specimens did not fail after 1000 hours. The ultimate strength of composite was not affected significantly by the fatigue tests whatever the sample was impact-damaged or not.
This work also confirms the effectiveness of the matrix self-healing process to protect the material against oxidation at 650°C. Investigation of lifetime of impact-damaged specimens at this temperature is in progress.

ACKNOWLEDGMENTS
The authors gratefully acknowledge DGCIS, Conseil Régional d'Aquitaine, DGAC, DGA, CNRS and Aerospace Valley for supporting this work in the frame of the program ARCOCE: "Arrière-Corps Composite Ceramique, Ceramic Matrix Composites for Exhausts", and particularly P. Diss (SAFRAN, SNECMA Propulsion Solide) for fruitful discussions.

REFERENCES

[1] R. T. Bhatt, S. R. Choi, L.M. Cosgriff, D. S. Fox, K. N. Lee, Impact resistance of environmental barrier coated SiC/SiC composites, *Mater Sci Eng A*, **476**, 8–19 (2008).

[2] R. T. Bhatt, S. R. Choi, L.M. Cosgriff, D. S. Fox, K. N. Lee, Impact resistance of uncoated SiC/SiC composites. *Mater Sci Eng A*, **476**, 20-28 (2008).

[3] S. R. Choi, Foreign object damage phenomenon by steel ball projectiles in a SiC/SiC ceramic matrix composites at ambient and elevated temperatures. *J. Am. Ceram. Soc.*, **91**, 2963–8 (2008).

[4] S. R. Choi, D. J. Alexander R. W. Kowalik, Foreign object damage in an oxide/oxide composite at ambient temperature. *J. Eng. Gas. Turb. Power*, 131, 021301-1–0212301-6 (2009).

[5] M. Virelli, Impact Behavior of a SiC/SiC Composite at an Elevated Temperature, High Temperature Ceramic Materials and Composites (HTCMC-7), editors W. Krenkel and J. Lamon, AVISO Verlagsgesellschaft mbHD-10117 Berlin, Germany, 519–530 (2010).

[6] K. Ogi, T. Okabe, M. Takahashi, S. Yashiro, A. Yoshimura, T. Ogasawara, Experimental characterization of high-speed impact damage behaviour in a three-dimensionally woven SiC/SiC composite, *Composites: Part A*, **41**, 489–498 (2010).

[7] V. Herb, Damage assessment of thin SiC/SiC composite plates subjected to quasi-static indentation loading, *Composites: Part A* (2010) doi:10.1016/j.compositesa. 2010.08.004

[8] I. Berdoyes, J. Thebault, E. Bouillon, Improved SiC/SiC and C/C materials applications parts. In: Proceedings of the European congress on advanced materials and processes, EUROMAT 2005, Prague, Czech Republic, 5–8 September 2005.

[9] P. Tan, L. Tong, G. P. Steven, Modelling for predicting the mechanical properties of textile composites—a review. *Composite Part A*, **28A**, 903–22 (1997).

[10] S. Momon, M. Moevus, N. Godin, M. R'Mili, P. Reynaud, G. Fantozzi, G. Fayolle, Acoustic emission and lifetime prediction during static fatigue tests on ceramic-matrix-composite at high temperature under air, *Composites: Part A*, **41**, 913-918 (2010).

[11] M. Moevus, D. Rouby, N. Godin, M. R'Mili, P. Reynaud, G. Fantozzi, G. Fayolle, Analysis of damage mechanisms and associated acoustic emission in two SiC$_f$/[Si–B–C] composites exhibiting different tensile behaviours. Part II: Unsupervised acoustic emission data clustering, *Comp. Sc. and Tech.*, **68**, 1250-1257 (2008).

[12] G. N. Morscher, 1998. Modal acoustic emission of damage accumulation in a woven SiC/SiC composite, *Comp. Sci. Tech.*, **59**, 687-697 (1999).

[13] E. Maillet, N. Godin, M. R'Mili, P. Reynaud, J. Lamon, G. Fantozzi, Indicators of the critical behavior of Ceramic Matrix Composites for rupture time prediction during fatigue tests at intermediate temperatures, *Composite Science and Technology*, submitted.

[14] C. Droillard, J. Lamon, Fracture toughness of 2D woven SiC/SiC CVI-composites with multi-layered interphases, *J. Am. Ceram. Soc.*, **79**, 849-858 (1996).

DETERMINATION OF ACOUSTIC EMISSION SOURCES ENERGY AND APPLICATION TOWARDS LIFETIME PREDICTION OF CERAMIC MATRIX COMPOSITES

E. Maillet, N. Godin*, M. R'Mili, P. Reynaud, J. Lamon, G. Fantozzi
INSA-Lyon, MATEIS CNRS UMR5510, F-69621 Villeurbanne, France
* Corresponding author. Tel: +33 472438073; Fax: +33 472437930; nathalie.godin@insa-lyon.fr

ABSTRACT

Ceramic Matrix Composites (CMCs) are anticipated for use in aeronautical engines. Expected lifetimes in service conditions that are of thousands of hours are unattainable during laboratory tests. The objective of this paper is to propose an Acoustic Emission (AE) based approach to lifetime prediction for CMCs. The energy of AE sources is extracted from the energy recorded during tests, with a view to obtain a proper measure of stress-induced damage during static fatigue at high temperature on woven $SiC_f/[Si-B-C]$ composites. The use of two AE sensors allows the evaluation of energy attenuation due to propagation distance, acquisition system and damage accumulation. The analysis of AE sources energy release in static fatigue reveals a critical evolution of local energy release in the vicinity of rupture location at 50% of rupture time. This criticality is assessed using the Benioff law and it is attributed to subcritical crack growth in fibers.

1 INTRODUCTION

Ceramic Matrix Composites (CMCs) are candidates for use in aeronautical engines owing to their low density and good mechanical properties at high temperatures. CMCs can experience large deformations ($\approx$1%) as a result of energy dissipation through multiple matrix cracking and deflection of cracks at fiber/matrix interfaces. Several approaches to describe the mechanical behavior of CMCs under tensile loading[1,2] have been proposed. Moreover, the Acoustic Emission (AE) technique has been considered in order to monitor microstructural changes and damage evolution in CMCs at ambient and intermediate temperatures[3-5].

In static fatigue at high temperature under low stresses, composite ultimate failure results from oxidation-activated crack growth in fibers and associated stress redistributions when fibers fail[6,7]. The development of the [Si-B-C] self-healing matrix allowed protection against oxidation. Above 550°C, under oxidizing atmosphere, boron trioxide and silicon carbide react producing a borosilicate glass that fills up the cracks. Expected lifetimes in service conditions are tens of thousands of hours, which can hardly be checked out using laboratory tests for practical reasons. Therefore, a real-time prediction of the remaining lifetime during tests is necessary. It requires the monitoring of damage evolution for which AE measurement is a suitable technique. In fact, the AE technique is based on the recording and analysis of transient elastic waves that are generated during material damage. It provides real-time data on initiation and evolution of damage in terms of location and mode.

The energy of an AE signal corresponds to a part of the energy released by the source. Therefore, it is a relevant measure of damage evolution. Momon et. al[8] studied the release of AE signals energy during static fatigue tests on woven $SiC_f/[Si-B-C]$ and $C_f/[Si-B-C]$ composites at intermediate temperatures. The coefficient of emission R_{AE} was defined. It allowed the identification of a characteristic time around 60% of rupture time. Moreover, the Benioff law[9], introduced at first in seismology, was applied. Experimental data recorded after the characteristic time fitted well this power law, indicating critical features of energy release prior to rupture. Recently[10], determination of the coefficient of emission was improved to allow real-time determination. In addition, the effects of attenuation due to propagation distance were eliminated. The use of equivalent energy allowed the study of local energy release in the rupture zone. The optimum circle method[11], developed in seismology, was applied. It showed that the energy release associated with AE sources located in a limited interval (10 to 20 mm) around the rupture point was well approximated by the Benioff law at a characteristic time equal to 55% of rupture time. This critical release of energy at a local scale was attributed to slow crack growth in fibers.

Gyekenyesi et. al[12] used acousto-ultrasonics in order to monitor the evolution of energy attenuation in SiC/SiC composites subjected to load/unload/reload tensile tests. A significant decrease of energy of the captured ultrasonic wave of over 50% was observed between undamaged and damaged states. The increase of energy attenuation was attributed to matrix cracks opening. Accurate measurement of damage evolution based on the energy of AE sources therefore requires the consideration of attenuation due to damage. To evaluate attenuation during test using acousto-ultrasonics usually requires calibration tests. Also, using a single sensor gives a measure of the relative evolution of attenuation. The present paper proposes a method for the evaluation of energy attenuation that uses the energy recorded from AE sources generated during material damage. The method is based on the calculation of the ratio of AE energy recorded for each source at both ends of the specimen using two sensors. Attenuation is evaluated from thousands of local attenuation measurements obtained for the AE sources which are detected. Finally, the analysis of the critical aspect of AE sources energy release prior to rupture is discussed for 7 static fatigue tests.

2 MATERIAL AND TESTING PROCEDURE

2.1 Material

The composite material was manufactured by SAFRAN Snecma Propulsion Solide. It was made of woven PyC coated SiC fibers (Hi-Nicalon, Nippon Carbon Ltd, Japan) and a multilayered [Si-B-C] matrix produced via chemical vapor infiltration. A seal coat protected the material. The fiber volume fraction was 35 to 40% and porosity was about 12 v%. Dog-bone shaped specimens were machined. Their total length and thickness were respectively 200 mm and 4.5 mm. The gauge section dimensions were 60 x 16 mm^2.

2.2 Mechanical testing

Fatigue tests were performed under a constant load. A pneumatic tensile machine was used. It has been designed to ensure high stability for long duration tests while reducing environment noise. Before loading, each specimen was heated up to the test temperature (450°C or 500°C) at a rate of 20°C/min. Then, after one hour at the test temperature, it was loaded up at a rate of 600 N/min to the test load selected in the range of 40 to 95% of the rupture load (indicated for each test in terms of stress as the ratio σ/σ_r). The rupture load (and rupture stress σ_r) had been determined during tensile tests under monotonous loading at room temperature. An extensometer was used for elongation measurement. The machine was equipped with a 25 kN load cell.

2.3 Acoustic Emission monitoring

The test configuration is presented on Figure 1. Hatched parts represent contact areas between specimen and water-cooled grips. Two sensors (micro80, Physical Acoustics Corporation) were positioned, 190 mm apart, on specimen. The sensors were placed in housings machined in the grips. A holding system with spring was used in order to maintain constant contact pressure between sensors and material throughout test. Medium viscosity vacuum grease was used as a coupling agent. Each sensor was connected to a preamplifier (gain: 40 dB, frequency range: 20-1200 kHz), which was connected to the data acquisition system (two-channel MISTRAS 2001, Physical Acoustics Corporation). The threshold was set to 32 dB in order to filter out signals from ambient noise. The acquisition parameters were set as follows: peak definition time 50 μs, hit definition time 100 μs and hit lockout time 1000 μs. For each AE signal, the following data were recorded: arrival time, stress and strain values, as well as signal energy. Wave velocity V (9500 m/s) was determined prior to testing using a pencil lead break procedure. Young's modulus decrease during the tests was monitored using unloading/reloading cycles performed every 6 or 12 hours and wave velocity V was corrected accordingly using coefficient Γ[13] defined as:

$$\Gamma = \sqrt{\langle E(t)\rangle / E_0}$$ (1)

where $\langle E(t)\rangle$ is the secant elastic modulus at time t and E_0 is the reference Young's modulus of undamaged composite. The locations of sources were determined as follows:

$$x(n) = \pm \frac{1}{2}.V.\left[t_j(n) - t_i(n)\right]$$ (2)

where $x(n)$ is the location of the nth source, $t_i(n)$ and $t_j(n)$ are the arrival times respectively at the closest sensor (i) and at the furthest one (j). $x(n)$ is negative if sensor 1 is triggered first and positive otherwise (Figure 1). Since two sensors were used, location was determined with respect to specimen longitudinal axis. Only those signals located in the gauge length (60 mm) were kept. Each source n was therefore described by the time at which the closest sensor was triggered $t(n)$, its abscissa $x(n)$ (ranging from -30 to +30 mm) and the recorded energy by two sensors $E_1(n)$ and $E_2(n)$.

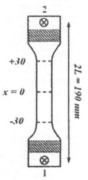

Figure 1. Test configuration and location of sensors 1 and 2
for static fatigue loading at high temperature

3 ENERGY OF ACOUSTIC EMISSION SOURCES

3.1 Recorded AE signal energy vs. source energy

It is generally accepted that the energy of an AE signal includes the energy released by the source at crack initiation. Various parameters affect recorded energy: distance of wave propagation, energy attenuation due to damage, coupling between sensor and material surface and sensor frequency response. Wave theory states that the energy of an acoustic wave decreases exponentially with the increase of propagation distance. Therefore, the following equation was proposed to describe the energy of recorded AE signals (for instance, at sensor 1) received from the source n:

$$E_1(n) = E_s(n).A_1.e^{-B(L+x(n))}$$ (3)

$E_s(n)$ is the energy released at source n in the form of elastic waves. Due to differences in coupling between sensor and material surface or in sensor frequency response, for a source located at equal distance, the sensors may record significantly different amounts of energy. Thus, A_i is the proportion of source energy that is recorded by sensor i. It is a characteristic of the sensor i. $L + x(n)$ is the distance of propagation from source n to sensor 1 (2L being the distance between sensors). The attenuation coefficient B is related to the propagation medium, which is subjected to changes due to damage evolution. Similarly, AE signal energy received at sensor 2 was expressed as:

$$E_2(n) = E_s(n).A_2.e^{-B(L-x(n))} \qquad (4)$$

3.2 Identification of attenuation parameters

To evaluate energy attenuation, the ratio of AE signal energies recorded at both sensors is calculated for each source n. For an easier identification of attenuation coefficient B, X(n) was defined as the natural logarithm of this ratio. From Equations 3 and 4, it comes:

$$X(n) = log\frac{E_1(n)}{E_2(n)} = log\frac{A_1}{A_2} - 2.B.x(n) \qquad (5)$$

A_1/A_2 represents the relative effect of acquisition that is for source n the fraction of source energy recorded by sensor 1 with respect to that recorded by sensor 2. The second term indicates the effect of propagation, which depends on source location x(n) and on attenuation coefficient B. Both parameters can be estimated from X(n) for various sources (1, 2, ..., n, ...) since X is a linear function of x. Because attenuation is expected to vary with damage, parameters estimation is performed at successive time intervals. For a given time interval, values of X(n) are observed in various space intervals in order to take into account uncertainties in AE sources localization.

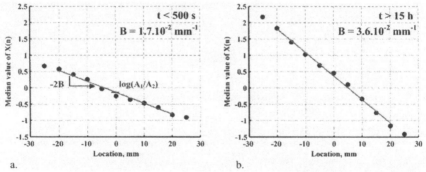

a.   b.

Figure 2. Median value of X(n) in every space interval vs. location and linear approximation a/ for the first 2000 AE sources (t < 500 s) and b/ for the last 2000 AE sources (t > 15 h) generated during the test at 500°C - σ/σ_r = 0.95

Estimation of attenuation constants was performed as follows. Attenuation was evaluated for a given time interval using the median values of X(n) in every space interval (width: 10 mm, overlapping: 5 mm). Each median value of X(n) corresponds to a few hundreds AE sources located in the same space-time interval. The median values of X(n) corresponding to the space intervals at both ends of the gauge length were discarded and a linear approximation was carried out. Figure 2a shows the median value of X(n) in every space interval vs. location for the first 2000 AE sources generated during the test performed at 500°C - σ/σ_r = 0.95. Likewise, Figure 2b shows the median value of X(n) in every space interval vs. location for the last 2000 AE sources generated during the same test. In both cases, the data points fit well a linear function (coefficient of determination R^2 greater than 0.98). This observation validates Equation 5 as well as the equation proposed for the recorded AE energy (Equations 3-4). Both attenuation parameters can therefore be determined, the coefficient of attenuation B from the slope of the linear fit and the ratio of A_i from its origin ordinate. In order to monitor the evolution of attenuation, the linear approximation is repeated in consecutive time intervals. Each time interval corresponds to 2000 AE sources. At every increment of 500 sources, a new time interval is considered. The overlapping is set in order to accurately monitor evolution of both attenuation parameters.

3.3 Method validation

Similar results were observed on every test and are reported here for the test at $500^{\circ}C - \sigma/\sigma_r = 0.95$. Figure 3 shows the distribution functions of $X(n)$ for every space interval during the first time interval (corresponding to the first 2000 AE sources). The black line corresponds to the values of $X(n)$ calculated for sources located in the interval [-5; +5] mm. The curves to the left represent the consecutive 10-mm wide intervals from +5 mm to +25 mm. Values of $X(n)$ decrease as the observed sources are further from sensor 1 (located at -95 mm). The curves to the right represent the intervals from -5 mm to -25 mm. In this case, values of $X(n)$ increase as the observed sources are closer to sensor 1. Considering different values of sources location results only in a translation of the distribution of $X(n)$. Therefore, it is relevant to use the median value of $X(n)$ to characterize its distribution in a given interval.

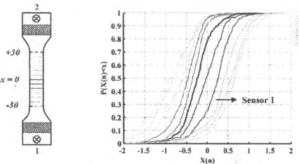

Figure 3. Distribution functions of $X(n)$ in the different 10-mm wide space intervals during the first time interval (2000 AE sources: t < 500 s)

For the space intervals located at the gauge length boundaries, changes in the overall shape of the distribution can be observed. These changes are attributed to a significantly lower number of sources available to define the distribution. To evaluate the effect of population size, the distribution function of $X(n)$ was analyzed for the space interval [-5; +5] mm during the first time interval. A number N of values was randomly selected to plot the distribution function. It was observed that the effect of population size only appears for N lower than 150. Therefore, the required number of sources located in the entire gauge length per time interval was set to 2000 sources. In the case of homogeneous activity throughout the gauge length (as observed during the early stages of damage), this corresponds to more than 300 sources per interval, which leads to a good description of the distribution of $X(n)$. However, later in the tests, space intervals located at both ends of the gauge length ([-30; -20] mm and [+20; +30] mm) displayed often an inhomogeneous activity with a number of sources smaller than 150 for several time intervals during the tests. Thus, these intervals were discarded.

The value chosen for the space interval width (10 mm) was validated by an observation of the effect of interval width on the distribution function of X(n). Considering the sources located during the first time interval (first 2000 sources), the distribution functions of X(n) were plotted for a space interval centered at x = 0 with various width values (ranging from 2 to 60 mm). It appears that distribution functions of X(n) are well superimposed with width values of 2 to 10 mm. This is mainly attributed to uncertainties on sources localization that make observations in the smallest interval (±1 mm) equivalent to those in the 10-mm wide interval. Further increase in interval width (from 10 to 60 mm) results in an increase in standard deviation of X(n). Therefore, the width value of 10 mm was selected in order to maximize the number of sources per interval without affecting the observed distribution.

3.4 Determination of AE sources energy

The ratio of A_i indicates which sensor recorded the greatest amount of AE sources energy. When the ratio of A_i is on average greater than 1, sensor 1 is the reference sensor. The following expression for source energy $E_s(n)$ can be derived from Equation 3:

$$E_s(n) = E_1(n).C_1.e^{+B(L+x(n))}$$ (6)

When the ratio of A_i is on average lower than 1, sensor 2 is the reference sensor. The expression of the source energy $E_s(n)$ becomes:

$$E_s(n) = E_2(n).C_2.e^{+B(L-x(n))}$$ (7)

According to Equation 6, C_1 should be equal to $1/A_1$. However, A_1 cannot be determined because there are 2 equations (E_1 and E_2) and 3 unknowns (A_1, A_2 and B). Therefore, only the ratio A_1/A_2 can be obtained. Based on the evolution of the ratio of Ai during tests (Figure 6, discussed in the next section), the value of C_1 was estimated as follows:
- With sensor 1 as the reference sensor:
  Before the extremum value of ratio of A_i: $C_1 = 1$
  After the extremum value of ratio of A_i: $C_1 = \dfrac{(A_1/A_2)_{extremum}}{(A_1/A_2)}$ (8)

- With sensor 2 as the reference sensor:
  Before the extremum value of ratio of A_i: $C_2 = 1$
  After the extremum value of ratio of A_i: $C_2 = \dfrac{(A_1/A_2)}{(A_1/A_2)_{extremum}}$ (9)

4   TOWARDS LIFETIME PREDICTION FROM THE ANALYSIS OF AE SOURCES ENERGY

4.1 Release of AE sources energy under static fatigue loading

Figure 4 represents the cumulative AE sources energy under static fatigue for the test at 500°C - $\sigma/\sigma_r = 0.67$. The energy released during initial loading was not plotted. It accounts for more than 90% of the total energy and was attributed to matrix cracking. A significant energy release is observed at the beginning of static fatigue. The activity then decreased until approximately 50% of rupture time. Finally, the energy release exhibited acceleration prior to rupture. Consistent observations were performed on the 7 studied fatigue tests. In this article, we discuss the applicability of the Benioff law to model energy release prior to rupture.

4.2 Critical aspect of sources energy release

In a previous paper[10], modeling of energy release prior to rupture using the Benioff law[9] was detailed. The optimum circle method[11] was also applied to determine the optimum space interval around the rupture point and optimum time when the energy release is well approximated by the Benioff law. The method is briefly outlined below.

The Benioff law[9], first introduced in seismology, describes the energy release as a power law of the remaining time before rupture:

$$\Omega(t) = \sum_{n=1}^{N(t)} \sqrt{E_s(n)} = \Omega_r.[\,1 - \frac{\psi}{1-\gamma}.(t_r - t)^{1-\gamma}\,] \qquad (10)$$

where Ω is the so-called cumulative Benioff strain. $N(t)$ is the number of AE sources generated prior to time t. Ω_r is the cumulative Benioff strain at rupture time t_r, γ and Ψ are constants.

The optimum circle method[11], also developed in seismology, is used to assess the applicability of the Benioff law. In addition, it determines which AE sources should be considered in order to achieve the best approximation. Two parameters are optimized: d corresponding to the half-width of the space interval around the rupture point and t_s, time after which the Benioff law is used to approximate the energy release. To determine the optimum values of both parameters, one has to consider every possible pair of values of d and t_s. Each pair of values leads to different energy release, corresponding to the AE sources located in the space interval of ±d around the rupture point during the time interval $[t_s; t_r]$. Two approximations are carried out on the energy release resulting from each pair of values of d and t_s: the Benioff law (Equation 10) and a linear approximation. The c-value is the ratio of the root mean square error of the approximation by the Benioff law over that of the linear fit. When the c-value is lower than 1, there is a positive contribution of the Benioff law since the approximation error is lower than that of a linear fit. It is a relative validation of the relevance of the approximation by the Benioff law. Therefore, to ensure quality of the approximation, only c-values lower than 0.5 are considered to be relevant. The optimum values d* and t_s* correspond to the lowest c-value.

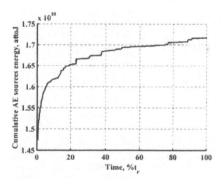

Figure 4. AE sources energy release under static fatigue. Test at 500°C - $\sigma/\sigma_r = 0.67$

5 RESULTS AND DISCUSSION

5.1 Identification of attenuation parameters

Attenuation was characterized using AE data recorded during 7 fatigue tests on woven SiC_f/[Si-B-

C] composites carried out at 450°C and 500°C. Testing conditions and corresponding rupture times are given in Table I. Attenuation constants were estimated as indicated above. To ensure the relevance of every value of attenuation coefficient B and ratio of A_i, the coefficient of determination R^2 was calculated for every approximation. When a linear approximation had a value of coefficient of determination R^2 lower than 0.95, the associated values of both parameters were discarded. Values of R^2 lower than 0.95 were not observed in every test, and they appeared only for time intervals either at the beginning or at the end of tests. They were attributed to heterogeneous acoustic activity throughout the gauge length.

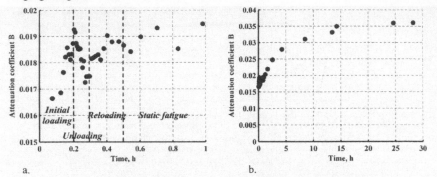

a.   b.

Figure 5. Test at 500°C - $\sigma/\sigma_r = 0.95$. Attenuation coefficient B vs. time
a/ During the first hour and b/ Throughout the fatigue test

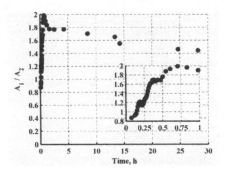

Figure 6. Test at 500°C - $\sigma/\sigma_r = 0.95$. Ratio A_1 / A_2 vs. time

Figure 5 shows the evolution of attenuation coefficient B during the test carried out at 500°C - σ/σ_r = 0.95. The initial value corresponding to the first 2000 AE sources is around $1.7.10^{-2}$ mm^{-1}. A slight increase is observed during initial loading up to maximum load at 0.2 hour. The specimen was then unloaded in order to monitor damage caused by initial loading. It results in a decrease of attenuation coefficient B down to nearly its initial value. Static fatigue loading began around 0.5 hour and a steep increase of attenuation coefficient B is observed up to 15 hours when the value reaches a plateau ($3.6.10^{-2}$ mm^{-1}). Figure 6 shows the evolution of A_1/A_2 during the above test. If both sensors had recorded the same amount of sources energy, the ratio of A_i would be equal to 1. However, its value was around 0.8 at the beginning of test, which indicates that for a same source equidistant from both sensors, sensor 1 receives about 80% of the energy received by sensor 2. This effect can be mainly attributed to differences in coupling between sensors and material. It is in good agreement with results of a previous paper[10] on the distribution of AE signal energies at both sensors for the AE sources located in an area centered on x = 0. It is confirmed here by taking into account the AE sources located in the entire gauge length. If the values of A_1 and A_2 were only dependent on the acquisition system, they would remain constant during test. However, a significant increase of ratio of A_i is observed during the first 45 minutes of test followed by a decrease until rupture. Previous local damage may affect wave propagation leading to A decrease. A could also decrease due to previous damage accumulated outside of the gauge ([-30; 30] mm, see Figure 1). During the tests, AE activity remained rather diffuse throughout the gauge. Therefore, changes of ratio of A_i could be attributed to damage occurring outside of the gauge length causing additional energy attenuation. In the present case, the ratio of A_i is greater than 1. Therefore, at first, more damage appears between the upper limit of the gauge length and sensor 2 (Figure 1). Then, the decrease of ratio of A_i means that damage growth in the lower part of the specimen becomes predominant. Using this parameter, the sensor that records the greatest part of AE sources energy is identified. In the present case, since the ratio of A_i is greater than 1 during most of the test, sensor 1 will be chosen as the reference sensor. After 45 minutes of test, the decrease of ratio of A_i indicates a decrease of the amount of sources energy recorded by sensor 1. It is mainly attributed to changes of the propagation medium between the lower part of the studied zone and sensor 1.

The 6 other tests exhibited a reproducible evolution of attenuation coefficient B with a significant increase between 10 and 50% of the rupture time (up to a factor of 2) and a plateau value beyond. Significant changes of ratio of A_i were also observed which justified the correction factor C introduced in Equation 6. For each test, the AE source energy of every source n generated at time t(n) was calculated using Equation 6 with values of both attenuation parameters characterized in the time interval closest to t(n).

5.2 Critical aspect of AE sources energy release

The optimum circle method was carried out for every test. Five values of d were considered: 5, 10, 15, 20 mm and d_{max} corresponding to the entire gauge length. A value of t_s was taken every hour in time interval of 10 to 90% of rupture time t_r. Figure 7 shows the evolution of the c-value vs. t_s for the five values of d for the test carried out at 450°C - σ/σ_r = 0.71. The minimum c-value appears clearly around 40% of rupture time with d equals to 5 and 10 mm. For values of d greater than 10 mm, no optimum is observed. Thus, for this test the best approximation by the Benioff law is achieved when considering AE sources located in an interval of ±10 mm around rupture point beyond 40% of rupture time. The very low c-value (0.28) indicates that when considering the optimum values d* and t_s* (10 mm and 40% of t_r respectively), the approximation using the Benioff law induces an error more than three times lower than that induced by a linear approximation.

Optimum values of t_s* and d* (leading to the minimum c-value) are reported in Table I. The Benioff law was relevant on average after 50% of rupture time when considering AE sources located in a limited space interval of ±14 mm around the rupture point. This indicates the critical aspect of local energy release around the rupture point. It also indicates that another damage mechanism becomes predominant locally after 50% of rupture time. The critical aspect of energy release was therefore attributed to subcritical crack growth in fibers.

These results are in good agreement with those of our previous paper[10] where only attenuation due to propagation distance in the undamaged material was taken into account. A characteristic time around 55% of rupture time was identified, after which the energy release was attributed to subcritical crack growth in fibers. The present paper shows that the attenuation induced by damage evolution occurred before 50% of rupture time. This phenomenon had no influence on the energy release in the second half of fatigue tests. However, it remains necessary to take into account attenuation due to propagation distance within the material in order to consider a local energy release around the rupture point.

Table I. Testing conditions, rupture times and optimum values of t_s^* and d^*.

Temp., °C	σ/σ_r	Rupture time t_r, h	t_s^*, %t_r	d^*, mm
500	0,95	42	62	20
500	0,91	43	58	10
500	0,79	136	29	20
500	0,67	195	55	10
500	0,44	480	41	15
450	0,71	372	39	10
450	0,62	591	68	10
Average			50	14
Standard deviation			14	5

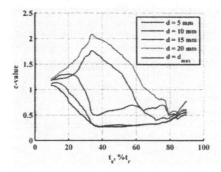

Figure 7. Test at 450°C - $\sigma/\sigma_r = 0.71$. Evolution of the c-value vs. t_s for the 5 values of d

6 CONCLUSION

This paper presents a method for the determination of AE energy attenuation based on AE energy of sources generated by material damage during fatigue tests. The effects of energy attenuation due to propagation distance, acquisition and damage accumulation are eliminated in order to accurately define the energy of AE sources. The energy release prior to rupture under static fatigue exhibits a critical evolution locally (±15 mm around the rupture point) at 50% of rupture time regardless of the applied stress level. It is attributed to slow crack growth in the SiC fibers.

The applicability of the Benioff law to model the energy release associated with fibers failures offers a possible tool for lifetime prediction of SiC/SiC composites under static fatigue. Therefore, future work will focus on the calibration of the Benioff law in order to use it for lifetime prediction. Moreover, the characterization of energy attenuation will be studied with a view to damage monitoring. The method proposed in this paper can be used in real time and it allows the detection of a significant increase of attenuation during early phases of damage. Future work will aim at determining the contribution of each damage phenomenon on energy attenuation.

ACKNOWLEDGMENTS

The authors gratefully acknowledge SAFRAN Snecma Propulsion Solide and the CNRS for supporting this work as well as G. Fayolle for helpful comments.

REFERENCES

[1]J. Lamon, A micromechanics-based approach to the mechanical behavior of brittle matrix composites, *Compos Sci Technol*, **61**, 2259-72 (2001).

[2]G. Fantozzi, P. Reynaud, Mechanical hysteresis in ceramic matrix composites, *Mat Sci Technol A*, **521-522**, 18-23 (2009).

[3]M. Moevus, D. Rouby, N. Godin, M. R'Mili, P. Reynaud, G. Fantozzi, G. Farizy, Analysis of damage mechanisms and associated acoustic emission in two SiC/[Si-B-C] composites exhibiting different tensile curves. Part I: Damage patterns and acoustic emission activity, *Compos Sci Technol*, **68**, 1250-57 (2008).

[4]M. Moevus, N. Godin, M. R'Mili, D. Rouby, P. Reynaud, G. Fantozzi, G. Farizy, Analysis of damage mechanisms and associated acoustic emission in two SiC/[Si-B-C] composites exhibiting different tensile curves. Part II: Unsupervised acoustic emission data clustering, *Compos Sci Technol*, **68**, 1258-65 (2008).

[5]S. Momon, N. Godin, P. Reynaud, M. R'Mili, G. Fantozzi, Unsupervised and supervised classification of AE data collected during fatigue test on CMC at high temperature, *Comp: Part A*, **43**, 254-60 (2012).

[6]W. Gauthier, J. Lamon, Delayed failure of Hi-Nicalon and Hi-Nicalon S multifilament tows and single filaments at intermediate temperatures, *J Am Ceram Soc*, **92**, 702-709 (2009).

[7]O. Loseille, J. Lamon, Prediction of lifetime in static fatigue at high temperatures for ceramic matrix composites, *J Adv Mat Res*, **112**, 129-140 (2010).

[8]S. Momon, M. Moevus, N. Godin, M. R'Mili, P. Reynaud, G. Fantozzi, G. Fayolle, Acoustic emission and lifetime prediction during static fatigue tests on ceramic matrix composite at high temperature under air, *Comp: Part A*, **41**, 913-18 (2010).

[9]C. G. Bufe, D. G. Varnes, Predictive modeling of the seismic cycle of the Greater San Francisco Bay Region, *J Geophys Res*, **98**, 9871-83 (1993).

[10]E. Maillet, N. Godin, M. R'Mili, P. Reynaud, J. Lamon, G. Fantozzi, Analysis of Acoustic Emission energy release during static fatigue tests at intermediate temperatures on Ceramic Matrix Composites: towards rupture time prediction, *Compos Sci Technol*, Accepted manuscript, DOI: 10.1016/j.compscitech.2012.03.011 (2012).

[11]D. D. Bowman, G. Ouillon, C. G. Sammis, A. Sornette, D. Sornette, An observational test of the critical earthquake concept, *J Geophys Res*, **103**, 24359-72 (1998).

[12]A. L. Gyekenyesi, G. N. Morscher, L. M. Cosgriff, In situ monitoring of damage in SiC/SiC composites using acousto-ultrasonics, *Comp: Part B*, **37**, 47-53 (2006).

[13]G. N. Morscher, A. L. Gyekenyesi, The velocity and attenuation of acoustic emission waves in SiC/SiC composites loaded in tension, *Compos Sci Technol*, **62**, 1171-80 (2002).

NONDESTRUCTIVE EVALUATION OF THERMAL BARRIER COATINGS BY OPTICAL AND THERMAL IMAGING METHODS

J. G. Sun
Argonne National Laboratory
Argonne, IL 60439

Thermal barrier coatings (TBCs) are widely used to improve the performance and extend the life of combustor and gas turbine components. As TBCs become prime reliant, it becomes important to determine their conditions nondestructively to assure the reliability of the components. Nondestructive evaluation (NDE) methods may be used to assess TBC condition for quality control as well as to monitor TBC degradation during service. Several NDE methods were developed for these applications, including laser backscatter, mid-IR reflectance (MIRR), and thermal imaging based on a multilayer analysis method. Both laser backscatter and MIRR are optical imaging methods and have been investigated for TBC health monitoring. The thermal imaging method is new and can measure TBC's thermal properties with high accuracy. Because TBC properties change with life, a model for TBC health monitoring based on the property change is proposed. This paper describes these NDE methods and presents preliminary experimental results related to TBC health monitoring.

INTRODUCTION

Thermal barrier coatings (TBCs) have been extensively used on hot gas-path components in gas turbines. In this application, a thermally insulating ceramic topcoat (the TBC) is bonded to a thin oxidation-resistant metal coating (the bond coat) on a metal substrate. TBC coated components can therefore be operated at higher temperatures, with improved performance and extended lifetime [1,2]. Because TBCs play critical role in protecting the substrate components, their failure (spallation) may lead to unplanned outage or safety threatening conditions. Therefore, it is important to determine TBC conditions nondestructively to assure their reliability. Nondestructive evaluation (NDE) methods may be used to assess TBC condition for quality control of as-processed coatings as well as to monitor TBC degradation during service. This study is focused on NDE monitoring of TBC degradation during TBC life.

TBC degradation normally starts from initiation of small cracks at the TBC/bond coat interface. These cracks then grow and link together to form delaminations which eventually cause TBC spallation or failure. Although qualitative NDE methods can detect large-scale damages such as delaminations, quantitative methods are necessary to determine damage progression during the entire TBC life cycle. Various methods have been developed to monitor TBC degradations. Electrochemical impedance spectroscopy [3] has been investigated to examine the change in TBC microstructure (and cracking) and phase transformation, and was found to be useful for TBC life prediction. Photo-luminescence piezo-spectroscopy has been widely used to determine the stress accumulation at the interface and the exposure temperature of the coating material, which may be coupled with models to predict TBC life [4,5]. These spectroscopy methods are usually used in laboratory experiments. For field applications, fast and large-area imaging NDE methods are necessary.

In this study, two optical and one thermal imaging NDE methods were evaluated for TBC health monitoring. The optical methods are laser backscatter and mid-infrared (IR) reflectance

(MIRR). Laser backscatter is an optical scanning method developed at Argonne National Laboratory (ANL) for detection of subsurface flaws in ceramic materials [6,7]. MIRR is an optical imaging method developed at NASA for TBC health monitoring [8]. Both optical methods detect the increased reflection/scattering of an incident light from the progressive cracking within the TBC volume and at the interface during TBC degradation. The thermal imaging method explores the thermal property change for TBC health monitoring. This is feasible because TBC thermal properties, specifically the thermal conductivity, exhibit characteristic changes during TBC lifetime. With the development of a multilayer analysis method [9] that can accurately measure TBC thermal properties, a TBC life prediction model based on TBC thermal conductivity was proposed. These optical and thermal imaging NDE methods were evaluated based on preliminary experimental results for TBC samples that were thermal cycled to various lifetimes.

OPTICAL IMAGING METHODS FOR TBC HEALTH MONITORING

Laser Backscatter Method

The laser backscatter method is based on the cross-polarization detection of the scattered light from the subsurface of a translucent material [6]. When a linearly-polarized laser light is incident on a translucent material such as a TBC, the total backscattered light consists of surface reflection and subsurface backscatter. However, the surface reflection typically has no change in its polarization state while the subsurface scatter has a significant change. Because of the polarization difference, the cross-polarization backscatter detection can selectively measure only the subsurface backscatter from the internal of the material while filtering out the surface reflection. This method can therefore be more sensitive to detect subsurface microstructure change, such as cracking within the TBC coating and at interface. The current ANL system uses a HeNe laser at 633nm wavelength and performs scanning for image construction; however, direct imaging at other wavelengths is possible. This method has been investigated for TBC health monitoring during isothermal heat-treatment testing and preliminary results for pre-spall prediction were obtained for EB-PVD and APS TBCs [7].

Mid-IR Reflectance (MIRR) Method

MIRR is an optical imaging method developed by Dr. Eldridge of NASA for TBC health monitoring [8]. In this method, a steady-state infrared light source is used to illuminate the TBC surface and the total reflection, including those from the TBC surface, TBC volume, and cracks near topcoat/bond coat interface, is imaged by an infrared camera at 4 μm infrared wavelength. Because optical penetration for TBCs is at maximum in mid-infrared wavelengths (3-5 μm), MIRR has higher sensitivity to detect TBC degradation and cracking near the topcoat/bond-coat interface. Correlations have been determined between the MIRR data and the crack/delamination progression that is related to pre-spall condition [8].

In the NASA MIRR system, a SiC IR emitter coupled with a parabolic mirror is used to provide collimated illumination to the TBC samples, and imaging is obtained at the 4-μm wavelength with a narrow-band filter [8]. A simplified system was built at ANL, as illustrated in Fig. 1. This system utilizes a standard IR heating lamp coupled with a sanded-glass plate to provide uniform illumination, and imaging is taken in the entire wavelength band (3-5 μm) of the IR camera. The wide-band imaging greatly improves the detection sensitivity so a weaker IR illumination can be used, which leads to reduced sample heating and improved accuracy. A disadvantage is the inclusion of part of an OH absorption band near 3 μm [8]. However, because the infrared sensor (indium antimonide) has a low response near 3 μm wavelength, only ~50% from the maximum at 5 μm, it is estimated that a large reflectance change of 10% at 3 μm due to OH absorption will only result in ~1% in the IR camera reading. In addition, the high transmittance of TBC coating within the 4-5 μm range [8] will further limit the OH absorption effect when detecting reflectance change from cracks near interface. Nevertheless, OH absorption is a factor for detection accuracy and should be studied further.

TBC Samples and Experimental Results from Laser Backscatter and MIRR

A set of thermal-cycled TBC samples (sample courtesy of Dr. Eldridge of NASA) was used to evaluate the laser backscatter and MIRR methods. It consists of 5 EB-PVD TBC samples that were cycled up to 75% life (100% life corresponds to failure). All samples have a dimension of 1-inch diameter and a TBC coating thickness of ~0.13mm. Figure 2 shows a photograph of the samples. It is noted that some small spallations existed at the edges of the 50% and 75% TBC samples, which were probably induced by handling.

Figure 3a shows the scan images of the 5 EB-PVD TBC samples using the laser backscatter method. The scatter intensity is generally uniform on each TBC surface, although surface markings (contaminations) in the 0% sample caused reduction in local intensity. The intensity is higher at the edge areas with subsurface TBC delamination, indicating the detection of the delaminations. The average scatter intensity is plotted as a function of TBC life in Fig. 3b; the intensity data seem not very sensitive within the lifetime range of these TBCs. In previous studies [7], however, it was found that the backscatter intensity increases significantly near the end of TBC life. Therefore, further study is necessary to evaluate TBCs in later stage of their life.

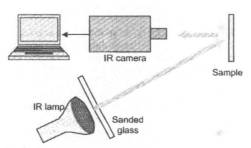

Fig. 1. Schematic of MIRR system setup at ANL.

Fig. 2. Photograph of 5 thermal-cycled EB-PVD TBC samples (% life is indicated).

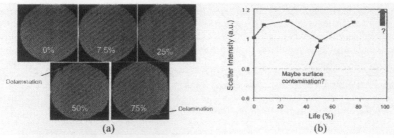

Fig. 3.   Laser backscatter (a) images and (b) intensity as function of TBC life for TBC samples.

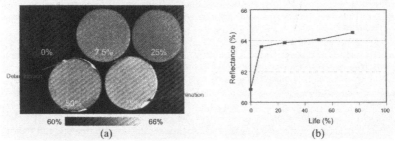

Fig. 4.   MIRR reflectance (a) image and (b) value as function of TBC life for TBC samples.

Figure 4a shows the MIRR image of the 5 EB-PVD TBC samples. MIRR also detected the surface markings (and contaminations) and small delaminations near TBC spalled areas. In addition, the reflectance value increases with TBC life, as shown in Fig. 4b. The reflectance has a large increase in the early stages of TBC life, but the change becomes smaller in later TBC life. The large initial reflectance change could be due to the difference in sample thickness or surface condition (contamination); the exact cause was not further investigated. This result is consistent to that in the literature [8], indicating that MIRR measurement within a wide-wavelength band (in 3-5μm) has similar detection sensitivity as the standard method.

Both laser backscatter and MIRR are intensity-based optical imaging methods that can be used for fast and remote imaging of large surface areas and their imaging results can be easily scaled to produce quantitative data. Because laser backscatter measures only the internal scattering from the TBC material and the interface, while MIRR also includes surface reflection, laser backscatter could be more sensitive to detect the interface cracking that is the major damage mode in TBC degradation. However, optical penetration within TBC is very weak at the 633nm wavelength used in the current laser backscatter experiments, by at least a factor of 5 compared to that at the 4μm wavelength for MIRR [8]. This results in a lower detection sensitivity for crack development at the interface by laser backscatter. If a longer wavelength was used, the detection sensitivity of the laser backscatter method should be improved. Nevertheless, optical penetration is a limiting factor for all optical methods. It was shown that the detection sensitivity for MIRR, which is already operated near the optimum wavelength, is usually limited to a coating thickness of ~0.3mm for APS TBCs [8]. In addition, optical intensity signal is also affected by TBC thickness (due to volume scattering) as well as surface contamination. Because of these difficulties, other NDE methods such as thermal imaging are needed for TBC health monitoring. In particular, thermal imaging is not limited by TBC thickness (e.g., from 50 μm to 1mm for typical TBCs; can be easily extended outside this range).

THRMAL IMAGING METHOD FOR TBC HEALTH MONITORING

The change of thermal conductivity with TBC life has been studied by many authors. In early exposure times, TBC undergoes a sintering process, resulting in a denser material and an increased conductivity (an interesting observation was that sintering does not affect MIRR reflectance [8]). As exposure time increases, the initiation and growth of microcracks causes a gradual reduction of TBC conductivity. When delamination occurs in the later stage of the TBC life, a significant conductivity decrease can be observed in the delaminated regions. This characteristic change in TBC conductivity with TBC life may be used to monitor TBC degradation condition and predict TBC life. Such approach for TBC health monitoring apparently has not been attempted before likely because of the lack of a NDE method that can accurately measure TBC conductivity. TBC conductivity can be measured by several methods. The most reliable and commonly used method is laser flash, which is normally a destructive method because it uses stand-alone TBC coating samples (sometimes with specially-prepared TBC samples with substrate) [10,11]. Although a large number of TBC thermal conductivity data is available, they are mostly for the intrinsic property of the TBC coating material without the contribution from the interface degradation inside a real TBC system. Therefore, these data are not suitable for TBC life prediction. What is needed is a NDE method that can measure the thermal conductivity of a TBC system (i.e., TBC coating on substrate), so TBC degradation and cracking at the interface is included within the measured data. Such a NDE method has been developed recently: the multilayer analysis method based on one-sided flash thermal imaging [9]. Preliminary studies showed that it has high measurement accuracy and repeatability.

For TBC health monitoring, it is necessary to establish a model based on a characteristic "damage" parameter. There are two major mechanisms affecting TBC conductivity: one is the increase of the intrinsic conductivity of the coating material due to continued sintering, and the other is the decrease of the TBC system conductivity due to the progressive cracking at the interface. Because the conductivity decrease due to cracking is the major TBC degradation mechanism, a damage parameter based on the conductivity decrease may be used as an indicator of the TBC degradation status. To define this parameter, the intrinsic TBC conductivity increase due to sintering needs to be determined.

Recent studies [12,13] showed that the increase of TBC thermal conductivity k due to sintering may be correlated with the Larson-Miller parameter (LMP):

$$\ln(k / k_0) = a + b\text{LMP}, \tag{1}$$

where LMP is defined as:

$$\text{LMP} = T(\ln t + C), \tag{2}$$

k_0 is the initial conductivity (at t=0 for as-spayed coating), T is an average exposure temperature, t is time, and a, b, and C are fitting constants. This correlation indicates that if a TBC is not damaged during its service life its conductivity should increase monotonically with time (or life). This correlation therefore determines the intrinsic thermal property of the coating material and sets an upper bound for the TBC coating system. Any damage inside the coating or at the interface (cracking and delamination) should reduce the TBC-system conductivity. Because the measured thermal conductivity includes the contributions from the intrinsic property as well as all damages, the difference between the intrinsic conductivity and measured conductivity is an indicator of the damage level in the TBC system. Based on this assumption, we can define the conductivity difference Δk as a TBC damage parameter:

$$\Delta k = k - k_{exp}, \qquad (3)$$

here k is the intrinsic TBC conductivity determined from Eq. (1) and k_{exp} is measured TBC conductivity (e.g., using the thermal imaging method). It is apparent that Δk is related to the thermal exposure temperature and time through the LMP defined in Eq. (2), both are important parameters affecting TBC lifetime.

In practical application, it is necessary to establish an approach to determine the intrinsic TBC conductivity k, which may be different for coatings with different composition and microstructure as well as under different sintering conditions. Therefore, the correlation Eq. (1) should be determined for each TBC system. To do this, we note that a TBC system is normally not damaged during its early life, so measured conductivity should be equal to the intrinsic conductivity, i.e., $k = k_{exp}$ or $\Delta k = 0$. This approach may then establish the relationship Eq. (1)-(3) for each TBC system from experimental data. The remaining issue involves with how to determine a criterion for Δk that corresponds to TBC failure, which is not simple because of the distributed nature of TBC damage as shown in the following example. Although more validations are necessary, development of this model may establish another fundamental approach to understand TBC damage process based on an important and most-studied TBC property [14,15]. The following example illustrates a preliminary evaluation of this model.

Figure 5 shows measured thermal conductivity of four EBPVD TBC samples using the thermal-imaging multilayer analysis method [9]. These samples were thermal cycled to various percentage of TBC life (sample courtesy of Mr. A. Luz, Imperial College London). The 100% TBC life is defined as when the total area of TBC spallation exceeds 20% of the surface area. As shown in Fig. 5, the as-processed TBC conductivity was generally uniform and low (1.8 W/m-K). At 33% life, TBC conductivity was still uniform, but at a higher value due to sintering. At 67% life, while the majority of the TBC surface had uniform conductivity, many small regions showed lower conductivity. These low-conductivity spots should correspond to the locations with microcracking. At 100% life, although TBC spallation occurred in a large surface area (the gray-color region), most TBC still remained. The conductivity value was high in some remaining TBC areas, comparable to those at 33% and 67% TBC life, and very low in other areas (below the low conductivity bound of 0.8 W/m-K used in the calculation). These low conductivity regions were all located at the edges of the remaining TBC, and they had been confirmed to be interface delamination.

Figure 6 shows the average TBC conductivity evolution that would be predicted by the above model, Eqs. (1)-(3). Because of the lack of samples, the correlation Eqs. (1)-(2) cannot be verified, so is only illustrated in the figure assuming that the correlation would follow experimental data in the early lifetime (say, up to 33%). The decease of measured conductivity in later lifetimes was due to TBC damage. Figure 6 also shows measured conductivity values at some damaged areas (two small regions at 67% life and the delaminated region in 100% life). Because the conductivity in damaged regions is much lower, it would be necessary to consider how to include them in the calculation of average TBC conductivity and the damage indicator Δk. Nevertheless, this result demonstrates that TBC conductivity, measured by the thermal imaging NDE method, can be used to monitor TBC degradation with life.

Fig. 5.   Measured thermal conductivity images of EBPVD TBC samples.

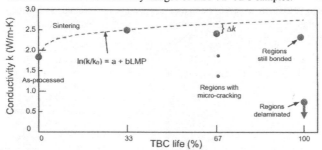

Fig. 6.   Measured thermal conductivity as function of TBC life for EBPVD TBC samples.

CONCLUSION
        Health monitoring and lifetime prediction are important issues for reliable utilization of TBCs in advanced turbine engines. In this study, two optical imaging methods, laser backscatter and mid-IR reflectance (MIRR), and a thermal imaging method were evaluated for these applications.   For the optical methods, the evaluation used a set of 5 TBC samples that were thermal-cycled to various percentages of life. Experimental results showed that both methods may detect delaminations below the thin TBC layers. The MIRR reflectance was found to increase monotonically with TBC life, while the laser backscatter intensity was not very sensitive to the range of TBC lifetime investigated in this study. However, because all optical methods are limited by optical penetration depth, which is ~0.3mm for typical APS TBCs, thermal imaging method that is not limited by typical TBC thicknesses is needed especially for thicker TBCs. With the recent development of a multilayer analysis method that can accurately measure TBC thermal properties, it is now possible to use the TBC thermal conductivity change for TBC health monitoring of real engine components. This approach is based on an established correlation of the intrinsic TBC conductivity change due to sintering and the measured TBC conductivity to define a conductivity difference parameter as an indicator of the TBC damage level. Experimental results were presented to illustrate this approach.   Although more validation is necessary, development of this model may establish another fundamental approach for TBC health monitoring and lifetime prediction.

ACKNOWLEDGMENT
        The author thanks Dr. J. Eldridge for valuable discussions related to the MIRR method. This work was sponsored by the U.S. Department of Energy, Office of Fossil Energy, Advanced Research and Technology Development/Materials Program.

REFERENCES
1. US National Research Council, National Materials Advisory Board, "Coatings for High Temperature Structural Materials," National Academy Press, Washington, DC, 1996.
2. A. Feuerstein and A. Bolcavage, "Thermal Conductivity of Plasma and EBPVD Thermal Barrier Coatings," Proc. 3rd Int. Surface Engineering Conf., pp. 291-298, 2004.
3. F. Yang and P. Xiao, "Nondestructive evaluation of thermal barrier coatings using impedance spectroscopy," Int. J. Appl. Ceram. Technol., Vol. 6, pp. 381-399, 2009.
4. X. Wang, R.T. Wu, and A. Atkinson, "Characterisation of residual stress and interface degradation in TBCs by photo-luminescence piezo-spectroscopy," Surface & Coatings Technology, Vol. 204, pp. 2472–2482, 2010.
5. C. Rinaldi, L. De Maria, and M. Mandelli, "Assessment of the Spent Life Fraction of Gas Turbine Blades by Coating Life Modeling and Photostimulated Luminescence Piezospectroscopy," Journal of Engineering for Gas Turbines and Power, Vol. 132, 114501, 2010.
6. J.G. Sun, W.A. Ellingson, J.S. Steckenrider, and S. Ahuja, "Application of Optical Scattering Methods to Detect Damage in Ceramics," in Machining of Ceramics and Composites, Part IV: Chapter 19, Eds. S. Jahanmir, M. Ramulu, and P. Koshy, Marcel Dekker, New York, pp. 669-699, 1999.
7. W.A. Ellingson, R.J. Visher, R.S. Lipanovich, and C.M. Deemer, "Optical NDT Methods for Ceramic Thermal Barrier Coatings," Materials Evaluation, Vol. 64, No. 1, pp 45-51, 2006.
8. J.I. Eldridge, C.M. Spuckler, and R.E. Martin, "Monitoring Delamination Progression in Thermal Barrier Coatings by Mid-Infrared Reflectance Imaging," Int. J. Appl. Ceram. Technol., Vol. 3, pp. 94-104, 2006.
9. J.G. Sun, "Thermal Property Measurement for Thermal Barrier Coatings by Thermal Imaging Method," in Ceramic Eng. Sci. Proc., eds. S. Mathur and T. Ohji, Vol. 31, no. 3, pp. 87-94, 2010.
10. H. Wang and R.B. Dinwiddie, "Reliability of laser flash thermal diffusivity measurements of the thermal barrier coatings," J. Thermal Spray Techno., Vol. 9, pp. 210-214, 2000.
11. J.G. Sun, "Thermal Conductivity Measurement for Thermal Barrier Coatings Based on One- and Two-Sided Thermal Imaging Methods," in Review of Quantitative Nondestructive Evaluation, eds. D.O. Thompson and D.E. Chimenti, Vol. 29, pp. 458-469, 2009.
12. Y. Tan, J.P. Longtin, S. Sampath, and H. Wang, "Effect of the Starting Microstructure on the Thermal Properties of As-Sprayed and Thermally Exposed Plasma-Sprayed YSZ Coatings," J. Am. Ceram. Soc., Vol. 92, no. 3, pp. 710–716, 2009.
13. D. Zhu and R.A. Miller, "Thermal Conductivity and Elastic Modulus Evolution of Thermal Barrier Coatings under High Heat Flux Conditions," J. Thermal Spray Technol., Vol. 9, no. 2, pp. 175-180, 2000.
14. J.G. Sun, "Development of Nondestructive Evaluation Method for Ceramic Coatings and Membranes," Proc. 23rd Annual Conference on Fossil Energy Materials, Pittsburgh, PA, May 12-14, 2009.
15. F. Cernushci, P. Bison, S. Marinetti, and E. Campagnoli, "Thermal diffusivity measurement by thermographic technique for the non-destructive integrity assessment of TBC coupons," Surf. Coat. Technol., Vol. 205, pp. 498-505, 2010.

VISUALIZATION OF INTERNAL DEFECTS IN CERAMIC PRODUCTS BY USING A UT PROBE ARRAY

Yoshihiro Nishimura and Takayuki Suzuki
National Institute of Advanced Industrial Science and Technology
1-2-1 Namiki, Tsukuba,
Ibaraki 305-8564, Japan

Katsumi Fukuda and Naoya Saito
Tokyo National College of Technology
1220-2, Kunugida, Hachioji,
Tokyo 193-0997, Japan

ABSTRACT

The possibility of inspecting defects in ceramic materials using a UT probe array was studied to investigate the reliability of ceramic products. Large structural components made of many small blocks are likely to include internal fatal defects, particularly in joints, and their fracture response is much faster and more drastic than that of metal materials. Therefore, non-destructive testing (NDT) is necessary to detect internal defects[1]. A UT probe array can scan the inside of a sample and reconstruct images of internal defects. However, this process takes a long time. Braconnier and Hirose proposed methods to reconstruct an image of internal defects more quickly. In the present study, samples with inserted defects were prepared to simulate defects in ceramics, and data was acquired to reconstruct images. An array probe with a large aperture was also developed to inspect defects deep within large ceramics samples (longer than 100mm) and applied to a large ceramic sample and an aluminum sample. To evaluate visualization using the UT array probe, some configurations of delay time, noise reduction, and image reconstruction methods were tried, and their influences were examined in order to construct an NDT system for inspecting large ceramic products.

INTRODUCTION

Ultrasonography is used to image the inside of structural components using ultrasonic probes for non-destructive testing (NDT) to certify the safety and soundness of industrial products and the infrastructure of our society (e.g., aircraft, railroads, and recreational equipment). Ceramics (e.g., silicon nitride (SN)) have excellent heat resistance and corrosion resistance; however, internal defects significantly influence their strength and fracture toughness. Larger the components are more likely to have internal fatal defects. Therefore, it is unreasonable to manufacture a large component in one body. Instead, large structural components are made of small defect-free or similar block units (Figure 1).

Figure 1. Large ceramic structure made of small block units

However, such components are likely to include fatal defects in joints, and their fracture response is much faster and more drastic than that of metal materials. Therefore, ultrasonography is necessary to detect internal defects in joints. Conventional UT imaging is conducted with a point focusing probe. Inspecting large components takes a long time because the probe employs a small beam diameter and a small sampling interval to derive a high-resolution image. An alternative is a probe array, by which ultrasonic waves can be focused on any position in a sample by controlling the delay time for each element to pulse into the sample. Electronically scanning the entire internal volume of a sample enables reconstructing a 3D image with less time. However, this system cannot use high frequency due to mutual interference between adjacent elements. Braconnier[2] and Nakahata and Hirose[3,4] proposed methods to address this problem. They did not focus each pulse at all positions in the sample but sent a plane wave instead, to receive the reflected wave and calculate reflection intensities at all positions with numerical focus in a computer. Braconnier constructed 3D images using a synthetic aperture; Nakahata and Hirose constructed 3D images using reverse-conversion based on the response corresponding to the pupil function of the probe in the frequency space. These methods are very effective but require extensive computation. Moreover, it is difficult for the probe array to compensate for the characteristics of each probe[5,6]. Considering these problems, the present study seeks to construct a system for visualizing internal defects.

RECONSTRUCTION PRINCIPLES
　　　　Ordinary phased-array methods, such as linear scanning and sector scanning (Figure 2), adjust the delay time of each element, focus an ultrasonic wave on the measurement region in the specimen, and scan only one or two lines. The images are blurred in regions other than the scanning focus lines. Many pulses are required to scan the entire region for better resolution, and these pulses create numerous reflections within the sample. The measuring system must wait for the reflecting pulses to decay before transmitting the next pulse, and this process takes a long time.
　　　　Therefore a method using one pulse was employed, as suggested by Braconnier[1] (Figure 3). This method pulses all the elements simultaneously, generates a plane wave (Figure 3(a)), and receives the wave reflected through each element. The reflection intensity calculated at any position on which the ultrasonic wave focuses in the entire measured region and reconstructed an image (Figure 3(b)). Then the refracting angles calculated, considering Snell's law, to reconstitute the image (Figure 3(c)). The image could therefore be reconstructed instantly in the entire measured region.

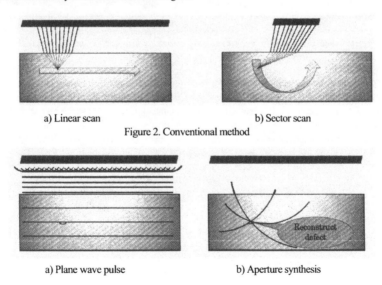

a) Linear scan　　　　　　　　　　　　　b) Sector scan
Figure 2. Conventional method

a) Plane wave pulse　　　　　　　　　b) Aperture synthesis

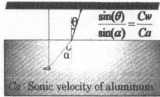

$$\frac{\sin(\theta)}{\sin(\alpha)} = \frac{Cw}{Ca}$$

Ca: Sonic velocity of aluminum

Cw: Sonic velocity of water

c) Consideration of Snell's law

Figure 3. Flash focus principles

APERTURE SYNTHESIS

The relationship between array probe and sample in Figure 4 are demonstrated. Let $\vec{x}_2 = (x_2, z_2)$ be the coordinate of an internal defect in the sample and $u(\vec{x}_2, t)$ be the wave received at $\vec{x}_2$. This wave is a plane wave.

$$u(\vec{x}_2, t) = \int \frac{f\left(t - \frac{|\vec{x}_1 - \vec{x}_2|}{c}\right)}{|\vec{x}_1 - \vec{x}_2|} \exp(-\lambda |\vec{x}_1 - \vec{x}_2|) dx_1 = f\left(t - \frac{|z_2|}{c}\right) \exp(-\lambda |z_2|) \quad (1)$$

Let $\mu(x_2, z_2)$ be the reflection coefficient at $\vec{x}_2$ and $v(x_3, t)$ be the wave received at $\vec{x}_3 = (x_3, 0)$. The attenuation coefficient λ is set at 0 to simplify the calculation.

$$v(x_3, t) = \int \mu(x_2, z_2) \frac{f\left(t - \frac{|z_2| + \sqrt{(x_3 - x_2)^2 + z_2^2}}{c}\right)}{\sqrt{(x_3 - x_2)^2 + z_2^2}} dx_2 dz_2 \quad (2)$$

The above equation is rewritten as follows, by defining

$$g(x_3 - x_2, z_2) = \frac{f\left(t - \frac{|z_2| + \sqrt{(x_3 - x_2)^2 + z_2^2}}{c}\right)}{\sqrt{(x_3 - x_2)^2 + z_2^2}} \quad (3)$$

$$v(x_3, t) = \int \mu(x_2, z_2) g(x_3 - x_2, z_2) dx_2 dz_2 \quad (4)$$

Because Eq. (4) includes convolution integrals, the reflected wave can be calculated simply by performing a Fourier transformation. The reflection coefficient $\mu^*(x_2, z_2)$ can be calculated from time series of wave data $v(x_3, t)$ received at each piezo element using Eq.(5).

$$\mu^*(x_2, z_2) = \sum_{x_3}^{n} v\left(x_3, \frac{\sqrt{(x_3 - x_2)^2 + z_2^2}}{c}\right) \quad (5)$$

The images are reconstructed using the above theory.

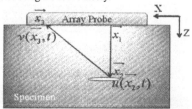

Figure 4. Array probe and specimen

EXPERIMENT EQUIPMENT
    NDT equipment with a high-precision 3D-scanner (Figure 5) was developed and applied to a prepared sample. This NDT equipment was used to position the array probe over the sample and acquire the time series of echo data; The probe array scanned all of the inside of sample. The derived data were converted to an internal section image of the sample.

Figure 5. Experiment equipment

    Figure 6 presents the array probes developed to inspect large ceramic samples; Table 1 lists the specifications.

a) Array probe 32ch   b) Array probe 64ch
Figure 6. Array probe

Table 1. Specifications

Material:	PZT	Configuration:	Rectangle
Channels:	32 or 64	Frequency:	5[MHz]
Probe dimensions		25×25×50 or 45×45×120mm	
Transducer dimensions		0.4×13 or 13×45mm	
Radius of curvature dimension		R = 50 or 400mm	
Clearance of transducer		0.1×13 or 0.2×45mm	

MEASUREMENT OF DEFECTS IN ALUMINUM BLOCK SAMPLES
    This technique was validated using three aluminum block samples, which were easy to machine. Figure 7 presents the profiles and dimensions of these samples, and Figure 8 presents reconstructed images. The wavelength of ultrasound is 1.2mm. Distances between defects were 1, 2, or 4mm. The upper surface of the defect was reproduced successfully; however, multiple lines appeared at the position of the defects' upper

surfaces because the original spike pulse included several periods of oscillation. The distances between defects were reproduced well.

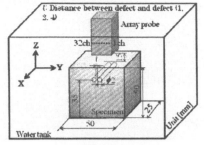

Figure 7. Measurement model

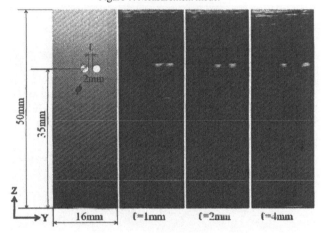

Figure 8. Reconstructed images

MEASUREMENT OF SMALLER DEFECTS IN ALUMINUM BLOCK SAMPLES

Attempted to inspect smaller and deeper defects in the sample than in the former measurement. It became difficult to detect deep defects because attenuation of ultrasonic signals increases with depth. The dimensions of defects are small compared to the wavelength of ultrasound. Figure 9 depicts the profiles and dimensions of the samples, and Figure 10 plots the measured time series of signal data derived through ch6.

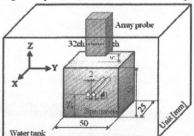

Figure 9. Measurement model

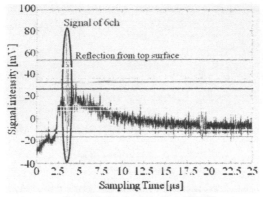

Figure 10. Received signal at 400MHz

The signal reflected from the specimen surface was easily detected; however, it was difficult to identify the signal reflected from a defect and the bottom of the specimen. Generally, each element of a probe array is far smaller than a single probe, and the reflected signal is weak. Therefore, signals measured by a probe array are much noisier than those measured by a single probe. Thus, the arithmetic mean was carried out to reduce white noise in the measured signal (Figure 11). Then the signal reflected from a defect and the specimen bottom, as well as the signal reflected from the top surface were confirmed.

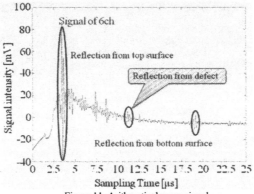

Figure 11. Arithmetical mean signal

It is difficult to identify the signal reflected from a defect due to noise (e.g., multiple reflections). Thus, the signal reflected from a defect was identified by comparing it with that of a sample without a defect. The arithmetic mean result of the sample without a defect is plotted in Figure 12. This indicates a noise signal. Noise was decreased by subtracting signals without a defect from signals with a defect (difference method). Using these data, images were reconstructed by aperture synthesis (Figure 13). The upper surface of the defect was reproduced successfully, and the distance between defects was reproduced well.

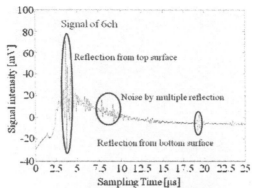

Figure 12. Arithmetical mean signal of no defects

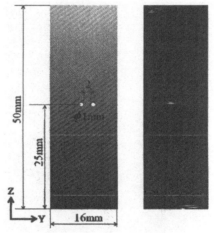

Figure 13. Reconstructed images

MEASUREMENT OF A SLIT DEFECT IN A LARGE ALUMINUM SAMPLE

Next, an aluminum sample with a slit defect at the bottom (Figure 14) was prepared and measured. The acoustic velocity of aluminum is 6300m/s, and that of ceramic (SN) is 9000m/s to 12000m/s. The acoustic velocity of the prepared sample was measured to be 10800m/s. A 150mm depth of a defect in aluminum corresponds to 90mm. Therefore, a 150mm-long aluminum block sample was used. Images were reconstructed by aperture synthesis (Figure 15). The signal used for reconstruction was processed by the method of finite differences. In addition, distance correction was used for reconstruction. In spite of its depth, the upper surface of the defect was reproduced successfully.

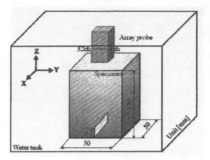

Figure 14. Measurement model

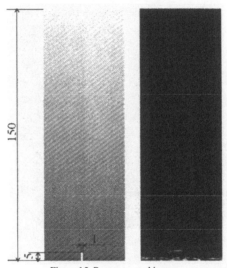

Figure 15. Reconstructed images

MEASUREMENT OF A SLIT DEFECT IN A LARGE CERAMIC SAMPLE (Si$_3$N$_4$)

A ceramic sample (SN) with a slit defect at the bottom (Figure 16) was prepared and measured. The ceramics were Si$_3$N$_4$. Their profiles and dimensions are indicated in Figure 16, and the measured time series of signal data derived through ch6 is plotted in Figure 17. The signal reflected from the specimen bottom was detected; however, it is difficult to identify the signal reflected from a defect. The influence of white noise was great, as with aluminum. Therefore the arithmetic mean was carried out to reduce white noise in the measured signal (Figure 18). The signal reflected from a defect was confirmed. However, the signal reflected from the specimen surface could not be observed due to a dead band, which formed when the pulse was transmitted early in the sampling period because each element simultaneously transmitted and received pulses. Figure 19 is the image of internal defect reconstructed by aperture synthesis. Distance correction was used for reconstruction. The image reconstructed for ceramics was similar to that reconstructed for aluminum.

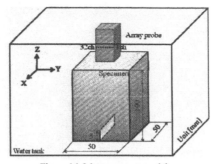

Figure 16. Measurement model

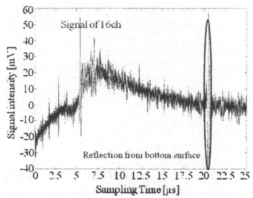

Figure 17. Received signal at 400MHz

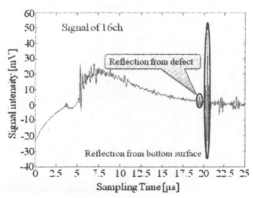

Figure 18. Arithmetical mean signal

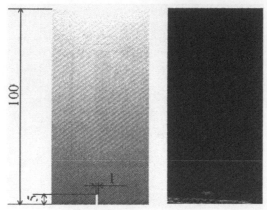

Figure 19. Reconstructed images

MEASUREMENT USING EVEN-ODD ELEMENT PULSING

To avoid a dead band, even-odd pulsing was performed (Figure 20). Even-odd pulsing first pulses even-numbered elements and simultaneously receives signals reflected from odd-numbered elements, and then pulses odd-numbered elements and receives signals reflected from even-numbered elements. The result is plotted in Figure 21. The signal reflected from the specimen surface could be obtained, and the dead band was reduced.

These data were used to reconstruct an image by aperture synthesis (Figure 22). A result similar to that of usual pulse imaging was obtained. Therefore, even-odd pulsing is effective for detecting internal defects.

Figure 20. Even-odd element pulsing

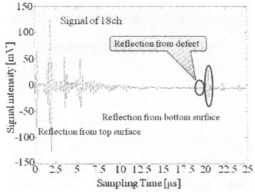

Figure 21. Averaged signal derived by even-odd element pulsing

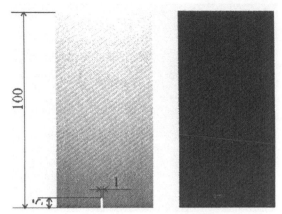

Figure 22. Images reconstructed from signals derived by even-odd element pulsing

CONCLUSION

Results of experimenting to visualize internal defects in ceramic products revealed the following.

1) Signals reflected from sample defects, surfaces, and bottoms were observed for all pulsing patterns.
2) White noise has a large influence but can be controlled by signal processing.
3) Images of a cylinder defect were reconstructed, the defect was reproduced, and the distance between defects was reproduced well.
4) A dead band formed during full-element pulsing but was controlled during even-odd element pulsing.
5) The image reconstructed in even-odd element pulsing was similar to that reconstructed in usual pulsing.
6) Therefore, the technique used here is an effective technology for detecting internal defects.

ACKNOWLEDGMENTS

This work was supported by the New Energy and Industrial Technology Development Organization (NEDO) and the Ministry of Economy, Trade and Industry as part of the project for the Innovative Development of Ceramics Production Technology for Energy Saving.

REFERENCES

[1]E. Sato, M. Shiwa, K. Yamashita, A. Sato, M. Koshirae, and K.Kitami: Method of Automatic Ultrasonic Inspection by Phased ArraySystem for Graphite Ingot: JAXA Research and Development Memorandum, JAXA-RM-05-006, (2006).
[2] Dominique.Braconnier, High-speed ultrasonic test using array probe, JSNDI proceedings, pp.35-38 (2006).
[3]K.Nakahata, S.Hirose, The study of the qualities of the reconstructed by array probe, JSNDI proceedings, p.39 (2006).
[4]K.Nakahata, M.Hirohata, S.Hirose: "Flaw Reconstruction from Scattering Amplitude using Full-waveform Sampling and Processing(FSAP)", Journal of JSNDI, 59, .6, pp.277-283 (2010).
[5]Y.NISHIMURA, A.SASAMOTO, T.SUZUKI, The study of 3D image reconstruction using EMAT, JSDNI proceedings, pp.171-176 (2006).
[6]Y. Nishimura, T. Suzuki, N. Kondo, H. Kita and K. Hirao: "Study of Defect Inspection in Ceramic Materials Using UT an X-Ray" , JSAEM Studies in Applied Electromagnetics and Mechanics, 14,pp.145-146 (2011)

EVALUATION OF CERAMIC MATERIALS AND JOINTS USING UT AND X-RAY

Yoshihiro Nishimura and Takayuki Suzuki

National Institute of Advanced Industrial Science and Technology

Advanced Manufacturing Research Institute, AIST Tsukuba EAST 1-2-1 Namiki Tsukuba Ibaraki, 305-8564, Japan

Naoki Kondo and Hideki Kita

National Institute of Advanced Industrial Science and Technology

Advanced Manufacturing Research Institute, AIST Chubu 2266-98 Anagahora Shimo-Shidami Moriyama-ku Nagoya 463-8560, Japan

ABSTRACT

Inspection methods to certify the reliability of large structural ceramic components composed of small blocks were studied. Ceramic blocks jointed with a silicon sheet or with glass were examined by UT and X-Ray. Silicon sheet or glass joints could be distinguished from other parts of component by UT more easily than by X-Ray. Fractures may originate in joints. More and larger defects were detected in non-joint parts and area than expected. It is important to develop technology to inspect joint layer deep within the sample. A new joint sample was developed, and its joint layer was inspected. It was found to have fewer defects in joints. Special probes were developed to inspect defects deep (95mm or deeper) within large components and were applied to large component samples. Defects of 0.5 mm were detected successfully.

INTRODUCTION

Ceramic materials are thought to be unreliable for manufacturing large structural components because they are likely to have fatal defects inside and their fractural responses are much faster and more drastic than those of metal materials. "Stereo Fabric" is a concept by which a large ceramic product can be assembled from many small blocks.[1,2] Its goal is to produce large, heat-resistant containers, pipes for the metal industry, large tables for the precision-electronics industry. Figures 1 (a) and (b) illustrate samples assembled into a large table

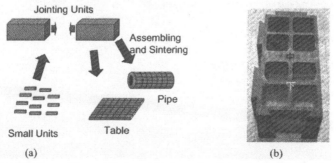

Figure.1. (a) Concept of Stereo Fabric and (b) Example of jointed units (block's dimension: 40 mm x 40 mm x 40 mm).

Of course, each small block must be verified to be defect-free for structural use. In particular, the joints between blocks must be defect-free. However, joints are known to be more likely to fracture than other parts of ceramic products. Non-destructive inspections of conventional ceramic components have been conducted by SEM and optical micro-scope. However, these inspection methods cannot to be applied to large structural components. X-Ray and UT are convenient non-destructive testing (NDT) methods, but they are not common in the field of ceramic development. Generally, micro X-Ray CT is used to inspect internal voids but not narrow cracks. However, it is inadequate for high-resolution measurement of inspected samples larger than 50 mm (e.g. stereo fabric products), due to geometrical limitations. In addition, UT has been used in the steel industry for many years. However, the acoustic velocity of Silicon Nitride (SN) is 9,000-12,000 m/s. For UT in water, the focal length of ceramic materials is one sixth to one seventh of that in water; therefore a probe whose focal length is 600 mm is required to inspect defects of SN to a depth of 100 mm. Thus, Special considerations are required for UT inspection of ceramics.

INSPECTION OF CONVENTIONAL SN JOINTS USING UT AND XRAY

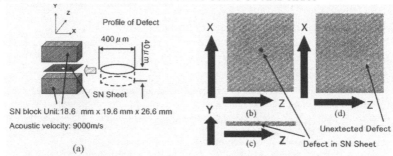

Figure 2. (a) Dimensions of SN sample and (b), (c), (d) Micro X-Ray/CT image.

Figure 2.(a) depicts an SN sample jointed with a silicon sheet with a hole defect in the middle and indicates the dimensions of the blocks(18.6 mm x 19.6 mm x 26 mm), The silicon sheet is made of SN. The SN samples and silicon sheet were sintered into one body. Figures 2. (b) and (c) present images taken by micro X-Ray/CT. Some unexpected defects were found (e.g. in addition to the planned hole defect in Figure 2. (d))

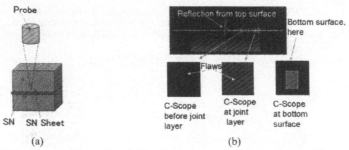

Figure 3. (a) UT probe and SN sample and (b) A-Scope image and C-Scope images (40 mm x 40 mm) of SN sample at 50MHz.

Figures 3 (a) and (b) present the SN sample illustrated in Figures 2 (a) and (b) and the images observed by UT. The focus is on the joint layer in this inspection. Reflection from the joint layer could be observed. The inserted defect could be observed by UT as well as by micro X-Ray/CT. SN blocks were carefully prepared to be defect-free; however, some unexpected defects were found in the SN block body in places other than the joint (Figure 3).UT detected more unexpected defects in the SN body

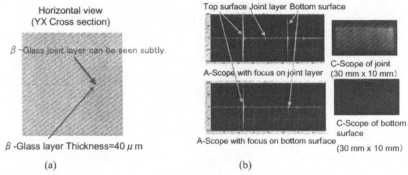

(a)   (b)

Figure 4. (a) Transmission X-Ray image and (b) A-Scope image and C-Scope images of SN sample at 50 MHz.

Figures 4 (a) and (b) present horizontal views of a transmission X-Ray image and a A-scope and C-scopes images of the SN sample jointed with β-glass. Figures 5 (a) and (b) are images of the SN sample jointed with a silicon sheet. An SN sample was formed from two blocks (19 mm x 45 mm x 45 mm). These samples were much larger than those depicted in Figure 2 and 3. Reflection from joint could not be observed without the focus on the joint layer.

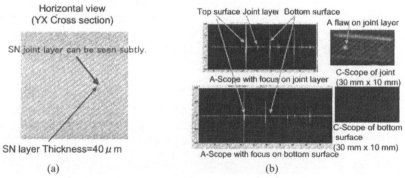

(a)   (b)

Figure 5. (a) Transmission X-Ray image and(b) A-Scope image and C-Scope images of SN sample at 50 MHz.

The β-glass joint in Figure 4 (a) was more visible than the SN joint in Figure 5(a). However, no defect in the SN sample jointed with a silicon sheet was observed. Both the reflections from joints in Figures 4 (b) and 5(b) could be seen more clearly than their transmission X-Ray images when their focal points were the sample's joint, not their bottom edges. The inserted defect was detected (Figure 5(b)).

EVALUATION OF SN-SN JOINT DEVELOPED FOR LARGE COMPONENTS
Special considerations are required for inspection of large ceramics component. Micro-Xray/CT cannot be used for them. Figure 6 depicts a system consisting of an acoustic point focal lens and an inspected ceramics sample.

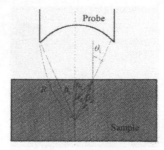

Figure 6. Acoustic lens and sample.

The focal point in the ceramics sample is shallower than that expected in water because the acoustic velocity in ceramics far exceeds that in water. The acoustic wave in this measurement focuses on a line or a point. Here, h_1 is the depth of the focal point calculated from the geometry of the probe, and h_2 is calculated from Snell's law. ($c_s / c_w = \sin\theta_2 / \theta_1$) The dependency of the ratio h_2 / h_1 on the incidence angle is calculated using equation(1) and plotted in Figure 7 (a). The equivalent distance L travelled by an acoustic wave from the acoustic lens to a focal point in the ceramics sample is calculated using equation (2) and the shortened equivalent travel length $c \Delta t$ is plotted in Figure 5 (b). Here, the acoustic velocity c_w of water is 1500 m/s and that c_s of SN is 9000 to13000 m/s. The critical incidence angle is calculated to be 8 degrees.

$$\frac{h_2}{h_1} = \frac{\tan\theta_1}{\tan\theta_2} \qquad (1)$$

$$\frac{L}{h_1} = \frac{R}{h_1} - (1 - \frac{1}{n^2})\frac{1}{\cos\theta_1} \qquad (2)$$

The aperture angle θ_2 / θ_1 of the acoustic lens used in this measurement is less than 5 degrees. Here, if h_1 is 100mm, the focus line in the sample is 0.77mm long. The acoustic wave focuses mostly on a point, and the depth ratio h_2 / h_1 is 1/8 when the UT inspections of the SN sample are conducted in water. The focal length of a UT probe for detecting a defect at depths of 80 mm in an SN sample must exceed 640 mm. Furthermore, the diameter of the piezoelectric element must exceed 25 mm to meet the condition for a short distance sound field.

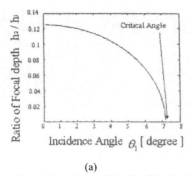

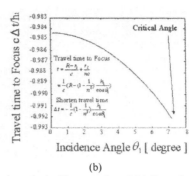

(a)   (b)

Figure 7. (a) Dependence of h_2 / h_1 of the focal depth on the incidence angle and (b) Dependence of the equivalent travel distance to the focus on the incidence angle.

UT equipment to inspect internal defects at a depth of more than 80 mm was prepared. (Figure 8) Its specifications are listed in Table 1. Derived wave forms were converted into time series of data by Acqiris DP210, whose maximum sampling rate is 2 GHz.

Figure 8. UT equipment with a 3D-scanner.

Table 1. 3D-Scanner's specifications.

Scan Speed	1 line/s (max 1000 mm/s)
Precision (X, Y, Z)	0.5 μ m
Maximum image resolution	8192 x 8192
Measurable Area	360 mm x 310 mm x 80 mm

Figure 9 indicates the dimensions of an SN-SN joint sample prepared by the Stereo Fabric Research Association. Two SN blocks were sintered into one body under high pressure without a silicon sheet or anything else between blocks. The measured acoustic velocity was 9300 m/s. Figures 10 (a) and (b) present a C-scope image acquired in the Y direction with a point focusing probe at the depth of 10 mm, whose focal length is 75mm and whose frequency is 15 MHz, and an image acquired in X direction. The focal length in this sample was calculated to be 12 mm and was thought to be long enough to inspect the joint layer in the Y direction. The A-scope image acquired at the X-Y position of a void defect in Figure 10 (a) is presented in Figure 11 (a). A void defect with a diameter exceeding 1 mm was found below the joint layer in Figure 11 (a). Moreover, reflection from the joint layer was observed in it. B-scope of Figure 10 (a) is presented in Figure 11 (b). The joint layer cannot be seen in Figure10.(b); however, a joint layer can be found in Figure 11 (b).

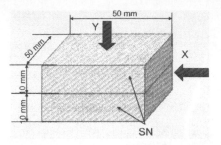

Figure 9. Dimensions of SN sample.

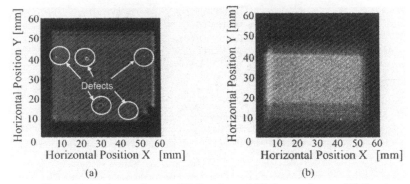

(a)   (b)

Figure 10. (a) C-scope image of Y direction and (b) C-scope image of X direction.

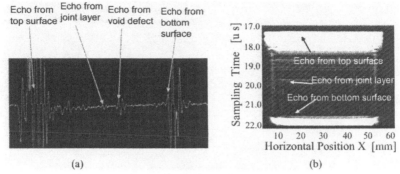

(a)   (b)

Figure11. (a) A-scope image at defect and (b) B-scope image of Figure 10 (a).

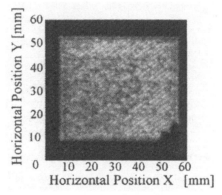

Figure 12. C-scope image of joint layer.

Figure 12 presents a C-scope image of the joint layer. No significant void defects could be found in the joint layer. There were fewer void defects in the joint layer than in SN blocks..

EVALUATION OF DEFECTS AT DEEP POSITIONS OF A LARGE SN SAMPLE

Inspection of defects deep within a sample is important to certify the health condition of large assembled components. A point focusing probe with a focal length of 600 mm and operation frequency of 5 MHz (Figure 13 (a)) was developed and applied to samples made of aluminium and SN.

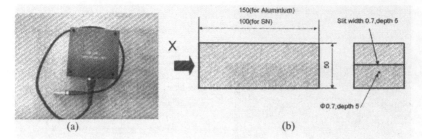

(a)   (b)

Figure 13. (a) Probe for inspection of deep defects and (b) Sample and dimensions.

Figure 13 (b) depicts samples and their dimensions. The SN sample and aluminium samples have a slit (1mm wide and 5 mm deep) and a hole (1mm diameter and 5mm deep) at the bottom edge. The SN sample was SUN-1 supplied by Nikkato, Inc., and had an acoustic velocity of 10800 m/s. The focal depth of the aluminium sample was 143 mm, and that of SN sample was 83 mm.

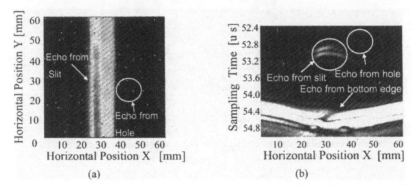

(a)   (b)

Figure 14. (a) C-scope image and (b) B-scope image including a slit and a hole defect.

Figures 14 (a) and (b) present the B-scope and C-scope images of aluminium sample acquired by a UT probe around the slit in the X direction. The slit was 1mm wide; however, it was blurred to 10 mm.in Figures 14. (a) and (b). The hole was be blurred in the C-cope image in Figures 14. (a) and (b) as well but its intensity was very low.

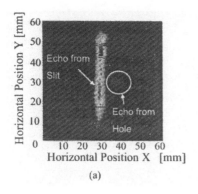

(a)

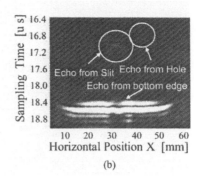

(b)

Figure 15. C scope image and (b) B scope image including a slit and a hole defect.

The SN sample was inspected and their C-scope and B-scope images are presented in Figures 15. (a) and (b). The slit and the hole were detected successfully.

CONCLUSION

UT and X-Ray were used to inspect a conventional ceramics joint layer bonded with β-glass or a silicon sheet. The joint layer as well as inserted defects and unexpected defects could be observed. A sample with joint developed for a large structure component was prepared, and its joint layer image was acquired by a long focus UT probe. Reflections from the joint layer itself could be observed but not particularly large void could be found. Moreover, a special UT probe with a focal length of 600 mm was applied to two kinds of large samples, metal and ceramics. It could inspect sample defects at depths exceeding 145 mm for metal and 95 mm for ceramics.

ACKNOWLEDGMENTS

This work was supported by the New Energy and Industrial Technology Development Organization (NEDO) and the Ministry of Economy, Trade and Industry as part of the project for the Innovative Development of Ceramics Production Technology for Energy Saving.

REFERENCES

[1]N. Kondou, Y. Nishimura, T. Suzuki, H. Kita, Evaluation of Jointed Silicon Nitride by X-ray Computed Tomography: J. of the ceramic society of Japan, 118-12, 1192-1194 (2010)

[2]Y. Nishimura, N. Kondou, H. Kita, T. Suzuki: Non-destructive Evaluation of Ceramics Material: Proc. of 7th Annual Conference at Omaezaki, Shizuoka by Japan Society of Maintenology, (July 13,2010)

INVESTIGATION OF NON-DESTRUCTIVE EVALUATION METHODS APPLIED TO OXIDE/OXIDE FIBER REINFORCED CERAMIC MATRIX COMPOSITE

Richard E. Johnston[1], Martin R. Bache[1], Mathew Amos[2], Dimosthenis Liaptsis[2], Robert Lancaster[2], Richard Lewis[2], Paul Andrews[3] and Ian Edmonds[3]

[1] Materials Research Centre, College of Engineering, Swansea University, Singleton Park, Swansea SA2 8PP, United Kingdom
[2] TWI NDT Validation Centre (Wales), ECM [2], Heol Cefn Gwrgan, Margam, Port Talbot, SA13 2EZ, UK
[3] Rolls-Royce plc, P.O. Box 31, Derby, DE24 8BJ, United Kingdom

ABSTRACT

Ceramic matrix composites may contain numerous internal features that could potentially affect their mechanical performance during service application. Therefore, potential non-destructive investigation techniques for the evaluation of aluminium oxide/aluminium oxide ceramic matrix composites (CMC) are presented. Several different non-destructive evaluation (NDE) methods were assessed for their capability to identify the internal structure on varying scale lengths. Internal features of interest included artifacts from the manufacturing process such as porosity, delamination and microcracking. Artificial defects of varying size were also deliberately introduced within the CMC during the manufacturing phase, to establish a threshold of the resolution for each technique. The NDE methods selected included X-ray computed tomography, 2-D radiography, flash thermography, immersion ultrasonic and resonance testing. This allowed for the ranking of each NDE method and recommendations for their future cost effective employment.

INTRODUCTION

Ceramic matrix composites (CMCs) are currently being developed for various high temperature applications, including advanced gas turbine engines. The potential use of CMC materials is due to their relative high strength/toughness and resistance to chemical reactions at high temperatures and their relative low density[1-2]. CMC materials are fabricated by several processing methods, which ultimately determine their mechanical and thermal properties. The fibre architectures of these composites vary and can be planar, two or three dimensional. Because of the complex microstructures and multiple processing steps involved during fabrication, CMC components could contain flaws or defects of various sizes and shapes[3].

Most of the discrete defects take the form of delamination and large voids as well as microscopic flaws such as interface debonding, cracks and porosity within the matrix. These flaws combined with various damage modes including environmental degradation and dynamic mechanical loading may initiate mechanical failures. In order to encourage the expanded application of ceramic-matrix composites, the use of suitable non destructive evaluation (NDE) techniques is critical to effective process control and reliable operation in service.

This paper details an extensive investigation carried out on oxide-oxide CMC materials, utilising ultrasonic immersion testing, pulsed thermography, 2D digital radiography, 3D computed tomography and resonance testing, to characterise internal structure and recognize the presence of deliberately embedded artifacts and surface damage.

EXPERIMENTAL PROCEDURE AND RESULTS

Figure 1 is a schematic representation of a CMC panel that was manufactured with embedded artifacts to assess the detection capabilities of various NDE techniques. The 200mm x 200mm panel was manufactured with twelve 0-90° plies of mullite fibres within an alumina matrix with a total thickness at 3.25mm. PTFE tape was cut to specific sizes and inserted into the layup during manufacturing at different depths to simulate delaminations. The PTFE tape is burnt out during sintering to produce a void of a given thickness. The sizes of the model defects were: $8mm^2$, $6mm^2$, $3mm^2$, $2mm^2$ and $1mm^2$. The spatial distribution of the defects and their respective depth are illustrated in Figure 1.

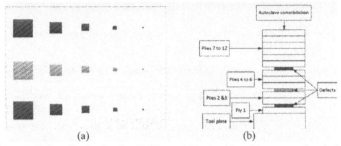

(a)   (b)

Figure 1. Defect distribution for CMC panel (a) plan view (b) cross sectional view

Ultrasonic Immersion Testing

In the ultrasonic immersion technique, the CMC panel containing the artificially induced defects was submerged into a 3-axis, X-Y scanning immersion tank (General Electric, KC-250) that can operate at frequencies up to 25Mhz. A double through transmission technique was used to scan the sample. The ultrasonic wave was propagated through the CMC sample and then reflected from a glass plate positioned below. The ultrasonic transducer had a centre frequency of 2.25MHz. A schematic diagram of the immersion set up and a photograph of the experimental apparatus are shown in Figure 2.

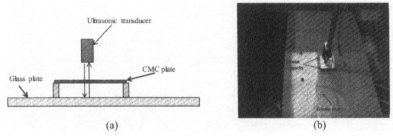

(a)   (b)

Figure 2. Ultrasonic immersion testing of CMC panel (a) schematic diagram (b) experimental set up

Figure 3 illustrates a typical reflected ultrasonic signal together with the monitoring gates. There are two ultrasonic reflected signals: the initial reflection from the top surface of the panel and reflection from the glass plate. Three monitoring gates were defined to measure the amplitude of the reflected signals: CMC top surface, CMC material volume and glass plate top surface.

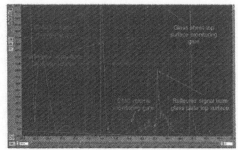

Figure 3. A-scan display of the reflected signals and inspection monitoring gates

Figure 4 shows the amplitude C-Scan result from the CMC panel inspected from the autoclave consolidation side (which results in a relatively rough surface profile compared to the tool side). The image was reconstructed using the amplitude response from the gate monitoring the reflected ultrasonic wave from the top surface of the glass plate after the wave had propagated through the CMC. The amplitude data demonstrates the intensity of the reflected signal with respect to Figure 3:

- Red corresponds to higher reflected signal amplitude, i.e. lower attenuation
- Blue/black corresponds to lower reflected signal amplitude, i.e. higher attenuation

In Figure 4 it is apparent that the PTFE tape induced void, embedded in the material to simulate delaminations, induces high levels of attenuation and therefore lower reflected amplitude. The majority of the embedded defects were detected. Furthermore, bulk density variations were defined across the area of the panel.

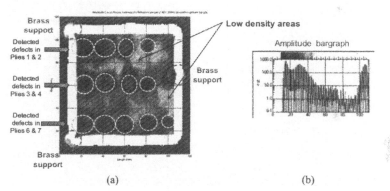

(a)   (b)

Figure 4. Ultrasonic immersion results from CMC panel inspected from autoclave consolidation side (a) amplitude C-scan image (b) amplitude bar graph.

Pulsed Thermography Testing

A popular technique used to detect defects and delaminations in organic composite materials is pulsed thermography, employing a thermal excitation source to rapidly heat the surface of the material. A highly sensitive infra-red camera then monitors the surface temperature during cooling. Any sub-surface flaws will affect heat transfer into the material resulting in temperature variations at the surface that change with time. The system used two flash lamps, mounted within a hood along with a Stirling Engine cooled medium wave infrared camera providing a 300mm square field of view. The flash lamps were triggered and the camera recorded a series of thermal images of the surface. The data were subsequently processed to facilitate review with a suite of analysis tools. Figure 5 shows a typical pulsed thermography inspection set-up.

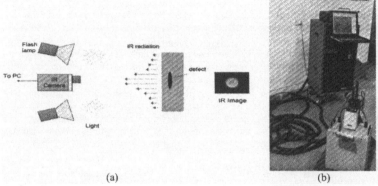

(a)   (b)

Figure 5. Pulsed thermography testing (a) schematic diagram (b) experimental set up

The optimised parameters applied to the CMC panel were 30Hz flash frequency and 45 seconds exposure time. A number of tools provided by the commercial thermography software (Mosaic) were used to analyze the data. Initially, the CMC sample was inspected from the autoclave consolidation side that has a rougher and relatively uneven surface profile compared to the tool side. Some defects were identified ($8mm^2$, $6mm^2$ and $4mm^2$) between plies 6 and 7 when inspecting from this orientation due to surface condition and depth of defects that are located between plies 6 to 12. Subsequently, the same parameters were applied while inspecting from the tool side with a smoother surface and the defects are located between plies 1 to 7. The thermographic results from the sample inspected from the tool side can be seen in Figure 6. As the defects were positioned at different depths in the panel, the thermography data are presented from different time frames.

The rate of dissipation through the material was captured by the infrared camera and analysed by the software. Figure 6 demonstrates that all fifteen simulated defects were detectable. Therefore, the pulsed thermography technique can detect delaminations of 0.03mm thickness at a depth of 1.6mm within the laminate from the tool plate face.

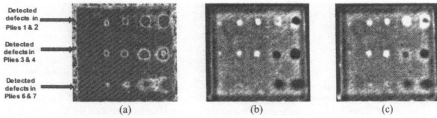

Figure 6. Pulsed thermography data sampled when inspected from tool surface at different stages of exposure (a) 0.77 sec (b) 2.94 sec (c) 3.5 sec.

2D Digital Radiography

Various radiographic parameters were assessed under 2D digital radiographic inspection to optimise sensitivity according to the thickness of material. Relatively low kV settings improved the contrast sensitivity and increasing the frames per image reduced noise thus improving the signal to noise ratio (SNR). Computed tomography (CT) inspection cannot be optimised in the same way due to a number of limitations. Firstly, CT reconstruction requires full X-ray penetration of the component throughout 360° of rotation. To ensure the X-ray beam penetrates the longest path requires high kV settings and reduces the sensitivity achieved. In addition, the number of frames integrated is normally reduced (compared to 2D inspection) to increase the scanning speed.

The CMC panel with artificial PTFE induced defects was subjected to digital radiography, sampling the central region of interest only. The optimised inspection parameters, making it possible to define the internal fibre architecture within the matrix, are provided in Table I.

Table I. 2D Radiography Inspection Settings

Parameter	Settings
kV	50
uA	75
Source to detector distance	1060mm
Source to sample distance	750mm
Filter	None
Exposure time	40millisecs
Frames per image	128

Despite enabling the internal fibre architecture to be resolved, X-ray radiography failed to identify any of the embedded defects. This is due to the limitations of conventional radiography for detecting planar defects, i.e. if the x-ray beam is orientated perpendicular to a planar feature (such as a delamination) then no overall density or thickness change exists in the material. The CMC panel was subsequently machined into three rectilinear sections to allow computed tomography inspection.

3D X-Ray Computed Tomography

Computed tomography is a radiographic based imaging method that produces 2D cross-sectional and 3D volumetric images of the object under inspection. Characteristics of the internal structure of an object, such as dimensions, shape, internal defects, and density, are readily available from CT images. Figure 7 illustrates the concept of a simple CT scanner. Here, an X-ray source generates an X-ray beam that propagates through the object under inspection. The resulting intensity of the X-ray beam is measured using an X-ray detector. This process is repeated at many angles to obtain a stack of 2D X-ray projections. A filtered back projection (FBP) reconstruction algorithm is then applied to the data to compute the attenuation coefficients of the material, either as a 3D volume or a cross sectional slice. Finally, the attenuation coefficients are assigned a grey value which relates directly to the material density.

The micro-focus X-ray system used for the experiments was capable of resolving detail down to approximately 15 microns, with geometric magnifications up to 160X. The system incorporates a 5-axis manipulator, and offers real-time radiography fed back to a charge coupled device (CCD) camera coupled with an image intensifier.

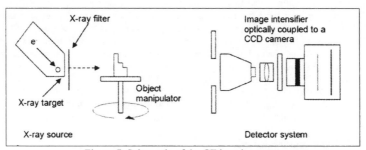

Figure 7. Schematic of the CT imaging setup

Measures were implemented to obtain sufficient image quality for the identification of defects on a micro scale within the sample material. The first measure involved stacking the samples together to reduce the aspect ratio. This allowed the penetrated material path lengths to remain similar for all projection angles. The result was an improvement in image quality by reducing noise levels caused by over exposed regions and image artefacts such as beam hardening. The second measure was to inspect the samples in sections at a higher magnification. The use of higher magnifications with a micro-focus source increased the resulting spatial resolution by both improving the geometric sharpness and increasing the number of detector elements that the object was projected over.

After sectioning, the CMC panel was subdivided into three strips, each with five of the different sized square artificial PTFE film defects embedded within. The film defects are characterised as: Type 1 – located between plies 1 and 2, Type 2 – located between plies 3 and 4 and Type 3 – located between plies 6 and 7. In each case, a total of 2 CT scans were completed, each 52mm in length with overlap between the top and bottom sections. The key CT inspection parameters are provided in Table II.

Table II. CT inspection parameters

Parameter	Settings
kV	100
uA	130
Focal spot	5 Micron
Detector type	CCD with 12-bit image intensifier
Target material	Tungsten (W)
Number of projections (x-ray images)	1440
Source to detector distance	1060mm
Source to sample distance	395mm
Rotation angle step	0.25 degrees
Voxel resolution	0.0432mm
Filter	1mm steel
Reconstruction algorithm	Filter back projection
Frames per image	32

Figure 8 is provided for guidance on the sample stacking geometry within the reconstructed CT images.

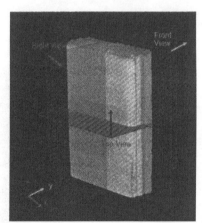

Figure 8. 3D view of the top section, slice orientation of sample illustrating cross sectional viewing planes (right view x, front view y, top view z).

Figure 9 displays the three sections of the CMC panel in a top view perspective. The 3mm^2 square PTFE defect is visible in each of the specimens, and this was the case for the 4 other defects when observed in a top view aspect (the largest PTFE film defect measuring 8mm^2 square was the most evident in the three specimens, and this is shown in Figure 10). This was also the case when the 15 different sized defects are observed in a front view (example given in Figure 11a) and a right view aspect (Figure 11b).

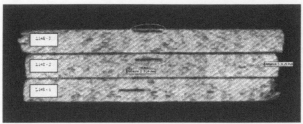

Figure 9. Top section - top view perspective. The mid-sized defect (3mm^2 square) in each of the three samples is highlighted.

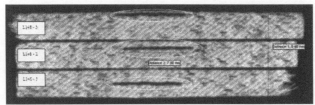

Figure 10 Bottom section - top view perspective. The largest defect (8mm^2 square) in each of the three samples is highlighted.

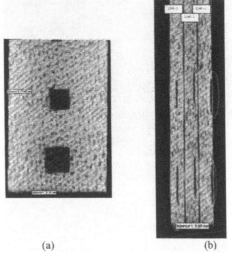

(a)   (b)

Figure 11. (a) Bottom section - front view (b) Bottom section - right view. The 6mm^2 and 8mm^2 PTFE defects in each of the three samples are highlighted.

The CT images demonstrate that the PTFE type induced defects were of a significantly lower density than the surrounding material, and the method clearly resolved the artefacts of varying size ($1mm^2 - 8mm^2$). The PTFE type induced defects resulted in delaminations with thickness of 0.18mm. However, the CT radiographic technique was not able to detect delaminations of 0.03mm in thickness in CMC samples with thinner induced defect material (i.e. release film).

Resonance Testing

Resonance inspection techniques are also widely adopted for the non-destructive testing of organic composite materials. A probe generates low frequency acoustic energy (typically 100 - 300 KHz) into the body of the composite material. The acoustic energy causes the target material to resonate and the acoustic impedance of the material is detected. This characteristic makes this technique ideal for thin samples and in particular multilayer composite materials. The electrical impedance of a resonance transducer is coupled to the acoustic impedance of the sample through a small amount of couplant. The transducer frequency is selected to be most sensitive to small inconsistencies in the density of material beneath the transducer. As the acoustic impedance of a composite is altered by defects such as delaminations, this form is easily detected by the resultant change in electrical impedance. Changes in material properties due to other defects such as inclusions can also be detected but the signal response will be different in both phase and magnitude.

Resonance mode inspection relies on a comparison between "good and bad" materials as the output is qualitative rather than quantitative. The challenge in this approach was to adjust the sensitivity such that spurious signals did not result from the operator's handling of the transducer. The technique typically involves nulling of the equipment such that the signal response lies at the cross within a target area on the output screen (Figure 12a). This is performed while the transducer is placed over a proven good area of the sample. The transducer is then lifted off the sample to provide a "lift-off" signal. This signal is adjusted to give an output at the far right hand side of the display grid and on the horizontal cross hair (Figure 12b).

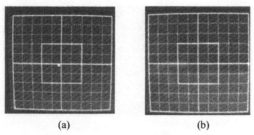

(a)   (b)

Figure 12. Resonance equipment calibration
(a) Nulling of transducer (b) Lift off signal from transducer

The next stage involved the inspection of various types of defects encountered in the material of interest. For each defect type a signal response was recorded on a second display grid. Figure 13a represents the response from one of the small PTFE defects embedded in the original CMC panel. During the inspection the signal response was maintained within an alarm zone represented by a box at the display grid centre (Figure 13b). Any response that moved towards the signal represented in Figure 13a would trigger an alarm to indicate the defect. The specific position and magnitude of the signal enabled a comparison of the defect with those detected in known samples.

The inspection of the CMC panel was performed by a manual raster scan across the sample surface. All the embedded defects indicated a change in signal response. However, due to the trade-off to maintain the signal within the alarm area during probe movement, although the smallest artificial defects of 1x1mm were observed they did not trigger an alarm. All other defects were detected and the response from different depths could be ascertained. As an example, the signal response from a 3mm x 3mm defect, which was located 1.625mm deep (6 plies) is shown in Figure 13c, can be differentiated from a 3mm x 3mm defect which was located 0.81mm deep (3 plies) as shown in Figure 13d. The difference in depth is represented by a difference in signal magnitude and a variation in the phase of the signal (the rotational position on the display grid).

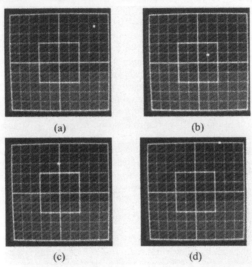

(a)   (b)

(c)   (d)

Figure 13. Inspection resonance results from (a) known 2mm x 2mm defect (b) defect free material (c) 3mm x 3mm defect at 1.625mm depth (d) 3mm x 3mm defect at 0.81mm depth

Resonance testing was proven to be a viable method of detecting artifacts of size $2mm^2$ to $8mm^2$ within this CMC at depths down to 1.62mm. This could be a useful low cost technique when applied to a manufacturing application, providing rapid results comparing produced CMCs to a known 'good' example. If a critical artifact threshold was determined, a 'go-no-go' manufacturing procedure could be implemented.

CONCLUSIONS

Ultrasonic immersion proved an effective and sensitive technique for the detection of delamination style defects and density variations in oxide-oxide CMC materials. A low frequency transducer and double through transmission utilising a reflective glass plate provided the optimum resolution of the internal features of interest.

Pulsed thermography also proved to be a rapid and effective method capable of detecting all of the different scale artificial delaminations. However, this technique was more effective when the sample was inspected from the tool side that offers a relatively smooth and consistent surface.

Furthermore, the artificially induced defects are located deeper when the sample is inspected from the autoclave consolidation side.

CT inspection offers the greatest resolution and through advanced imaging software provides the inspector with detailed, high precision three dimensional detail. However, the relatively high cost and more limited equipment availability may restrict this inspection technique to high performance applications. The 2D digital radiographs were not able to detect delamination type defects due to the limitations of conventional radiography for detecting planar defects.

Resonant mode inspection may be favoured as a simple approach to investigate the internal structure of ceramic matrix composites. The technique requires small amounts of couplant and results in clear differentiation in signal responses from both defect depth and defect size. The inspection trial resulted in all but the smallest artificial defects being detected and successfully interpreted. However, it is difficult to differentiate and characterize the defect type using this technique.

ACKNOWLEDGEMENTS

The current research has been conducted under Technology Strategy Board / Engineering and Physical Science Research Council, contract TS/G000484/1. Umeco (formerly Advanced Composites Group Ltd.) and Birmingham University are acknowledged for their joint role in processing the current CMC materials.

REFERENCES

[1] A. G. Evans, and R. Naslain, High-Temperature Ceramic-Matrix Composites I: Design, Durability and Performance, *Ceram. Trans.*, **57**, (1995a).

[2] A. G. Evans, and R. Naslain, High-Temperature Ceramic-Matrix Composites II: Manufacturing and Materials Development, *Ceram. Trans.*, **58**, (1995b).

[3] J. G. Sun, C. M. Deemer, W. A. Ellingson, and J. Wheeler, NDE Techologies for Ceramic Matrix Composites: Oxide and Non-Oxide, Material Evaluation, **64**, 52-60 (2006).

[3] J. Stuckey, J. G. Sun, and W. A. Ellingson, Rapid Characterization of Thermal Diffusivity in Continuous Fiber Ceramic Composite Components, Nondestructive Characterization of Materials VIII , 805-810, (1998).

[4] J. G. Sun, D. R. Petrak, T. A. K. Pillai, C. Deemer, and W. A. Ellingson, Nondestructive Evaluation and Characterization of Damage and Repair for Continuous Fiber Ceramic Composite Panels, *Ceram. Eng. Sci. Proc.*, **19 [3]**, 615-622 (1998b).

[5] J. G. Sun, Analysis of Quantitative Measurement of Defect by Pulsed Thermal Imaging, Review of Quantitative Nondestructive Evaluation, **21**, 572-576 (2001).

[6] J. G. Sun, S. Erdman, and L. Connolly, Measurement of Delamination Size and Depth in Ceramic Matrix Composites Using Pulsed Thermal Imaging, *Ceram. Eng. Sci. Proc.,* **24 [4]** , 201-206 (2003).

[7] W. A. Ellingson, and C. Deemer, Nondestructive Evaluation Methods for CMC (defect characterization), Department of Defense Handbook – Ceramic Matrix Composite, MIL-17-5, 100-108 (2002).

[8] M. F. Zawrah and M. El-Gazery, Mechanical Properties of SiC Ceramics by Ultrasonic Non-Destructive Technique and its Bioactivity, **106**, 330-337 (2007).

[9] K. Jeongguk, and P. K. Liaw, The Nondestructive Evaluation of Advanced Ceramic-Matrix Composites, Journal of Minerals, **50**, 1-13 (1998)

Wear, Chipping, and Fatigue of Ceramics and Composites

EDGE CHIP FRACTURE RESISTANCE OF DENTAL MATERIALS

Janet B. Quinn, George D. Quinn, and Kathleen M. Hoffman
Paffenbarger Research Center
American Dental Association Foundation
Stop 854-6, NIST
Gaithersburg, MD 29899

ABSTRACT
   The edge chipping test was used to measure the fracture resistance of twenty-four materials including many dental restoration ceramics. Materials tested included feldspathic and leucite porcelains, aluminas, yttria-stabilized zirconias, indirect filled resin-matrix composites, a filled and an unfilled resin denture material, glass, glass ceramics, and human dentin. The effects of variations in some of the experimental procedures were explored.

INTRODUCTION
   The edge chipping test is used to evaluate the resistance of brittle materials to flaking near an edge as shown in Figure 1.[1-22] Edge chip resistance may be correlated to fracture toughness for many materials, but the fracture processes are not the same. Chips are formed by advancing an indenter or stylus into a material near an edge. The force required for chip formation is recorded as a function of the distance from the edge. The greater the load application point distance from the edge, the greater is the force that is needed to create the chip. The shape of the chip is relatively independent of the material tested. For many glasses and ceramics, the force – distance trend is linear and follows equation 1:

$$F = T_e d \qquad (1)$$

where F is the chipping force and d is the distance from the edge. The slope of the fitted line (usually in terms of force in Newtons per millimeter distance) describes the susceptibility of the material to edge chipping and has been defined as the *edge toughness, T_e* (N/mm) or the *edge chip resistance, R_e.*[23] Dimensional analysis shows that these equivalent to N·mm /mm^2, or an energy per unit area. The slope of the line which goes through the origin (force and distance = 0) also may be interpreted as the force necessary to create a chip at a distance of 1 mm. Sometimes a linear trend analysis is used, but with a non-zero intercept. On the other hand, there are some examples where a power law function seems to fit better:

$$F = A d^n \qquad (2)$$

where n is a constant, usually from 1 to 2. A is a constant, and may be interpreted as the force to create a chip a distance 1 mm away from an edge.
   Originally developed to study hard metals at the National Physical Laboratory in London in the 1980s,[1-4] the edge chip method has been adopted by other groups to study a range of materials.[5-22] Chipping is a common failure mode in many brittle material applications, especially dental restorations and some of the fractures physically resemble *in-vivo* failures. The edge chip test has been used with dental restorative materials in several studies.[5,7,8,10,17-20] It even has been used to evaluate the chip resistance of human dentin.[21] A preliminary test method standard for technical ceramics is currently under review by the European community.[23] Unfortunately, divergences in technique, analysis, and reporting of results have emerged. For example, sometimes the chip resistance ("edge strength") is

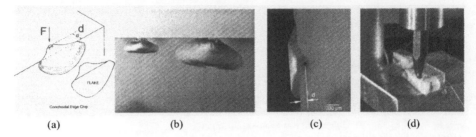

(a)	(b)	(c)	(d)

Figure 1. Edge chips. (a) shows a schematic of the edge chip process. The indenter applies an increasing force F at a distance d away from the edge until a piece spalls off. (b) shows a side view of chips in glass. (c) shows a top view of a chip in human dentin. (d) shows a close-up of the edge chipping machine with the indenter in contact with a human tooth. (c) and (d) from ref. 21.

expressed as a <u>force</u> in N derived from the linear trend, but arbitrarily normalized to a distance of 0.5 mm.[19,20] Indenters have included the original Rockwell C diamond indenters,[1-4] Knoop and Vickers pyramidal indenters, or the sharp-tip diamond 120° conical indenters that we favor. The latter is nearly identical to a Rockwell C indenter. Both have a 120° conical shape, but the Rockwell C indenter has a rounded tip with a 200 μm radius.

Our current sharp conical indenters have a tip radius less than 5 μm. Many people use the Rockwell, Vickers or Knoop diamond indenters since they are readily available. The drawback is that they are expensive and can wear out or break. The sharp 120° conical indenters we use are obtained from simple diamond scribers. They are durable, inexpensive, and expendable. This indenter also has the advantage that the indenter footprint geometry is self-similar irrespective of the force applied or the depth of indentation. This is unlike the rounded-tip Rockwell C indenter, whereby at low loads only the relatively blunt tip is engaged. The contact geometry changes to include the rounded radiused sides and even the conical straight sides as loads increase for greater edge distances. Another variation in edge chipping testing involves the edge distance measurement. One group[13] has advocated measuring distance not from the point of load application, but from the distance of the furthest most damage of a chip scar. This approach seems unwise to us. All the analytical fracture mechanics models to date are based on the point of load application. The furthest most damage distance depends upon the extent of indenter penetration and may change drastically with depth of penetration with a rounded tip Rockwell indenter. "Overchipping," described later, is also a problem.

Most edge chipping studies have applied force perpendicular to the edge, but Quinn and Ram Mohan[9] studied the effect of varying the indentation angle onto pieces with perpendicular edges. Danzer et al.[11,12] explored the effect of testing specimens with non-perpendicular edges.

The work presented herein presents a summary of most of our edge chipping results from 1998 to now. It includes data from a new study to evaluate the edge chip resistance of dental restorative materials. Chipping is a common failure mode for dental restorations. The experimental procedures and analyses have evolved somewhat since the method was first applied by one of us (JBQ) in the late 1990s to dental materials. Our experimental procedures become more refined. We have investigated the effects of using alternative indenter geometries, expanded the range of materials evaluated, and compared power law versus linear fits to the force - distance data.

MATERIALS

Over a dozen dental materials including leucite and feldspathic dental porcelains, a glass ceramic, a glass-infused alumina, and a 3Y-TZP (3 mol % yttria-stabilized tetragonal zirconia polycrystal) were evaluated previously. Details and full characterization of these materials are in the original reference that is the PhD thesis[5] by one of us (JBQ). Excerpts from this work were published only in a few conference proceedings and short papers.[7,8,10] Therefore, for the reader's convenience, these early results are summarized here for comparison to our newer results.

We have recently evaluated a newer commercial dental grade 3Y-TZP (Lava, 3M-ESPE)[a] with a grain size of 0.5 μm. The material is typically fabricated in the form of milling blanks for CAD/CAM machining. Bend bars (3 mm x 4 mm x 45 mm) from an earlier study were tested.

New experiments have been done with 3 mm x 4 mm x 28 mm bend bars of the feldspathic dental porcelain (Mark II, Vita). A detailed characterization of this material, its microstructure, fracture origins, and fracture toughness is in Ref. 24. This fine-grained, relatively strong porcelain chips very consistently and we are starting to use it as a quasi-reference material in our work.

A commercial indirect filled resin-composite was also evaluated. It was received in the form of conventional milling block blanks (Paradigm, 3M-ESPE). The company literature listed the material as containing 0.85 mass fraction spherical sol gel derived particles comprising nanocrystalline zirconia in silica. The spherical particles averaged 0.6 micrometers in size, but had a broad size distribution. They are in a highly cross-linked polymeric matrix of bis-GMA and TEGDMA. Additional details on this material, including hardness, flexural strength, fracture toughness measurements, and comprehensive fractography of strength limiting flaws are in Ref. 18.

Chipping of porcelain veneers applied to zirconia or metal crowns is sometimes a clinical problem. To evaluate the sensitivity to edge chipping, layered specimens for edge chipping experiments were made by laminating metal alloy of 0.65 mass fraction gold and 0.26 mass fraction palladium (Ultracrown SF, Dentsply, Ceramco) with a feldspathic veneer porcelain (Ceramco 3) with 0.30 volume fraction leucite. 3Y-TZP zirconia substrates (Cercon) were veneered with a feldspathic porcelain (Ceramco PFZ) with no leucite. Additional details are in Ref. 17.

Twelve human teeth were tested in a study of the effects of calcium hydroxide endodontic treatments on the fracture resistance of human dentin.[21] There were indications in the literature

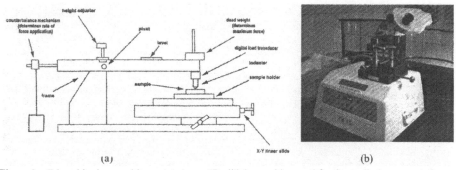

(a) (b)

Figure 2. Edge chipping machines. (a) shows "Rudi" the machine used for the preliminary work from 1998 to 2000. (b) shows the CK10 machine used for work after 2002.

---

[a] Commercial products and equipment are identified only to specify adequately experimental procedures and does not imply endorsement by the authors, institutions or organizations supporting this work, nor does it imply that they are necessarily the best for the purpose.

that the calcium hydroxide treatments makes the teeth more susceptible to fracture. Four subsets of extracted teeth with different treatments including a control untreated set were created, with each subset including one incisor, one premolar, and one molar. A pulpectomy was performed on each tooth and the canal sealed. The teeth were mounted in polymethylmethacrylate (PMMA) resin and then sectioned and polished to expose dentin near the canal. Additional details are in Ref. 21.

Our most recent work has been with a traditional denture material (Blueline DCL, Ivoclar-Vivadent, Liechtenstein). This denture material is a cross-linked PMMA. In an effort to improve the wear resistance and long-term aesthetics, the manufacturer has introduced alternative filled resin-matrix composites with micro- and nano-sized inorganic fillers. The manufacturer provided us with several grades of these composites in the form of rectangular 13 mm x 8 mm x 3 mm wear test type blocks. It was not clear whether the edge chip test would work on such softer denture materials.

EXPERIMENTAL PROCEDURE

In the preliminary 1998-2000 work by one of us (JBQ),[5,7,8,10] an edge chipping machine shown schematically in Figure 2a was constructed consisting of a rigid base and movable machinist X-Y stage that held test specimen pieces such as flat plates and bend bars. A lever beam with weights applied force to the beam and the indenter. The indenters were 120° conical indenters with tip radii between 10 μm and 50 μm radius. A load cell monitored the force application rates and the peak load at fracture to within (± 0.05 N). Force was gradually applied at approximately 1N/s. For most materials, when fracture occurred an audible ping was heard and the chip flew off the side of the specimen. Care was taken to ensure that chips were made far enough from each other so as to not interact. Edge distances were set approximately before each chip was made. The chip distances could not be precisely located with this apparatus since there was no microscope on the machine. Hence, edge distances were measured after the chips were made by inspecting the top specimen surface (see Figures 1a and 3a) with a stereoptical microscope with a precision X-Y stage with a resolution of 1 μm. Indentation divots and starburst cracks helped mark the indention site. Repeated measurements indicated the accuracy of the edge chip distances were about ± 5 μm in difficult cases. Additional details are in References 5 and 7.

In our work after 2002, we used a commercial edge chipping machine (Engineering Systems Model CK 10, Nottingham, UK) shown in Figures 2b and 1d. The 1000 N load cell readout resolution was 0.1 N and the load cell was certified to be accurate to better than 0.1 N. Load was gradually applied (displacement control, 1 mm/min for most materials, but 3 mm/min for the denture resin materials) until the chip fractured off the specimen. The peak load was recorded. We initially used the same testing protocol as described above: chips were made in the testing machine, but distances were measured afterwards in a stereoptical microscope. We changed this procedure in 2009. In addition to having a micropositioning X-Y stage with 0.01 mm resolution for the specimen, the CK10 apparatus had a moveable head with the diamond indenter and a locating microscope. This is like many common indentation hardness type machines. Hence, we could set the edge distance to a prescribed value prior to a test. This eliminated the need to interpret and measure distances with the stereoptical microscope afterwards. We continued to keep the chips far away from each other, but gradually we began to allow them to be closer spaced (still trying to eliminate overlap) in order to get more chips and data points per specimen. Some of our early (1998 – 2000) force - distance curves were based on as few as six chips, but we came to realize that more chips are better. The uncertainties in the edge toughness, T_e, estimates, which are based on the slope of the line through the data, are improved when more data is collected. We now typically make twenty to thirty-five chips.

We continued to use sharp 120° conical diamonds, but adopted tighter controls on the indenter angles and the tip radii. The indenters were fashioned from simple, inexpensive ($25 in 2012) hand scribers used to mark the surface of hard materials (model HS301-120, Gilmore Diamond Tools,

Attleboro, MA, USA). The aluminum handles are cut off and the diamond scribe end inserted into the edge chipping machine as shown in Figure 1d. Tip radii are 5 µm or less and included angles are within 1 degree of 120°. These are inspected frequently for tip damage and replaced when necessary. A limited number of experiments were done with Knoop and Rockwell C indenters on the dental resin materials. One series of experiments with sharp conical scribe indenters with angles of 90°, 100°, 120°, and 140° were done with the feldspathic porcelain. The CK10 machine had a break detection circuit that detected sudden drop off in force when a chip formed. This circuitry was reliable for most of the brittle ceramics, but not adequate with the softer resins. In the latter cases, a chip formed but did not always fly off. There was not a sudden load drop off. In such cases, the formation of a chip was observed by eye and the machine stopped manually. A bright light was shone on the polished specimen side surface to help detect when the chip popped off. It is essential to rigidly mount test pieces; otherwise they might tilt during loading or get nudged sideways by the indenter after a chip popped off. Testing a specimen with no fixative, or alternatively, held by double-sided sticky tape was unsatisfactory. Specimens are now waxed to a mounting plate.

In our early work, edge distances from as small as 0.05 mm to as large as 0.30 mm were used. Our dental material specimens were often small and we could not get many large chips. As our work progressed in the late 2000 s, it became apparent that chips at some of the shorter distances were difficult to obtain and measure, and the uncertainties expressed as a percent were larger. It also became evident that some of the force-distance data trends at the very short distances (0 mm to 0.10 mm) were slightly different than for the larger distances. This led to interpretations of slight nonlinearities in the force - displacement curves. For some of the materials, the nonlinearities disappeared once data was collected over a larger range of edge distances. Data is now collected routinely from 0.10 mm to 0.50 mm, which requires peak forces of 100 N to 400 N, depending upon the material.

Test specimens varied in size and shape from 3 mm x 4 mm cross section sized bend bar fragments to square blocks of 10 mm to 20 mm length per side. Specimens were at least several millimeters thick. The top indented surfaces and the side surfaces were ground and polished. Specimen preparation procedures varied with the materials, which ranged from hard ceramics such as alumina, zirconia, and porcelain, to glasses, to soft human teeth, to resinous materials and their filled composites for denture materials. In the preliminary work, the hard ceramic specimens were polished flat and parallel with diamond suspensions of grit sizes 45, 30, 15, 6, 3, 1 and 0.25 µm, respectively. Glasses and the resins were ground with progressively finer carbide papers from 300 to 4000 grit.

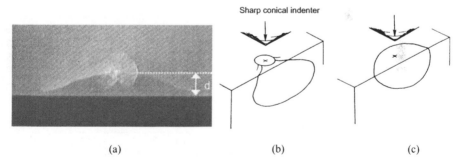

Sharp conical indenter

(a)   (b)   (c)

Figure 3. Illustrations of properly-formed edge chips (a) and (b). (a) is a top view of a chip in a leucite porcelain interrupted just prior to final pop off. It shows the starburst pattern of radial cracks around the indention site divot and the two dominant cracks that form the chip. (c) shows an overchip.

Table 1. Edge chip results.

Material	Machine[X]	Indenter[XX]	Distances (mm)	n†	Te, linear eq.1 (N/mm)[Y]	Power law eq. 2	Refs.
Glass, soda lime	CK10	120° sc	0 – 0.20	12	206 + 13	-	9
MGC[Z]　fine grain size	Rudi	120° sc	0 – 0.50	27	93 ± 5 [NZ]*	$F = 100\ d^{1.32}$	5,7,10
MGC[Z]　medium grain size	Rudi	120° sc	0 – 0.50	34	106 ± 4 [NZ]*	$F = 130\ d^{1.29}$	5,7,10
MGC[Z]　coarse grain size	Rudi	120° sc	0 – 0.50	23	131 ± 6 [NZ]*	$F = 158\ d^{1.28}$	5,7,10
Porcelain (Ceramco silver body)	Rudi	120° sc	0 – 0.35	10	111 ± 16[NZ]	-	5
Porcelain (Ceramco silver incisal)	Rudi	120° sc	0 – 0.30	15	129 ± 12 [NZ]	-	5
Porcelain Enamel (Ceramco Finesse)	Rudi	120° sc	0 – 0.30	19	94 ± 8 [NZ]	-	5
Porcelain (Ivoclar Empress I)	Rudi	120° sc	0 – 0.35	14	108 ± 6 [NZ]	-	5
Glass ceramic (Ivoclar Empress II)	Rudi	120° sc	0 – 0.35	10	188 ± 11 [NZ]	-	5
Infused alumina (Vita Inceram)	. Rudi	120° sc	0 – 0.20	10	273 ± 20 [NZ]	$F = 448\ d^{1.31}$	5,7
Sintered 99.9% alumina (Coors AD999)	Rudi	120° sc	0 – 0.20	6	296 ± 9	-	7,10
Silicon nitride (Norton, NC 132)	Rudi	120° sc	?	?	590 ± 20	-	10
Jasper (silicate mineral)	Rudi	120° sc	0 – 0.30	13	194 ± ? [NZ]	-	6
3Y-TZP　(St. Gobain, Prozyr)	Rudi	120° sc	0 – 0.10	9	534 ± 19 [NZ]	$F = 1430\ d^{1.42}$	5,7,10
3Y-TZP　(Dentsply, Cercon)	CK10	120° sc	0 – 0.10	21	414 ± 33	-	unpubl.
3Y-TZP　(3M-ESPE, Lava)	CK10	120° sc	0 – 0.50	25	601 ± 21	-	new
Leucite Porcelain (Vita Fortress)	CK10	120° sc	0 – 0.35	15	114.9 ± 11.6	$F = 220\ d^{1.47}$	unpubl.
Feldspathic porcelain (Vita Mark II)	Rudi	120° sc	0 – 0.45	7	87.7 ± 2.1 [NZ] 77.2 ± 4.3	$F = 106\ d^{1.28}$	5,7,10 unpubl.
Feldspathic porcelain (Vita Mark II)	CK10	120° sc	0 – 0.50	26	137.5 ± 5.3	$F = 168\ d^{1.20}$	new
Porcelain veneer for metal crowns	CK10	120° sc	0 – 0.55	29	-	$F = 440\ d^{2.00}$	17
Porcelain veneer for zirconia crowns	CK10	120° sc	0 – 0.55	27	-	$F = 407\ d^{2.00}$	17
Filled Resin composite (3M-ESPE, Paradigm)	CK10	120° sc	0 – 0.28	15	119.7 ± 13.0 171.5[NZ]	$F = 366\ d^{1.72}$	18 18
Filled Resin composite (3M-ESPE Paradigm)	CK10	120° sc	0 – 0.30	22	124.7± 14.2	$F = 433\ d^{1.90}$	new
PMMA denture material (Ivoclar, Blueline DCL)	CK10	120° sc Rockwell C Knoop	0 – 0.50 - 0 – 0.50	18 - 32	475 ± 44 did not chip 642 ± 108	- - -	new
Composite denture material (Ivoclar, Phonares I)	CK10	120° sc Rockwell C Knoop	0 – 0.50 0 – 0.50 0 – 0.50	25 17 32	249 ± 38 412 ± 35 554 ± 63	-	new
Human dentin, untreated	CK10	120° sc	0 – 0.50	7	219 ± 21	-	21
Human dentin, Ca(OH)₂ treated	CK10	120° sc	0 – 0.50	12	283 ± 20	-	21

[X]　Rudi denotes the self-made device shown in Figure 2a;
　　CK10 denotes the Nottingham CK10 edge chipper shown in Fig. 2b
[XX]　120° sc is a 120° sharp conical indenter.
*　Best linear fit, for the data above a force of 6 N

†　number of chips
[Y]　Uncertainties are ± standard deviation
[Z]　MGC = Machinable Glass Ceramic
[NZ]　Non zero intercept

The human teeth were sequentially polished with 240, 320, 450, and 600 grit silicon carbide papers. Side walls were carefully ground perpendicular to the top surface that was to be indented. Additional details are in Ref. 21.

One experimental problem that we encountered had to do with the detection of chip formation. In many of the brittle materials, chip formation was abrupt and easily detected. Interpretation of the event was unequivocal. In a few materials, however, chip formation was not so distinct. This was a problem with softer materials, with those that did not have a lot of stored elastic energy at the onset of critical fracture, or with those with rising R-curve behavior (i.e., increasing fracture toughness with increasing crack size). In such cases, the chip would form gradually and suddenly pop most of the way through, but not entirely off. If there was no sudden force drop off, the testing machine continued to press the indenter into the contact site causing additional cone-like cracks to form on the back side of the contact impression. These cracks then ran out to the side surface and created what we termed an "overchip" as shown in Figure 3c. Consequently, all chips are now inspected immediately after formation and overchips are rejected.

RESULTS

Table 1 shows the new and previously published results. Some of the specific force – distance sets data are shown and discussed below with an emphasis on the newer work. Edge toughness, T_e, was calculated from the slope of a regression fit of the force – distance data in accordance with eq. 3. The standard deviation of T_e was calculated from:[25]

$$s.\,dev.\,T_e \; = \; \left(\frac{residual\ mean\ square}{regression\ sum\ of\ squares} \right)^{1/2} T_e \; = \; \left(\frac{1 - R^2}{(n-2)\ R^2} \right)^{1/2} T_e \tag{3}$$

where n is the number of chips and R^2 is the correlation coefficient for the regression fit with zero intercept.

Figure 4 shows a comparison of early results[5,7] for a hip joint grade 3Y-TZP (Prozyr, St. Gobain), a dental glass-infused alumina (Inceram, Vita), a sintered 99.9% alumina (AD-999, Coors),

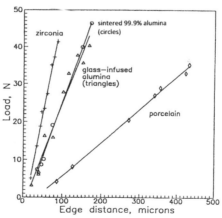

Figure 4. Data for a hip joint grade zirconia (3Y-TZP), glass-infused alumina, a sintered 99.9% alumina, and a dental feldspathic porcelain.

and a dental feldspathic porcelain (Mark II, Vita). Despite the limited number of data, the trends are clear. The two aluminas are very similar and fall in between the much more chip resistant zirconia and the more chip sensitive dental feldspathic porcelain. Data for the zirconia were only collected for very small edge distances, however. As noted above, close examination of our data collected in subsequent years over greater distances (0 mm to 0.50 mm) revealed that the short distance data (0 mm to 0.10 mm) did not seem to have "settled down" and might deviate from the overall trend. There seemed to be some non-linearities at short distances.

New plots for the indirect resin filled-composite, the same feldspathic porcelain, and a dental 3Y-TZP zirconia are in Figure 5. Although the trend line for the filled-composite is linear, a power law fit seemed to be an equally good fit for this material as discussed and shown in Ref. 18. The filled-composite and the feldspathic porcelain have similar edge chip resistances. The feldspathic porcelain is the same material used for the data in Figure 4, but the new curve is based on more data points. Table 1 has data for three different 3Y-TZP zirconias. The T_e values differ, but these data were collected over different distance ranges. One of the zirconia data sets, shown in Figure 5, is over a much broader range, giving a better estimate of T_e. The two other data sets in Table 1 were collected over very small (0 mm to 0.10 mm) distances such as shown in Figure 4. A better comparison would be if all three zirconias had data collected over a 0 full mm to 0.50 mm range.

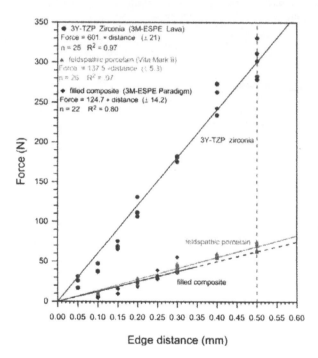

Figure 5. New data for three dental restorative materials by a 120° sharp conical indenter. The line slopes are the "edge toughness, T_e." The dashed vertical line shows the "edge strengths" at 0.5 mm.

Figure 6 shows very nonlinear results for the veneering porcelains on the zirconia and metal substrates.[17] Despite the fact that the porcelains were very different, the edge chip resistances were almost the same. The porcelains were tested while bonded to their substrates. The chips started in the porcelain veneers and for forces greater than 30 N, the chips penetrated through to the substrate-veneer interface. Interactions with the residual stresses at the interface probably caused the nonlinearity in these two cases. Figures 4 and 5 showed that porcelains usually have simple linear edge chip trends.

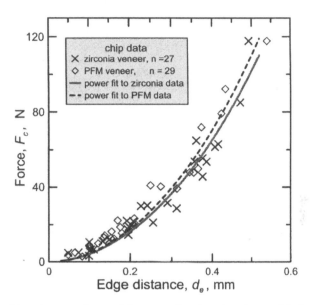

Figure 6. Edge chip resistances of a porcelain veneered to zirconia and a porcelain fused to metal (PFM). Ref. 17.

Figure 7 shows the results for two of the denture materials, including the cross-linked PMMA and one filled resin-matrix composite for comparison. In the case of the PMMA baseline material (Ivoclar, Blueline DCL), we had great difficulty making chips with the Rockwell C indenter. Large divots formed at the indentation sites. The material gradually bulged outwards from the surface and ruptured at the midpoint as shown in Figure 8a. Of eighteen chip attempts, only seven semi-valid chips formed, and they had extreme variability (a factor of 2) in peak force. We initially had similar difficulties with the sharp 120° conical indenter, but eventually obtained eighteen valid chips as shown in Figure 7a, but there also was large scatter. Chips were more easily formed in the PMMA when we switched to a Knoop indenter with its long axis aligned parallel to the specimen edge. Scatter became very large once distances exceeded 0.35 mm (Figure 7a). In contrast, Figure 7b shows that chips were formed with all three indenters in the more brittle filled-composite denture material (Ivoclar, Phonares I). The Knoop indenter required the greatest forces to make a chip. The sharp conical indenter required the least. The blunter Rockwell C indenter data was in-between.

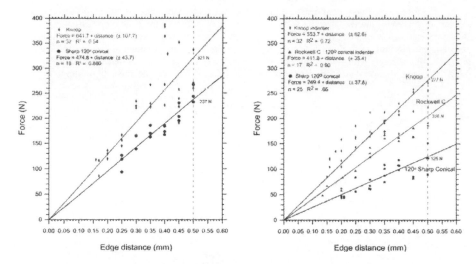

Figure 7. Edge chip resistance of denture materials. (a) is for the traditional denture material (Ivoclar-Vivadent, Blueline DCL), a highly cross-linked PMMA. (b) is for a more modern filled-composite (Ivolcar-Vivadent, Phonares I) that is more wear resistant. The line slopes are the "edge toughness, T_e." The dashed vertical lines help illustrate the "edge strengths" at 0.5 mm.

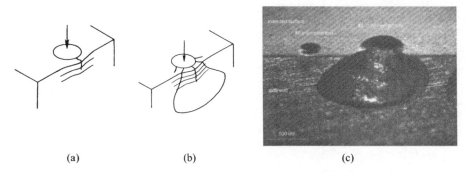

(a)   (b)   (c)

Figure 8. Chips were difficult to form in the PMMA denture material with the conical indenters. (a) shows how the material pushed out from the side. The thin wall membrane eventually ruptured. The filled-composites also bulged from the side, but chips popped off as shown in (b) and (c).

We further investigated the effect of using different indenter shapes by testing the feldspathic dental porcelain (Vita Mark II) with four sharp diamond scribes with different included angles. The results shown in Figure 9 indicate that the blunter the indenter angle, the greater the force needed to create chips. This strongly suggests that a wedging stress component contributes to chip formation.

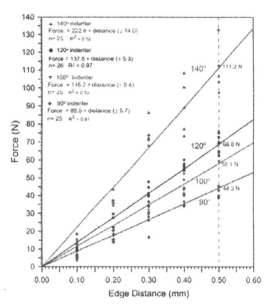

Figure 9. Data for the feldspathic porcelain collected with sharp conical indenters with different included angles. The dashed vertical line shows the "edge strengths" at 0.5 mm.

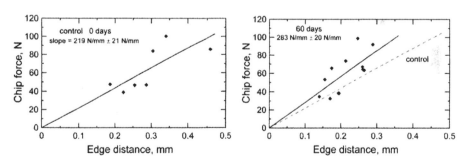

Figure 10. Edge chip resistance of human dentin. (a) shows the data for untreated and (b) for teeth exposed to 60 days of calcium hydroxide. (Ref. 21)

Figure 10 shows the results for the human teeth.[21] Only two of the four data sets are shown. While there was considerable scatter in all the data, greater forces were required to produce larger chips. There did not appear to be a trend among the different tooth types (e.g., incisors, premolars, molars). Some of the variability may be due to variations in the direction of dentinal tubules, the location within the dentin that fracture was initiated, the age of the patient/teeth, and experimental uncertainty, particularly the assessment of the initial contact point. Measurements of the edge distance were still being done with the stereoptical microscope for this work. The edge toughness values (± one standard deviation) were 219 (± 21) N/mm for the control group and 283 (± 20) N/mm for the group exposed to calcium hydroxide for 60 days. One-way ANOVA showed the calcium hydroxide treated values were significantly different than that for the control group using 95 % confidence intervals. The forces necessary to create edge chips were greater, on average, for the treated specimens than for the untreated controls. This was somewhat surprising since, as discussed in ref. 21, the treatment had been reported to increase the fracture sensitivity of teeth. Our data shows an opposite trend. A likely explanation was that the treatment caused softening (a change in hardness) of the dentin, so that more of the energy of the indenter was expended in deformation rather than brittle chipping. Additional details and data are in ref. 21.

DISCUSSION
As noted in the introduction, variations in procedures and interpretations are proliferating and are creating confusion. Different indenter types (sharp Knoop, Vickers, or conical or blunt Rockwell) make intercomparsion of data more difficult.[13] The edge toughness parameter T_e (the line slope) can vary dramatically with indenter type as shown in Figures 7b and 9. Morrell and Gant[4] and Gogotsi and Mudrik[13] have commented on the different responses to different indenters. One important reason for the differences stems from fact that edge chipping is a multistep process that entails:

a. Formation of a small divot (analogous to a hardness indentation)
b. Formation of short stable radial cracks.
c. Additional propagation (wedging?) of some of the cracks downward and parallel to the side surface.
d. Unstable crack propagation towards the side surface of one or a pair of cracks that cause a chip to pop off.

Different indenters may alter the initiation and propagation steps, and alter the ratio of energy expended in fracture versus deformation processes. Steps b and c may be sensitive to loading rate and environment (e.g., slow crack growth in glasses and ceramics, or relaxations in polymers and composites).

Early work[1-4,11,12] emphasized the relationship between edge toughness and traditional fracture parameters such as K_{Ic}, the fracture toughness; G_{Ic}, the critical strain energy release rate, or γ_f, the fracture surface energy. For the customary linear regression fit line through the force – distance data, most investigators interpret the slope as "edge toughness." It has units of force/distance which is dimensionally analogous to (force · length) / area or energy/area. Thus, it should not be surprising that the edge toughness sometimes correlates better with G_{Ic}, the critical strain energy release rate, or to γ_f, the fracture surface energy than with K_{Ic}, the fracture toughness.

Our experience with the human dentin, the indirect filled-composite, and the polymeric based denture materials revealed that considerable amounts of deformation may occur in softer materials. Chips usually did not occur in the polymethylmethacrylate (PMMA) denture material with the sharp conical and Rockwell C indenters. Even with the more chisel-like Knoop indenter, one had the sense that material was pried off the side and not chipped off. Clearly there are limits of applicability to the edge chip testing. One cannot chip warm butter.

A power law relationship between force and distance was a better fit in many instances. If the exponent is 2, then the constant A may be deemed an "*edge strength*" since the constant in the relationship (equation 2) has units of force/area, the same units as for stress. Otherwise, if n ≠ 2, A may be considered the force necessary to create a chip at 1 mm distance. Table 1 shows that most of the materials have values between 1 and 2. Morrell and Gant[4] also showed non-linear data for hard metals. Gogotsi et al.[13,14] had some non-linear data for zirconia ceramics. A recent indentation fracture mechanics model[26] for edge chip resistance supports a power law relationship, but only for an exponent of 1.5. That indentation model shares many of the shortcomings of the common indentation fracture mechanics models and it should not be surprising that actual behavior deviates from the model and that the claimed accuracy is no better than 25%.[26] Future work should explore why some materials seem to have a linear behavior while others have a power law trend. It is likely that deformation processes and/or the multistep chipping process cause the variations.

The work summarized here and in more detail in Ref. 21 was the first instance of the edge chipping test being applied to human dentin. There had been indications in the literature that calcium hydroxide treatments during endodontic treatments made dentin more susceptible to fracture. The edge chipping tests did not confirm this, but dentin is a relatively soft material and much of the energy of the indentation was expended in deformation. It would be interesting to apply the method to human enamel, the harder outer shell of a tooth.

An interlaboratory comparison project for edge fracture, edge flaking, edge chip, or edge strength resistances might be appropriate.

CONCLUSIONS

The data base of edge chipping results has been expanded to include a wide range of dental materials. Testing procedures have been refined. Different indenters give rise to different edge toughness values. Some materials follow a linear data trend, but others that have significant deformation (composites, layered structures) have power law trends.

ACKNOWLEDGMENTS

This work was made possible by a grant from NIH, R01-DE17983, and the people and facilities at the National Institute of Standards and Technology and the ADAF Paffenbarger Research Center. The authors wish to thank Dentsply-Ceramco, York, PA, USA; 3M-ESPE, St. Paul, MN, USA; and Ivoclar-Vivadent, Schaan, Liechtenstein for donating materials for this study. Tony Giuseppetti assisted with specimen preparation.

REFERENCES

1 E. A. Almond and N. J. McCormick, Constant-Geometry Edge-Flaking of Brittle Materials, *Nature,* 1, 53-55 (1986).
2 N. J. McCormick and E. A. Almond, Edge Flaking of Brittle Materials, *J. Hard Mater* 1 [1], 25-51 (1990).
3 N. J. McCormick, Edge Flaking as a Measure of Material Performance, *Metals and Materials,* 8, 154-156 (1992).
4 R. Morrell and A. J. Gant, Edge Chipping of Hard Materials, *Int J Refract Metals and Hard Mat* 19, 293-301 (2001).
5 J. B. Quinn, Material Properties of Ceramics for Dental Applications, PhD thesis, University of Maryland, Department of Materials Science and Engineering, College Park, MD, June, 2000.

6   J. B. Quinn, J. W. Hatch, and R. C. Bradt, The Edge Flaking Test as an Assessment of the Thermal Alteration of Lithic Materials, Bald Eagle Jasper, pp. 73 -86 in *Fractography of Glasses and Ceramics IV, Ceramic Transactions* 122, J. R. Varner and G. D. Quinn, eds., The American Ceramic Society, Westerville, OH, 2001.

7   J. B. Quinn, L. Su, L. Flanders, and I. K. Lloyd, "Edge Toughness" and Material Properties Related to the Machining of Dental Ceramics, *Mach. Sci. and Technol.*, **4**, 291-304 (2000).

8   J. B. Quinn and I. K. Lloyd, Flake and Scratch Size Ratios in Ceramics, pp. 55-72 in *Fractography of Glasses and Ceramics IV, Ceramic Transactions* 122, J. R. Varner and G. D. Quinn, eds., The American Ceramic Society, Westerville, OH, 2001.

9   J. B. Quinn and V. C. Ram Mohan, Geometry of Edge Chips Formed at Different Angles, *Ceram. Eng. Sci. Proc.*, **26** [2] 85-92 (2005).

10 J. B. Quinn, I. Lloyd, R. N. Katz, and G. D. Quinn, Machinability: What Does it Mean? *Ceram. Eng. Sci Proc.*, **24** [4] 511-516 (2003).

11 R. Danzer, M. Hangl, R. Paar, How to Design with Brittle Materials Against Edge Flaking. *6th International Symposium on Ceramic Materials for Engines*, Arita, Japan, 658-62 (1997).

12 R. Danzer, M. Hangl, and R. Paar, Edge Chipping of Brittle Materials, pp. 43-55 in *Fractography of Glasses and Ceramics IV*, American Ceramic Society, 2001.

13 G. Gogotsi, S. Mudrik, and V. Galenko, Evaluation of Fracture Resistance of Ceramics: Edge Fracture Tests, *Ceram. Int.*, **33**, 315-320 (2007).

14 G. Gogotsi, V. I. Galenko, S. P. Mudrik, B. I. Ozersky, V. V. Khvorostyany, and T. A. Khristevich, Fracture Behavior of Y-TZP Ceramics: New Outcomes, *Cer. Int.*, **36**, 345-350 (2010).

15 G. Gogotsi, S. Mudrik, and A. Rendtel, Sensitivity of Silicon Carbide and Other Ceramics to Edge Fracture: Method and Results, *Cer. Eng. Sci. Proc.*, **25** [4], 237-246 (2004).

16 G. Gogotsi and S. Mudrik, Fracture Barrier Estimation by the Edge Fracture Test Method, *Ceram. Int.*, **35**, 1871- 1875 (2009).

17 J. B. Quinn, V. Sundar, E. E. Parry, and G. D. Quinn, Comparison of Edge Chipping Resistance of PFM and Veneered Zirconia Specimens, *Dental Materials,* **26** [1], 13-20 (2010).

18 J. B. Quinn and G. D. Quinn, Material Properties and Fractography of an Indirect Dental Resin Composite, *Dental Materials,* **26** [6], 589-599 (2010).

19 D. C. Watts, M. Issa, A. Ibrahim, J. Wakiga, K. Al-Samadani, M. Al-Azraqi, and N. Silikas, Edge Strength of Resin-Composite Margins, *Dental Materials*, **24**, 129-133 (2008).

20 K. Baroudi, N. Silikas, and D. C. Watts, Edge-Strength of Flowable Resin-Composites, *J. Dentistry*, **36**, 63-68 (2008).

21 E. R. Whitbeck, G. D. Quinn, and J. B. Quinn, Effect of Calcium Hydroxide on Dentin Fracture Resistance, *J. Res. NIST*, **116** [4], 743-749 (2011).

22 R. Morrell, Edge Flaking – Similarity Between Quasistatic Indentation and Impact Mechanisms for Brittle Materials, pp. 14-23 in *Fractography of Advanced Ceramics II, Key Engineering Materials* vol. 290, eds. J. Dusza, R. Danzer and R. Morrell, Transtech Publ. Switzerland, 2005.

23 European prestandard, prTS 843-9, Advanced Technical Ceramics – Mechanical Properties of Monolithic Ceramics at Room Temperature, Part 9: Method of Test for Edge-Chip Resistance, European Standard Committee TC 184, Brussels, 2009.

24 G. D. Quinn, K. Hoffman, and J. B. Quinn, Strength and Fracture Origins of a Feldspathic Porcelain, 28 [5] 502–511 (2012).

25 J. Neter, M. H. Kutner, C. J. Nachtsheim, and W. Wasserman, *Applied linear Statistical Models, 4th ed.*, Irwin Series, McGraw Hill, Chicago, 1996.

26 H. Chai and B. R. Lawn, A Universal Relation for Edge Chipping from Sharp Contacts in Brittle Materials: A Simple Means of Toughness Evaluation, *Acta Met,* **55**, 2555-2561 (2007).

HIGH PRESSURE SEAWATER IMPINGEMENT RESISTANCE OF LOW SILICA ALUMINUM OXIDES

Tim Dyer and Ralph Quiazon
Energy Recovery, Inc,
San Leandro, CA, USA

Mike Rodgers
Rio Tinto Alcan
Gardanne, France

ABSTRACT

High purity alumina is used to make energy recovery devices for seawater reverse osmosis (SWRO) desalination. These devices reduce energy consumption of large water factories by up to 60%. SWRO applications are unusual for ceramics since they are continuous processes that combine corrosive and abrasive fluid conditions with occasional cavitation energy.

Selecting a starting material for a product formulation is critical. High purity specialty alumina powders, with minimized soda and silica levels, have been developed for high performance ceramics. Available products included the now obsolete Alcan RAC45B, and a new alumina developed by Rio Tinto Alcan, P172IIPB which targets levels of <0.03% soda and <0.02% silica. Though similar products are available in the market, aluminas made from these raw materials perform differently in seawater impingement.

Super ground, >99.5% pure alumina powders, including Rio Tinto Alcan P172HPB, were processed into sample coupons using Energy Recovery Inc.'s (ERI) production spray dry, cold isostatic press, and sintering processes. These ceramic coupons, along with commercially procured ceramic reference samples, were subjected to a modified ASTM G134 – 95 cavitation test followed by structural characterization. After testing, surface roughness increased to 2.54 to 5.08 μm R_a. Cavitation damage appears to correlate with hardness, grain size, and fired density. Fired density data also reveals that the P172HPB should offer physical properties equal to or better than the other aluminas available. In general, raw materials with overall lower impurity levels appear to perform better in SWRO conditions.

INTRODUCTION

Isobaric seawater reverse osmosis (SWRO) energy recovery devices, such as ERI's PX Pressure Exchanger device shown in Figure 1, operate at over 96% efficiency and thus can save up to 60% of the energy required for SWRO over the life of the desalination plant. Having reliable materials of construction is critical to long term and trouble-free SWRO desalination processes. High purity aluminum oxide (alumina) is found in other select, severe pump and energy recovery device applications, however little information is published about the wear and corrosion performance of this material within desalination applications. SWRO applications are unusual since they combine corrosive and abrasive conditions with possible cavitation energy. Cavitation can occur in pumps, turbines, valves, couplings, bearings, internal flow passages, and other various components in a SWRO system. Advanced alumina ceramics are generally known to perform better than plastics or metals in a severe corrosive and abrasive environment such as a SWRO system.

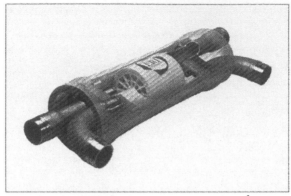

Figure 1.PX Rotary Isobaric Pressure Exchanger (≈1400 m³/day)

To evaluate long term SWRO service effects on aluminum oxide ceramics, the authors evaluated coupons manufactured using three similar, >99.5% pure, calcined and super-ground raw alumina powders. The powders used in this study, including Rio Tinto Alcan P172HPB, contained less than 200 ppm of silica and were received at a surface area measuring from 7 to 9 m²/g using the BET method. For reference purposes, samples of two high purity commercial aluminas were also evaluated. One of these reference samples consisted of LTD998; a commercial alumina ceramic manufactured approximately 5 years ago using the now obsolete Rio Tinto Alcan RAC45B alumina powder.

All ceramic material samples were evaluated for SWRO durability using high pressure, submerged water jet, impingement tests. The test and the apparatus used for this study are similar to those described in ASTM G134 – 95[1]. Impingement tests were completed by spraying water through a 0.508 mm sapphire orifice at 11032 kPa over a period of 6 hours with a 6.35 mm sample distance (also known as the stand-off distance). The sample and the nozzle for the cavitation jet remained submerged in filtered water. This paper will quantify ceramic surface wear using a precision stylus profilometer over the eroded sample area. Samples typically have an initial surface finish of 0.508-0.762 μm R_a. After testing, the surface finish increased to a range typically between 2.54 to 5.08 μm. Cavitation damage to the material is typically seen as multiple concentric rings. Material properties of the different alumina samples were also evaluated using four-point bending test (modulus of rupture), grain size, and Vickers hardness. The objective of this paper is to improve the general understanding of how different alumina ceramics withstand the severe conditions within a SWRO process stream.

EXPERIMENTAL
Cylindrical ceramic coupons manufactured using raw alumina powders described above incorporated into ERI's proprietary formulation. Following cold isostatic pressing (isopressing) and production sintering processes, impingement samples were diamond core drilled and mounted in epoxy. Reference commercial alumina samples were similarly core drilled out of larger fired samples and mounted in epoxy using the same process. To minimize the effects of surface roughness on impingement resistance, all samples were polished using resin-bonded diamond plates until a uniform 0.508-0.762 μm surface finish and was achieved. Surface roughness measurements were conducted using a Zeiss Surfcom 130A with a 2μm radius, 60° conical diamond stylus tip. Samples were then placed into the impingement tester, centered face-up beneath the orifice, and held in place with a two-jaw, v-block based collet. The nozzle assembly, which houses the orifice, was then adjusted until a

6.35 mm standoff distance was achieved between the sample face and the nozzle assembly. Tap water filtered to 10 μm was used to submerge the sample and nozzle assembly. Freshly filtered tap water was then pumped through the nozzle assembly until the water was observed streaming onto the sample surface. After being fully immersed the uploader valve (a throttling and bypass valve) equipped high pressure pump was switched on and pressure gradually increased to 11032 kPa where it remained steadily over the course of six hours. The complete test circuit is diagramed in Figure 2.

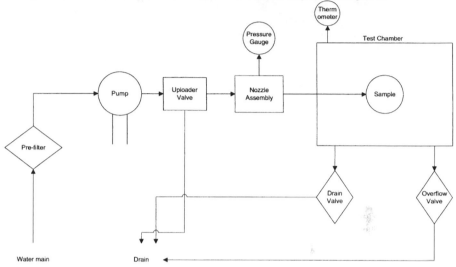

Figure 2. Test Circuit

After six hours of testing was completed, water from the test chamber was drained and the sample was removed from the collet. Each sample was then carefully centered under the profilometer stylus with the aid of light-emitting diode (LED) side-lighting to locate the center. An image of a sample mounted on epoxy, ready for profilometer measurement can be seen in Figure 3. Three measurements were taken on each sample, with three samples for each material.

Figure 3. Close-up image of a sample prior to impingement testing and post impingement testing

In addition to impingement testing, a quantity of 10, 7.5 to 8.0 mm diameter center-less ground rods of each material were produced and evaluated using 4-point bend testing to determine flexural strength. Flexural strength testing was compliant to ASTM C 1161-02c using a "b-configuration" fixture from Wyoming Test Fixtures. ERI uses a modified fixture retrofitted with tungsten carbide pins to minimize deformation and wear during strength testing. The fixtures were placed into an Instron 3366 10 kN universal testing machine prepared for compressive bend testing. Samples were diametrically measured and centered in the fixture before running the Instron Bluehill program.

Further material testing was completed on the alumina samples by performing Archimedes specific gravity measurements and Vickers hardness tests on the sintered alumina. Specific gravity of the ceramic coupons were measured using a high precision scale, the Mettler Toledo AB104-S/FACT with density kit via the Archimedes method.

To measure hardness ceramic coupons were mounted in epoxy and polished to a 1 μm surface finish with a Struers LaboPol System. Micro-hardness measurements were then taken using a Vickers indenter at 9.81 N force and holding for a period of 12 seconds on a Struers Duramin hardness tester.

RESULTS AND DISCUSSION

After six hours of impingement testing of high purity alumina, there was noticeable damage to the surface of each sample. As shown in Figure 3, the cavitation damage resulted in concentric rings, with a broad outer ring and a small pinhole in the middle.

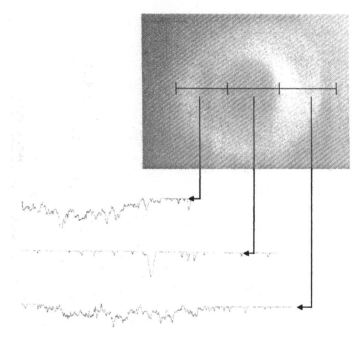

Figure 4. Typical area of surface roughness measurements on an impingement test sample

In comparing surface roughness measurements after impingement testing, it was found that a high specific gravity, high hardness, and small grain size sample provided the best impingement resistance. A summary of the results is shown in Figure 5.

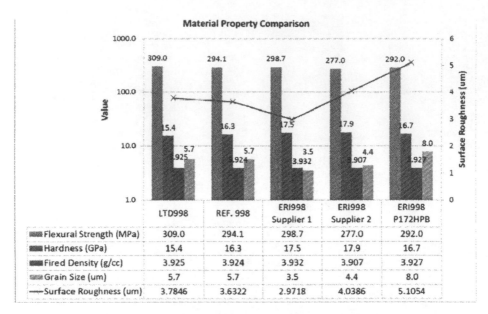

	LTD998	REF. 998	ERI998 Supplier 1	ERI998 Supplier 2	ERI998 P172HPB
Flexural Strength (MPa)	309.0	294.1	298.7	277.0	292.0
Hardness (GPa)	15.4	16.3	17.5	17.9	16.7
Fired Density (g/cc)	3.925	3.924	3.932	3.907	3.927
Grain Size (um)	5.7	5.7	3.5	4.4	8.0
Surface Roughness (um)	3.7846	3.6322	2.9718	4.0386	5.1054

Figure 5. Comparison of material properties and their relation to impingement resistance

The results of our cavitation testing suggest that increasing fired density and hardness result in improved cavitation resistance (minimum roughness change after impingement), however it is believed that the large grain size of experimental ERI998 P172HPB alumina led to the decrease in cavitation resistance of that particular sample. Once a large grain was pulled out from impingement of an air bubble on the sample surface, the void left behind becomes a site for more concentrated cavitation. We found this observation to be in agreement with a similar study published by Nour in 2007[2]. On the other hand, despite the high hardness of ERI998 Supplier 2, its' relatively low fired density and subsequent micro-porosity hampered resistance to impingement damage. From impingement testing, it can be concluded that high purity alumina materials exhibiting a combination of high fired density, high hardness, and smaller grain size perform the best for cavitation resistance.

By subtly changing the standard ERI998 formulation, another batch of ERI998 P172HPB was produced that yielded more exceptional results, as shown in Table I. The decrease in grain size coupled with slight increase in hardness and fired density suggests that this sintered alumina will perform exceptionally against cavitation. Impingement testing results for this revised P172 HPB alumina formulation is pending testing and not available for inclusion in this paper.

Table I. Properties of P172HPB Spray Dried With a Modified ERI998 Formulation

Flexural Strength (MPa)	280 ± 30
Hardness (GPa)	16.8
Fired Density (g/cc)	3.929
Grain Size (μm)	4.20

CONCLUSION

Impingement and physical properties data from PX devices and reference alumina ceramics coupons demonstrate that, though similar in composition, sintered aluminum oxide ceramics produced using different raw powders exhibit unique properties. Therefore, one must be careful when selecting a raw material for an application specific ceramic formulation. We have also determined that physical properties of the new Rio Tinto Alcan P172 HPB should be equivalent or better than other aluminas past or present when incorporated into the ERI product formulation. Materials with high hardness, small grain sizes, and high fired densities appear to resist surface cavitation damage better than others. Though extreme cavitation testing appears to be able to damage all alumina samples, those exhibiting the highest specific gravity and hardness, with small grain size faired the best; the trends agree with previous research done on the cavitation on stainless steels[3].

REFERENCES

[1]Standard Test Method for Erosion of Solid Materials by a Cavitating Liquid Jet, ASTM G134-95 (2007)
[2]W. Nour, U. Dulias, J. Schneider, K. Gahr, The Effect of Surface Finish and Cavitating Liquid on the Cavitation Erosion of Alumina and Silicon Carbide Ceramics, *Ceramics –Silikaty* (1) **30-39** (2007)
[3]G. Bregliozzi, A. Di Schino, S.I.-U. Ahmed, J.M. Kenny, H. Haefke, Cavitation Wear Behavior of Austenitic Stainless Steels with Different Grain Sizes, *Elsevier Wear,* Neuchâtel, Switzerland, Terni Italy, 2004.

WEAR BEHAVIOR OF CERAMIC/METAL COMPOSITES

M. K. Aghajanian, B. P. Givens, M. C. Watkins, A. L. McCormick, and W. M. Waggoner
M Cubed Technologies, Inc.
1 Tralee Industrial Park
Newark, DE 19711

ABSTRACT

Ceramic particle reinforced aluminum metal matrix composites (MMCs) have been well studied due to their utility in various structural and thermal applications. In addition, MMCs have value in various wear applications due to the presence of the hard ceramic phase within the ductile metallic matrix. Potential uses include slurry and dry erosion components in the mining, coal power and paper processing industries where ceramic wear components are not viable due to low fracture toughness (e.g., potential breakage if component is struck with large piece of debris such as a rock). The present work quantifies the wear properties of ceramic particle reinforced aluminum as a function of ceramic content and type. Results are compared to those for traditional unreinforced ferrous and aluminum-based metals; and with a traditional Al_2O_3 wear resistant ceramic. Also, correlation of results with microstructure and properties of the composites is made. Finally, the wear scars are examined by SEM to assess failure mode.

INTRODUCTION

Particulate reinforced metal-matrix composites are of interest due to their improved mechanical and thermal properties relative to unreinforced alloys. Discontinuous particulate reinforcement provides isotropic properties as compared to that of continuous-fiber-reinforced composites. Moreover, these materials are versatile and can be tailored for specific uses by altering processing conditions and/or raw materials. For instance, factors such as the alloy chemistry, the reinforcement shape (particulate, platelet, whisker, etc.), the reinforcement chemistry (Al_2O_3, SiC, etc.), the reinforcement loading, the processing method, post heat treatment, and cold work can have a significant impact on the structural behavior of the resultant composite. Such composites are now seeing widespread use in thermal management, precision equipment, and automotive applications where composition and microstructure are tailored to provide the desired mechanical and/or thermal properties [1-3].

Another area of interest for MMCs is in the field of abrasive wear resistance, as is encountered in the mining, coal power, paper, and powder processing industries. Metals perform poorly in abrasive wear applications because of low hardness (i.e., high hardness abrasive particles quickly erode relatively soft exposed metal surfaces). Ceramics have sufficient hardness to provide high wear resistance. However, ceramics are susceptible to catastrophic failure when exposed to impact due to low fracture toughness. Moreover, as a result of processing limitations, ceramics are not available in the large sizes and complex shapes needed in many industrial applications. Ceramic particle reinforced metals provide high abrasive wear resistance due to the presence of the hard ceramic phase, and provide toughness due to the presence of the metallic matrix. In addition, these materials offer more processing flexibility with respect to component size and shape (e.g., via casting). To this end, the present paper examines the abrasive wear resistance of SiC and Al_2O_3 particle reinforced aluminum MMCs.

TEST PROCEDURES

Physical and mechanical properties were measured with the test methods shown in Table I. Density was measured only once per material on the bulk billet of composite. Young's modulus was measured at three locations on each billet. Flexural strength, tensile strength and fracture toughness were each measured using five to ten samples. The mechanical tests utilized a Sintech universal test frame in conjunction with Test Works materials testing software. Microstructures were evaluated using a Leica D 2500 M optical microscope and the Clemex Vision PE imaging software. Wear scars were examined with a Joel JSM-6400 scanning electron microscope (SEM).

Table I: Test Methods

Property	Test Description	Test Procedure
Density	Water Immersion	ASTM B 311
Young's Modulus	Ultrasonic Pulse-Echo	ASTM E 494
Hardness	Knoop Indentation	ASTM C 1326
Tensile Strength	Cylindrical Dog-Bone	ASTM B 557M
Fracture Toughness	Four-Point Bend Chevron Notch - K_{Ivb}	ASTM C 1421

Wear resistance was evaluated using the ASTM G65A dry sand rubber wheel procedure. During this test, silica sand is introduced between the specimen and a rotting rubber wheel. The specimen undergoes constant erosive wear as the abrasive is pulled between the rubber wheel and the surface of the specimen. The G65A procedure is known as a low stress abrasion test, meaning that the abrasive sand is not fractured during the test.

RESULTS AND DISCUSSION

The present work examined particle reinforced MMCs with a wide range of reinforcement contents. The low reinforcement content composite (30%) was produced with a casting process [4-5], and the high reinforcement content composites were produced with a pressureless preform infiltration process [6-7]. Microstructures of representative MMC samples are provided in Figures 1 and 2. To eliminate the effect of particle size on wear performance, all composites were produced with nominally 40 to 45 μm ceramic particles. The exception was the 70% loaded Al/SiC MMC where a blend of 45 and 12 μm particles was needed to achieve the desired reinforcement content. In all cases, excellent wetting of the matrix alloy to the ceramic particles is seen, with metal filling all gaps, including between closely spaced particles.

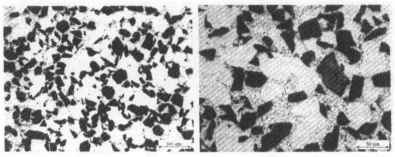

Figure 1: Optical Photomicrographs at Low and High Magnification of Cast Al/SiC-30p

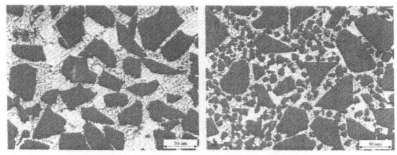

Figure 2: Optical photomicrographs of Al/SiC composites:
Left: Al/SiC-55p (45 μm particles)
Right: Al/SiC-70p (blend of 45 and 12 μm particles)

A summary of all MMC property and wear data is provided in Table II. As expected, the Al/SiC MMCs show increased Young's modulus, hardness and wear resistance, and decreased strength and fracture toughness, as the reinforcement content increases. Other than wear resistance, all property changes are gradual as the reinforcement content is changed. With wear resistance, a dramatic improvement is seen between 30 and 55% SiC, and only a modest improvement thereafter to 70% SiC. The high wear resistance of Al/SiC MMCs is consistent with literature data [8], however the non-linear response over a wide range of reinforcement contents is not previously discussed.

Compared to the Al/SiC-55p MMC, the Al/Al₂O₃-55p MMC has much higher strength and fracture toughness. This can be attributed to the difference in Al alloy matrix. The Al/SiC MMCs were made with an Al-Si alloy (nominally A356) to prevent Al-SiC reactions during processing. Presence of the brittle Si phase within the Al alloy leads to decreased toughness and elongation. The Al/Al₂O₃ MMC was produced with a ductile Al-Mg matrix alloy (nominally A518). Due to the lower hardness and stiffness of Al₂O₃ relative to SiC, the Al/Al₂O₃ MMC had lower Young's modulus, hardness and wear resistance as compared to the Al/SiC MMC with the same reinforcement content.

Table II: Summary of Property and Wear Data for MMCs

Composite	Density (g/cc)	Young's Modulus (GPa)	Knoop 500 g Hardness (kg/mm^2)	Ultimate Tensile Strength (MPa)	Fracture Toughness (MPa-m$^{1/2}$)	ASTM G65A Wear (vol. loss in mm^3)
Al/SiC-30p	2.78	125	165	317	15	516
Al/SiC-55p	2.95	200	340	280	11	78
Al/SiC-70p	3.01	270	650	230	10	33
Al/Al₂O₃-55p	3.36	180	285	412	16	177

Figure 2 plots the wear resistance results for the MMCs versus those for various traditional metals. The high reinforcement content Al/SiC MMCs show superior wear resistance to high hardness 4340 steel, and the Al/SiC-30p MMC shows similar performance to softer ferrous metals such as austenitic stainless steel and cast iron. The Al/Al₂O₃-55p MMC is similar in performance to hard steel.

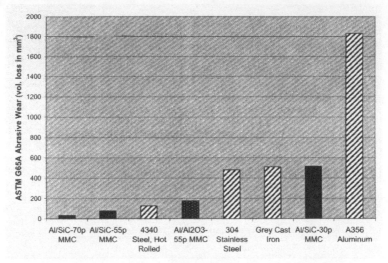

Figure 3: Wear Resistance of MMCs Compared to Traditional Metals

To understand the very high wear resistance of the high reinforcement content Al/SiC MMCs and the large drop-off in wear resistance as reinforcement content drops to 30%, the wear scars were examined by SEM. Figure 4 provides the wear scars for steel and aluminum and Figure 5 provides the wear scars for the Al/SiC MMCs. The metallic samples show the typical abrasive failure mode with uniform gouging of the surface. Both the aluminum and steel surfaces are flat and grooved, similar to a ground sample. The MMC surfaces appear very different after wear testing. In all cases, the samples are not flat, with plateaus of SiC standing proud above the aluminum matrix that is worn below the SiC particles. In particular, the 55 and 70% loaded samples show very well formed plateaus. It is postulated that once these plateaus form, the wear rate greatly slows as it is difficult for sand to reach the recessed aluminum matrix. The plateaus are much less pronounced in the 30% loaded sample, indicating why this sample shows significantly less wear resistance. It is postulated that the spacing between plateaus is too large to prevent continuous access to the recessed aluminum matrix by the abrasive media. Supporting these findings is work with particulate reinforced steels where the addition of hard particles was shown to both help (increased hardness) and hurt (addition of hard particle debris to system) wear resistance. It was stated that the reinforcement content needs to be high enough in such systems to counteract the negative effect of the hard debris particles that are formed [9].

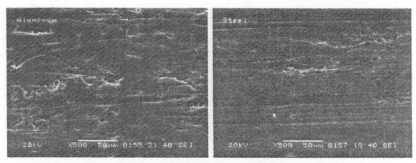

Figure 4: Wear Scars for Aluminum Steel (4340) Samples

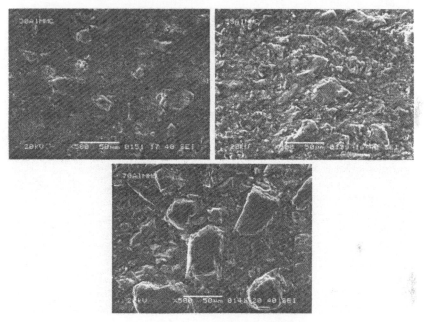

Figure 5: Wear Scars for Al/SiC MMCs

In abrasive wear applications where toughness is not required, Al_2O_3 tiles are often used (nominal fracture toughness of 4 MPa-m$^{1/2}$). Three different versions of 96% sintered Al_2O_3 wear tiles were purchased and tested in parallel to the MMC samples. Results are provided in

Table III. The data show a significant grain size effect, with abrasion resistance increasing with grain size. This is a common response in abrasive wear of ceramics, where reduced wear is obtained when grain boundary surface area is reduced (i.e., wear occurs fastest at grain boundaries). Comparing the data in Tables II and III, the wear resistance of high reinforcement content Al/SiC MMCs is similar to that sintered alumina, while providing greater than 2 times the fracture toughness.

Table III: Comparative Wear Data for 96% Sintered Al_2O_3

Grain Size	ASTM G65A Wear (vol. loss in mm³)
1.5 μm	67
7.4 μm	35
11.7 μm	11

SUMMARY

Particulate-reinforced Al/SiC and Al/Al_2O_3 MMCs with various particulate loadings were fabricated and tested for physical, mechanical and wear properties. In addition, wear resistance of traditional metals and Al_2O_3 ceramic was measured and compared to results for the MMC materials. From this work the following conclusions were made.

1. Al/SiC MMCs display increases in Young's modulus, hardness and wear resistance as SiC content increased. In particular, wear resistance of MMCs increases non-linearly with SiC content, and at 70% SiC is better than hard steel and similar to Al_2O_3 ceramic.

2. Production of Al/Al_2O_3 MMC does not require the use of a brittle Al-Si Al alloy, which yields a composite with higher strength and higher toughness than its equivalent Al/SiC counterpart. However, due to reduced hardness, this MMC provides less wear resistance than Al/SiC.

3. Examination of wear scars in the SEM shows very different failure modes for metals and MMCs. Metals wear in a uniform manner, whereas MMCs display a multi-level wear scar with proud plateaus of SiC and sub-surface erosion of the softer Al matrix phase.

REFERENCES

1. A. Evans, C. SanMarchi and A. Mortensen, *Metal Matrix Composites in Industry: An Introduction and a Survey* (Kluwer Academic Publishers, Norwell, MA, 2003).

2. S.V. Prasad and R. Asthana, "Aluminum Metal-Matrix Composites for Automotive Applications: Tribological Considerations," *Tribology Letters*, **17** (3) 445-53 (2004).

3. A. Mortensen and J. Llorca, "Metal Matrix Composites," *Annu. Rev. Mater. Res.*, **40** 243-70 (2010).

4. J. T. Burke, C. C. Yang and S. J. Canino, "Processing of Cast Metal Matrix Composites," *AFS Trans.*, **94-179** 585-91 (1994).

5. B. Givens, W. M. Waggoner, K. Kremer, and M. Aghajanian, "Effect of Particle Loading on the Properties of Al/SiC Metal Matrix Composites," in *Aluminum Alloys: Fabrication, Characterization and Applications II*, Yin et al. editors, TMS, Warrendale, PA (2009) 197-202.

6. M.K. Aghajanian, J.T. Burke, D.R.White and A.S. Nagelberg, "A New Infiltration Process for the Fabrication of Metal-Matrix Composites," *SAMPE Q,* **20** (4) 43-47 (1989).

7. M.K. Aghajanian, M.A. Rocazella, J.T. Burke, and S.D. Keck, The Fabrication of Metal Matrix Composites by a Pressureless Infiltration Technique," *J. Mater. Sci.* **26** 447-54 (1991).
8. R.L. Deuis, C. Subramanian and J.M. Yellup, "Abrasive Wear of Aluminum Composites – A Review," *Wear* **201** 132-44 (1996).
9. S. Ala-Kleme, P. Kivikyto-Reponen, J. Liimatainen, J. Hellman and S. Hannula, "Abrasive Wear Properties of Metal Matrix Composites Produced by Hot Isostatic Pressing," *Proc. Estonian Acad. Sci. Eng.* **12** (4) 445-54 (2006).

USE OF CERAMIC SLIDING SYSTEMS IN A PROTOTYPE GASOLINE PUMP WITH OPERATING PRESSURES OF UP TO 80 MPA

C. Pfister[1], H. Kubach[1], U. Spicher[1], M. Riva[2], M. J. Hoffmann[2]
[1] Institut fuer Kolbenmaschinen,
[2] Institute of Applied Materials – Ceramics in Mechanical Engineering
Karlsruhe Institute of Technology (KIT)
Kaiserstrasse 12, D-76131 Karlsruhe, GERMANY

ABSTRACT

The fuel consumption of modern combustion engines can be significantly reduced through the use of gasoline direct injection (GDI). As the time for mixture preparation in GDI engines is very short, the fuel has to be injected at high pressure into the combustion chamber. Modern injection systems provide an injection pressure of up to 20 MPa, whereas investigations performed at the Karlsruhe Institute of Technology have shown strong potential for reduced pollutant emissions by increasing the pressure to 80 MPa. However, the low lubricity of gasoline causes severe friction and wear in the high-pressure pump at fuel pressures above 20 MPa. The use of ceramic components in the sliding systems should help to overcome this limitation.

A three-piston radial pump based on ceramic sliding systems that delivers fuel at up to 80 MPa has been designed at Karlsruhe Institute of Technology. The prototype is based on the knowledge accumulated during former investigations performed on a single-piston pump. It is fitted with pressure and temperature sensors in each cylinder and a torque sensor on the driveshaft in order to measure its efficiency with various material combinations in its sliding systems. The results show that the use of silicon carbide or Sialon enable operation at injection pressures of up to 80 MPa with very good mechanical efficiency and low wear.

INTRODUCTION

Finite fossil energy resources as well as climate change lead to intensive research in combustion engines with the objective of reducing both fuel consumption and pollutant emissions. The gasoline direct injection enables the reduction of fuel consumption from 5% to 50%, depending on the engine operating point considered. This technology requires a high injection pressure in order to achieve good mixture preparation in the cylinder, as the time available for fuel evaporation before ignition is very short[1]. If the fuel does not completely vaporise before combustion, the particle emissions in the exhaust gas increase significantly. In order to accelerate the fuel vaporisation, the diameter of the injected fuel droplets must be decreased[2]. This can be achieved by increasing the fuel injection pressure[3]. Modern injection systems provide an injection pressure of up to 20 MPa whereas investigations performed at the Karlsruhe Institute of Technology have shown that the soot concentration in the exhaust can be significantly reduced by injecting fuel at pressures above 20 MPa[4].

However, increasing the delivery pressure of gasoline high-pressure pumps leads to new challenges as the lubricity of gasoline is very low, thus causing friction and wear issues. Experiments performed with commercial pumps have shown that conventional materials and coatings do not guarantee reliable operation of gasoline-lubricated high-pressure pumps when increasing the delivery pressure above 20 MPa[5]. As ceramics provide excellent properties regarding hardness and compressive strength, the use of ceramic components in the sliding systems of high-pressure pumps should help to overcome these difficulties. Materials such as silicon carbide and Sialon have been tested at Karlsruhe Institute of Technology in a single-piston model pump delivering fuel at up to 50 MPa[6]. The results have shown that very low friction coefficients and wear rates can be achieved with these materials.

In order to demonstrate the potential of ceramic sliding systems for high-pressure gasoline pumps, a new 3-piston radial-pump operating at up to 80 MPa has been designed. This pressure level represents a good compromise between the reduction of the particles emitted by the engine and a moderate pump driving torque. The prototype pump is fitted with several sensors in order to measure its efficiency and thus to assess the performance of the ceramic components in its sliding systems. More details about this new prototype are given in the next section.

PROTOTYPE PUMP

The prototype pump designed at the Karlsruhe Institute of Technology is a fuel lubricated three-piston radial pump based on the knowledge accumulated during former investigations on a single-piston pump[6]. This pump design is considered to be the best compromise between size, efficiency and cost[1]. Figure 1 gives a sectional view of the prototype with focus on the main sliding systems (piston/cylinder, cam/sliding shoe, and eccentric shaft/cam).

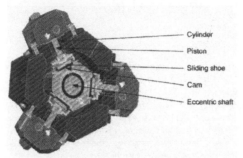

Figure 1. Sectional view of the prototype pump.

The kinematics of the pump is depicted in figure 2. For simplicity's sake, only one piston/cylinder group is depicted. The rotation of the eccentric shaft drives the cam in a circular translation. The cam pushes the piston upwards between 0 °CA and 180 °CA (TDC) via the sliding shoe. The piston moves backwards between 180 °CA and 360 °CA (BDC) via a retaining spring which is not shown in the figure. The sliding direction in the cam/sliding shoe system changes at 90 °CA and 270 °CA, thus causing the piston to rock in the cylinder. This minor effect is due to the small clearance (5 μm) between the piston and the cylinder, which serves to lubricate this sliding contact while maintaining low levels of fuel leakage.

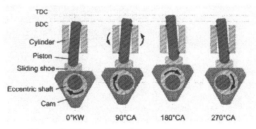

Figure 2. Kinematics of the prototype pump.

In contrast to the cam/eccentric shaft system, in which the relative sliding speed remains constant at constant pump rotation speed, the change of sliding direction in the piston/cylinder and cam/sliding shoe systems causes breakdown of the lubrication film and leads to increased friction and wear. Figure 3 shows the relative motion, sliding speed and contact pressure in the cam/sliding shoe system at a pump rotation speed of 500 rpm. The maximum contact pressure is proportional to the pump delivery pressure. At 80 MPa delivery pressure, the maximum contact pressure is approximately 14 MPa.

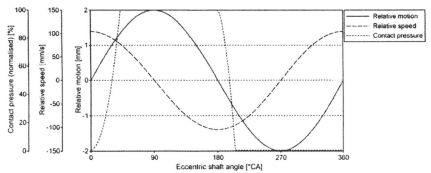

Figure 3. Relative motion, speed and contact pressure in the cam/sliding shoe system at 500 rpm.

The components of the cam/sliding shoe and piston/cylinder systems have been designed in order to allow easy replacement in the prototype (figure 4). The cam and the sliding shoes have diameters of 25 mm and 20 mm, respectively. The piston stroke length is 4 mm and its diameter is 8 mm, leading to a total pump displacement of 603 mm³.

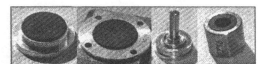

Figure 4. From left to right: sliding shoe, cam, piston, cylinder.

INVESTIGATED COMPONENTS

Materials
        For all the investigations, the material used for the cylinder is sintered silicon carbide (SSiC). The pistons are made of hardened bearing steel (equivalent to AISI 52100). Former tests on this sliding system have shown very good results regarding friction and wear for this material combination[7].
        Different ceramic materials have been investigated in the cam/sliding shoe system that is subjected to the most important tribological stresses in the pump: commercial SSiC (self-mated or in combination with bearing steel), two different Sialon ceramics and a Sialon-SiC composite. The Sialon and Sialon-SiC parts used for the tests have been manufactured at the Institute of Applied Materials – Ceramics in Mechanical Engineering of the Karlsruhe Institute of Technology. Sialon ceramics are derived from silicon nitride by substituting silicon with aluminium and nitrogen with oxygen. The possibility to stabilise two crystal structures offers the opportunity to design materials ranging from pure α-, mixed α/β-, up to pure β-Sialons. In general, α-Sialons are in the form of equiaxed grains with high hardness and good wear resistance but low fracture toughness, whereas β-Sialons have

elongated grains with high fracture toughness but relatively low hardness. To combine the advantages of both Sialon phases, α/β-Sialons have been developed over the past years[8]. For the application in the high-pressure pump, we chose a mixed Sialon with α/β-ratio of 60/40 and an α-rich Sialon with an α/β-ratio of 90/10 that have shown very good characteristics in tribometer tests[9].

In order to improve tribological properties of these materials further, another possibility that has been examined was to combine the qualities α/β-Sialon ceramics with the higher hardness, wear resistance and thermal conductivity of silicon carbide. Therefore Sialon-SiC composites with different amounts of SiC were densified by the field-assisted sintering technique (FAST) and characterised in tribological tests[10]. For the application in the high-pressure pump, a composite with 30 Vol.-% SiC particles has been chosen particles as best compromise between mechanical and physical properties.

The main characteristics of the investigated materials are listed below:
- Silicon carbide (EKasic F from ESK Ceramics):
Sinter type: sintered SiC
Hardness (Vickers): 2500 HV 10
Bending strength (S_0): 405 MPa
Weibull modulus: 7.2
Fracture toughness: 3.65 MPa.m$^{1/2}$
Elastic modulus: 430 GPa
Thermal conductivity: 120 Wm^{-1}K^{-1}

- SiAlON 60/40 (Karlsruhe Institute of Technology):
α/β-ratio: 60/40
Hardness (Vickers): 1850 HV 10
Bending strength (S_0): 904 MPa
Weibull modulus: 6.2
Fracture toughness: 6.5 MPa.m$^{1/2}$
Elastic modulus: 319 GPa
Thermal conductivity: 20 Wm^{-1}K^{-1}

- SiAlON 90/10 (Karlsruhe Institute of Technology):
α/β-ratio: 90/10
Hardness (Vickers): 1970 HV 10
Bending strength (S_0): 959 MPa
Weibull modulus: 12.3
Fracture toughness: 6.2 MPa.m$^{1/2}$
Elastic modulus: 321 GPa
Thermal conductivity: 20 Wm^{-1}K^{-1}

- SiAlON-SiC (Karlsruhe Institute of Technology):
α/β-ratio: 60/40, 30 Vol-% SiC
Hardness (Vickers): 2030 HV 10
Bending strength (S_0): -
Weibull modulus: -
Fracture toughness: 5.6 MPa.m$^{1/2}$
Elastic modulus: 345 GPa
Thermal conductivity: -

- Bearing steel
Type: 100Cr6 (equiv. AISI 52100)
Hardness (Vickers): 800 HV 10
Elastic modulus: 210 GPa
Thermal conductivity: 35 $Wm^{-1}K^{-1}$

Surface roughness and texture

The piston and the cylinder each have a surface roughness of approximately R_a = 0.05 μm. The surfaces of the cams and sliding shoes have been finely ground in order to reach a roughness of R_a = 0.1 μm, which is a good compromise between short run-in times and low friction coefficients[6].

A micro-texture has been tested on cams made of silicon carbide in order to reduce the pump friction losses. It is expected that this texture keeps the fuel in the sliding system on the one hand and stores the particles released by wear on the other hand. The texture consists of circular micro-dimples with a diameter of 60 μm, a depth of 10 μm and a surface ratio of 20 % (see figure 5). The texture parameters have been specially optimized at the Karlsruhe Institute of Technology for the present application[11]. Silicon carbide in combination with hardened steel has only been tested with textured cams as former tests have shown that the texture is necessary for this material pair to avoid adhesion in the sliding system[5].

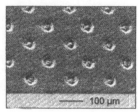

Figure 5. SEM picture of a textured cam (SSiC).

INVESTIGATION METHODS

Test cycle

The pump is driven by an electric motor on a test bench in a sequence of operating points in order to simulate real operation of a high pressure gasoline injection pump driven by the camshaft of a combustion engine. The fuel circulates in a closed loop. An electric feed pump delivers fuel with 0.5 MPa pressure to the high-pressure pump. The output pressure of the prototype is kept constant by a pressure regulation valve. The return flow is water cooled in order to maintain a pump inlet temperature of 20 °C.

The experiments focus on low rotation speed as the friction losses are higher in this part of the camshaft speed range[6]. During a test cycle, the pump delivery pressure is varied from 20 MPa to 80 MPa and the rotation speed from 300 rpm (idle camshaft speed of a combustion engine) to 1500 rpm (middle range camshaft speed). One cycle lasts 4 hours and is repeated four times to reach an amount of 900,000 revolutions, corresponding to a sliding distance of 7,200 meters in the cam/sliding shoe and piston/cylinder systems. The results presented in this paper were measured at 500 rpm and 1300 rpm, corresponding to a mean relative speed of 0.066 m/s to 0.173 m/s between the sliding parts. At 20 MPa pump delivery pressure, the contact pressure between the cam and the sliding shoe is approximately 3.5 MPa and increases linearly to up to 14 MPa at 80 MPa delivery pressure.

Some material pairs were not tested at low delivery pressure and high speed (figure 8 and 10) because of pressure regulation problems in the fuel circuit. However, these problems were not dependent on the performance of the investigated materials.

Fuels

All the tests have been performed with commercial gasoline with a research octane number of 95. This gasoline contains 5 % ethanol. As the proportion of ethanol is meant to increase in the future due to new legislation, complementary tests have been performed with nearly pure ethanol (99 %). The densities and dynamic viscosities of the fuels are given in table 1.

Table 1. Main characteristics of the fuels used for the investigations.

Fuel type	Density [-]	Dynamic viscosity [10^{-3} Pa.s]
Gasoline	0.739	0.65
Ethanol	0.794	1.2

Measured parameters

The prototype pump is fitted with pressure and temperature sensors in each cylinder. A temperature sensor measures the temperature of the fuel in the pump housing. In addition, the test bench is fitted with a torque sensor between the electric motor and the high-pressure pump. The pressures and temperatures in the fuel circuit as well as the output pump mass flow are also measured.

Analysis method

The pump driving torque and the cylinder pressures are measured at each operating point over 100 cycles with a resolution of 1 °CA. The results presented in this paper were measured after the two first test cycles, as it can be inferred from previous investigations that the sliding parts are already run-in at this stage[6].

It is possible to calculate the total pump efficiency with the mean values of the driving torque (M_d), the angular speed (ω), the intake and outlet pressures (p_o and p_i) and the measured fuel volumetric flow rate (V_{real}) via equation (1)[12].

$$\eta_{total} = \frac{\dot{V}_{real} \cdot (p_o - p_i)}{M_d \cdot \omega} \quad (1)$$

However, the total pump efficiency is not satisfactory for assessing the friction losses in the prototype, as it is not only influenced by the mechanical efficiency (η_{mech}) but also by the volumetric (η_{vol}) and hydraulic (η_{hydr}) efficiencies. The relation between these parameters is given by equation 2.

$$\eta_{total} = \eta_{vol} \cdot \eta_{hydr} \cdot \eta_{mech} \quad (2)$$

The volumetric and hydraulic losses are caused for instance by leakage between the pistons and cylinders and by the pressure losses in the pump valves. In order to evaluate the friction losses via the mechanical efficiency independently from the volumetric and hydraulic efficiencies, the cylinder pressure signals have been used. These signals enable the calculation of the theoretical required pump driving torque assuming that all friction coefficients in the sliding systems are equal to zero. The quotient of the mean values of the calculated ($M_{d,sim}$) and measured ($M_{d,mea}$) signals leads to the mechanical efficiency as shown in equation (3). Figure 6 gives an example of these signals at 80 MPa output pressure and a pump speed of 500 rpm.

$$\eta_{mech} = \frac{M_{d,sim}}{M_{d,meas}} \quad (3)$$

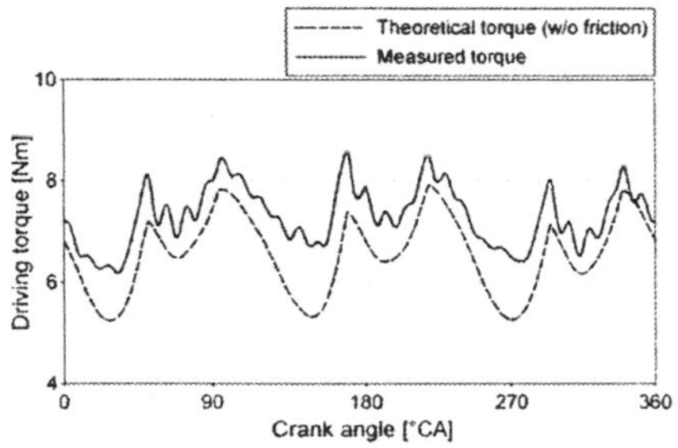

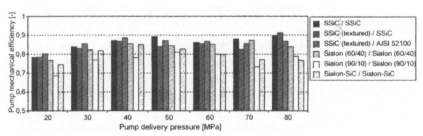

Figure 6. Example of theoretical and measured pump driving torque.

RESULTS

Mechanical efficiency with gasoline as delivered fuel
    Figure 7 shows the mechanical efficiency of the prototype pump operating at 500 rpm with various material pairs in the cam/sliding shoe system and gasoline as delivered fuel. The performance of the pump improves with increasing delivery pressure up to 50 MPa for all investigated material pairs. The results obtained in this lower delivery pressure range is pretty similar for all material combinations except for the Sialon 90/10, which shows higher friction losses. At higher pressure levels, self-mated silicon carbide (textured or not) performs the best with a mechanical efficiency of up to 0.9 at 80 MPa delivery pressure. The combination of textured silicon carbide in combination with AISI 52100 shows very stable performance with an efficiency of approximately 0.85. The lower mechanical efficiency measured with the Sialon pairs can be explained by the lower hardness when compared to silicon carbide on the one hand, but also by their lower thermal conductivity which prevents a good cooling of the sliding surfaces under high stress on the other hand.

Figure 7. Mechanical efficiency of the prototype pump at 500 rpm. Fuel: gasoline.

Figure 8 shows the mechanical efficiency of the prototype with the same material pairs as used in figure 7 at a rotation speed of 1300 rpm. All the investigated materials show almost the same performance over the whole pressure range with an increasing mechanical efficiency of 0.8 at 20 MPa to approximately 0.95 at 80 MPa. This improvement when compared to lower rotation speed (figure 7) can be explained by the better lubrication at higher mean sliding speed in the cam/sliding shoe system. At this speed, the light differences in hardness or in thermal conductivity of the investigated materials do not seem to have a significant influence anymore.

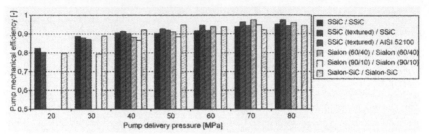

Figure 8. Mechanical efficiency of the prototype pump at 1300 rpm. Fuel: gasoline.

The total pump efficiency of the prototype operating at 20 MPa and 500 rpm with self-mated silicon carbide in the cam/sliding shoe system is approximately 0.7. Investigations performed at the Institut fuer Kolbenmaschinen with commercial pumps (Bosch HDP1 and Bosch CP1) have shown values of 0.2 to 0.45 at this operating point[5]. There is no commercially available gasoline pump operating at higher delivery pressure. The Bosch CP1 (designed for delivering diesel fuel at up to 135 MPa) was tested at 80 MPa, but only at 900 rpm and 1100 rpm with a total pump efficiency of 0.1 to 0.2. At lower speeds, the maximum pressure could not be reached because of increasing leakage between the pistons and the cylinders. At higher speed, the friction caused by the low lubricity of gasoline led to reduced motion of the pistons and very unstable operation. At the same rotation speeds, the prototype amounts a total efficiency of 0.5 to 0.6, thus confirming the benefit of ceramic sliding components.

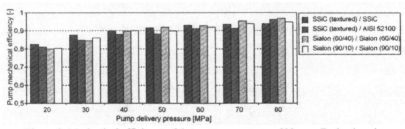

Figure 9. Mechanical efficiency of the prototype pump at 500 rpm. Fuel: ethanol.

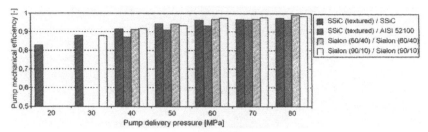

Figure 10. Mechanical efficiency of the prototype pump at 1300 rpm. Fuel: ethanol.

Figures 9 and 10 show the mechanical efficiency of the prototype with ethanol as delivered fuel and lubricant at 500 rpm (Figure 9) and 1300 rpm (figure 10). The self-mated silicon carbide and Sialon-SiC pairs were not tested with this fuel. There are no significant differences between the investigated material pairs. Even the sliding speed does not have a high influence on measurements, as the performance of the pump is only slightly improved with increasing speed. The mechanical efficiency of the prototype increases from approximately 0.8 at 20 MPa to 0.95 at 80 MPa for all material pairs. This enhancement in comparison to gasoline lubrication can be explained by the higher dynamic viscosity of ethanol (see table I).

SEM pictures of the sliding parts

Every cam and sliding shoe was analysed with a scanning electron microscope (SEM) in order to evaluate wear and to better understand the results presented in this paper. Figure 12 shows examples of SEM images taken from the middle of the cam after 16 hours of operation (12 hours in the case of Sialon-SiC composite). Because this region of the surface is always in contact with the sliding shoe (see figure 11), it is subjected to the harshest lubrication conditions.

Former investigations on a single-piston pump have already shown that there is no measurable wear with silicon carbide or with Sialon 60/40[6]. The same observation could be made with the present material in the prototype pump. Some cavities can be seen on the surface of the silicon carbide cam (black spots). These cavities exist before testing and even if they were considered as a material weakness, they keep fuel in the sliding system and improve lubrication as well as thermal conductivity, thus leading to good performance as shown in the previous diagrams. The texture that has been tested on silicon carbide enhances this effect with a higher volume for lubrication and storage of wear particles. In the case of silicon carbide in combination with bearing steel, the micro-dimples are filled with wear material but there is no steel layer on the surrounding contact surface as could be observed without texture[6]. As the first hours of test with this material pair have shown increasing mechanical efficiency over time followed by stable performance in the last hours, it can be inferred that the micro-dimples were filled during the first hours of the test, thus avoiding the adhesive action of the steel wear particles released during the run-in period. The grinding grooves are still visible on the surfaces of Sialon 60/40 and Sialon-SiC composite cams, which indicates very low wear. In former investigations with Sialon 60/40 under similar test conditions, the initial depth of the grooves enabled to calculate an approximate wear-rate of less than 10^{-8} mm³/Nm. Despite its higher hardness, Sialon 90/10 does not lead to better performance than Sialon 60/40 and cracks could be observed on the surface (figure 12). The low thermal conductivity of Sialon, combined with the lower fracture toughness of Sialon 90/10, could explain these observations.

Figure 11. Overview of the cam.
Arrow: sliding direction. Circles: extreme positions of the sliding shoe.

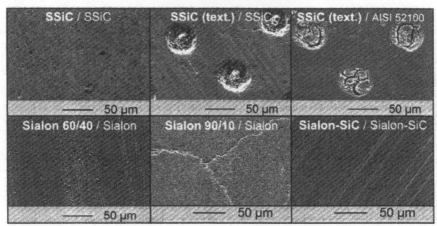

Figure 12. SEM pictures of the investigated cams.

TRANSFER INTO A COMMERCIALLY AVAILABLE HIGH-PRESSURE PUMP
     In order to demonstrate the feasibility of a compact and reliable high-pressure pump with ceramic sliding components, a commercially available pump has been modified and compared with its original design. As the housings of modern gasoline pumps are not designed for delivery pressures above 20 MPa, a Bosch CP1 capable of delivering diesel fuel at 135 MPa has been chosen. The original eccentric shaft bearings, the cams, and the cylinders have been replaced with silicon carbide (EKasik F from ESK Ceramics). The original eccentric shaft, the sliding shoes, and the pistons have been replaced with hardened bearing steel (AISI 52100, 800 HV10). In order to ensure good run-in and stable performance with silicon carbide in combination with bearing steel in the cam/sliding shoe system, the cam has been textured with micro-dimples. A sectional view of the modified pump is shown in figure 13.

Figure 13. Modified Bosch CP1 diesel pump
with ceramic components in the sliding systems.

Both pumps have been investigated with commercial gasoline in the same operating conditions as described earlier in this paper. Figure 14 shows the mechanical efficiency of the original and the modified Bosch CP1 at 500 rpm. The modified pump shows better performance at all investigated operating points with a mechanical efficiency increasing from 0.62 at 10 MPa output pressure to 0.87 at 80 MPa output pressure. Due to geometrical differences between the prototype pump and the modified CP1, the contact pressure between the cams and sliding shoes is higher for a given delivery pressure. At the highest delivery pressure level, this contact pressure is approximately 17 MPa, compared to 14 MPa in the previous prototype. The original CP1 shows mechanical efficiencies of approximately 0.45 at 10 MPa delivery pressure to nearly 0.65 at 40 MPa delivery pressure; it was not able to reach an output pressure of more than 60 MPa at 500 rpm. This can be explained by reduced piston motion due to high friction levels on the one hand, and increased leakage between piston and cylinder because of high wear on the other hand. Moreover, the investigations with the original CP1 have shown a 40% decline in efficiency within 9 hours of testing, thus confirming that conventional materials or coatings are not adequate for fuel-lubricated high-pressure gasoline pumps[5].

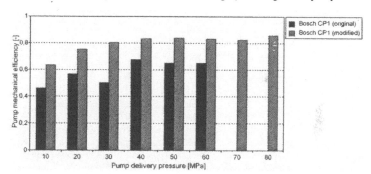

Figure 14. Mechanical efficiency of the original and the modified Bosch CP1 at 500 rpm.

The testing with the original CP1 was stopped after 8 hours (this corresponds to a sliding distance of 5200 meters in the piston/cylinder and cam/sliding shoe systems) as the mechanical and volumetric efficiencies of the pump were too low to continue the experiment. The modified CP1 was investigated for 12 hours (corresponding to a sliding distance of 6500 meters in the piston/cylinder and cam/sliding shoe systems). Figure 15 shows pictures of the cams, pistons and sliding shoes of both pumps after the test. Scratches are visible on the surfaces of the original CP1 components, thus confirming the high friction and wear levels that could be observed during the experiments on the test bench. The coating on the cam disappeared and has been partially transferred to the previously uncoated sliding shoe. In contrast, no surface deterioration could be observed on the components of the modified CP1. As already observed on the previous prototype, the surfaces are smoothed but the grinding grooves of the cams and the sliding shoes are still visible in some regions of the surfaces.

Figure 15. Pictures of sliding components of the original (top) and the modified (bottom) Bosch CP1 after investigation.

CONCLUSION

This paper proves that the use of ceramic components in the sliding parts of a high-pressure gasoline pump enables fuel delivery at pressures of up to 80 MPa with very low friction and wear levels. Sintered silicon carbide shows the best global performance regarding the pump mechanical efficiency but Sialon with an α/β-ratio of 60/40 gives the advantage of a higher bending strength and fracture toughness, thus allowing an easier integration in very compact pump designs. Sialon with an α/β-ratio of 90/10 seems inadequate for this application as it leads to higher friction losses and cracks could be observed on the surface of the investigated components. The lower pump efficiency observed with Sialons can be explained by their lower hardness and thermal conductivity which lead higher friction in the sliding systems. A good compromise can be reached with Sialon-SiC composite which combines high strength and high hardness as well as a higher thermal conductivity than pure Sialons.

Texturing the cam when using self-mated silicon carbide does not lead to a significant improvement at the operating points investigated in this paper. However, further investigations at lower pump rotation speeds have demonstrated a significant increase of the pump mechanical efficiency[6]. The texture also allows the use of silicon carbide in combination with bearing steel, thus enabling a simpler and more economical design than one that utilizes self-mated pairs. The integration of this material combination in a commercially available high-pressure pump leads to an increase of the mechanical efficiency of approximately 30% when compared to the original pump.

The use of a high-pressure pump capable of delivering fuel at up to 80 MPa could help to improve the mixture formation in gasoline direct injection engines. This could help to exploit the full benefit of this technology and thus to reduce the specific fuel consumption as well as the pollutant emission of combustion engines.

REFERENCES

[1] Spicher, U. et al.: Ottomotor mit Direkteinspritzung. Vieweg, ISBN 978-3-8348-0202-6, 2007.

[2] Anderson et. al: Understanding the thermodynamics of direct injection spark ignition (DISI) combustion systems: An analytical and experimental investigation. SAE-Paper 962018, 1996.

[3] Nauwerck, A., Pfeil, J., Velji, A., Spicher, U., Richter, B.: A basic experimental study of gasoline direct injection at significantly high injection pressures. SAE-Paper 2005-01-0098, 2005.

[4] Schumann, F., Buri, S., Kubach, H., Spicher, U., Hall, M.: Investigation of particulate emissions from a DISI engine with injection pressures up to 1000 bar". 19. Aachener Kolloquium Fahrzeug- und Motorentechnik, Aachen, 2010.

[5] Pfister, C, Kubach, H., Spicher, U.: Experimental investigations on a high pressure gasoline pump with ceramic sliding systems operating at up to 80 MPa. SIA International Conference and Exhibition, Strasbourg, 2011.

[6] Pfister, C, Kubach, H., Spicher, U.: Increasing the operating pressure of gasoline injection pumps via ceramic sliding systems. 35th International Conference and ex-position on Advanced Ceramics and Composites, Daytona Beach, 2011.

[7] Häntsche J. P., "Entwicklung und experimentelle Untersuchungen einer Hochdruckpumpe für Ottokraftstoff basierend auf ingenieurkeramischen Gleitsystemen", Logos Verlag Berlin, ISBN 978-3-8325-2464-7, 2009.

[8] Ekström T., Nygren M.: SiAlON Ceramics. Journal of the American Ceramic Society, 75, (1992), 259-276.

[9] Abo-Naf S.N., Dulias U., Schneider J., Zum Gahr K.-H., Holzer S., Hoffmann M.J.: Mechanical and tribological properties of Nd- and Yb-SiAlON composites sintered by hot isostatic pressing. J. Mater. Process. Tech., 183: 264-272, 2007.

[10] Schneider, J., Riva, M., Hoffmann, M. J.: Tribologische Charakterisierung von Sialon-SiC-Mischkeramiken im Hinblick auf den Einsatz in Kraftstoff geschmierten Gleitsystemen. Tribologie + Schmierungstechnik, 58-1, (2011), 22-26.

[11] Wöppermann, M., Zum Gahr, K.-H., Schneider, J.: SiC-Gleitkomponenten mit deterministischen, Texturen unter reversierender Beanspruchung in niedrigviskosen Flüssigkeiten. Tribologie-Fachtagung Gesellschaft für Tribologie, Göttingen, s. 25/1-11, 2008.

[12] Löhner, K. E.: Kolbenpumpen und Kolbenverdichter. Hrsg.: A. Kuhlenkamp. Wissenschaftliche Verlagsanstalt K. G. Hannover, 1948.

CONTACT INFORMATION

Dipl.-Ing. Christophe Pfister
Karlsruhe Institute of Technology (KIT), Institut fuer Kolbenmaschinen
Kaiserstrasse 12, 76131 Karlsruhe, Germany
Tel.: +49 721 / 608 48528 Fax: +49 721 / 608 48578
e-mail: christophe.pfister@kit.edu

ACKNOWLEDGEMENTS
This study is funded within the Collaborative Research Centre SFB 483 "High performance sliding and friction systems based on advanced ceramics" by the Deutsche Forschungsgemeinschaft, Federal Ministry of Education and Research, Germany.

DEFINITIONS/ABBREVIATIONS
- BDC: Bottom Dead Centre
- CA: Crank Angle (eccentric shaft angle)
- FAST: Field Assisted Sintering Technique
- GDI: Gasoline Direct Injection
- SEM: Scanning Electron Microscope
- SiC: Silicon Carbide
- SSiC: Sintered Silicon Carbide
- TDC: Top Dead Centre

MACHINABILITY STUDIES OF Al/SiC/B$_4$C METAL MATRIX HYBRID COMPOSITES USING PCD 1600 GRADE INSERT

Akshay Maheshwari [1*], E.N.Ashwin Kumar [2], and Anuttam Teja [3]

1, 2, 3- Under graduate students, Department of Mechanical Engineering, Sri Venkateswara college of Engineering, Pennalur, Sriperumbudur – 602 105, Tamil Nadu, India

ABSTRACT

Hybrid Metal Matrix Composites (MMC) (Al-SiC/B$_4$C) are widely used in aeronautical and automobile industries due to their excellent mechanical and physical properties. Due to the above properties these composites will slowly replace the conventional material application in aeronautical and automotive industries. However machining these composites is difficult because of the hard reinforcement of SiC and B$_4$C particles. Wear is high which reduces the life of the tool. Conventional and coated tool materials lose their life within a short period of time. This paper presents the experimental investigation on turning A356 matrix metal reinforced with 10 % by weight of silicon carbide (SiC$_p$) particles and 5% by weight of boron carbide particles (the particle size of the silicon carbide and boron carbide ranges from 20 microns to 50 microns) fabricated in house by stir casting method. Fabricated samples are turned on medium duty lathe of 2kW spindle power with poly crystalline diamond (PCD) inserts of PCD grade 1600 at various cutting conditions. Parameters such as power consumed by main spindle, machined surface roughness, and tool wear and tear are studied. Scanning Electron Microscope (SEM) images support the result. It is evident that surface finish and power consumed are good for PCD grade 1600 at higher cutting speed also tool wear is strongly dependent on abrasive nature of hard reinforcement particles.

INTRODUCTION

Metallic matrix hybrid composites have found considerable applications in aerospace, automotive and electronic industries [1] because of their improved strength, stiffness and increased wear resistance over unreinforced alloys [2]. However, the final conversion of these composites into engineering products is always associated with machining, either by turning or by milling. A continuing problem with hybrid MMCs is that they are difficult to machine, due to the hardness and abrasive nature of the reinforcing particles [2,3]. The particles used in the MMCs are harder than most of the cutting tool materials. Most researchers report that diamond is the most preferred tool material for machining MMCs and hybrid composites [4-9] Research on machining MMCs is concentrated on the study of cutting tool wear and wear mechanism [1-3,10,15,16]. Researchers investigated the performance of polycrystalline diamond in machining MMCs which contained aluminum oxide fiber reinforcement. They compared the tool life of cemented carbide with PCD and concluded that sub-surface damage is greater with cemented carbide than that of PCD tools.

Lane [4] studied the performance of PCD tools of different grain size. He reported that, PCD tools with a grain size of 25µm are better able to withstand abrasion wear than tools with grain size 10 µm. He also reported that further increases in the grain size do not have any influence on the tool life but cause significant deterioration in the surface roughness.

The works carried out by Andrews et al. [11] characterized the wear mechanisms of PCD and CVD diamond tools in the machining of MMCs. The conclusions can be applied for design of better diamond tools and optimization of the machining process. There are few research works in hybrid polymer composites. But only a limited number of papers are available in hybrid metal matrix composites. In view of above machining problems, the main objective of the present work is to investigate the influence of cutting parameters on surface finish and power consumption. The results are analyzed to determine the best machining parameters. Tool flank wear was studied by using the best parameter for time duration of 30 minutes and also to study the tool wear pattern.

EXPERIMENTAL PROCEDURE

Cylindrical bars (60mm D x 330mm L) having 10% of SiC particles and 5% of B$_4$C on matrix of Al 356, were fabricated using stir casting method. The bars were turned on self centered three jaw chuck, medium duty lathe of spindle power 2 KW. Fig -1 shows the microstructure of 10% SiC and 5% of B$_4$C particles (of size ranging from 20 microns to 50 microns) reinforced work piece. Table -1 shows the chemical composition of the work piece for experimentation. Parameters such as power consumed by main spindle were measured using a digital wattmeter (make-Nippon Electrical Inst.Co, Model 96x96–dw 34 Sr.No:070521485 CTR 5A/415 V AC F.S 4 KW). The machined surface was measured at three different positions and the average surface roughness (R$_a$) value was taken using a Mitutoyo surf test (Make-Japan –Model SJ-301) measuring instrument with the cutoff length 2.5 mm.

According to Taguchi method, three machining parameters are considered as controlling factors (cutting speed, feed rate and depth of cut) and each parameter has three levels. Table -2 shows the machining parameters and tool insert specification. Table – 3 shows the experimental data for PCD 1600 grade insert on machining A356/10%SiC/5%B$_4$C

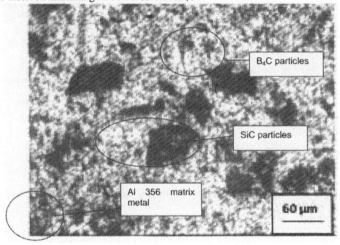

Fig-1 Microstructure of Al 356 reinforced with 10% SiC and 5 % B$_4$C

Figure 2 shows the experimental set up with Aluminium hybrid composite, consisting of 15% SiC and 5% B₄C particles reinforced composite

Fig –2 Experimental set up with Al /Hybrid composite

Table1.Chemical composition of Al-SiC(20p)-MMC

Type of Hybrid MMC	Reinforcement	%SiC	%B₄C	%Si	%Mg	%Fe	%Cu	%Mn	%Zn	%Ti	%Al
Particulate MMC	SiC and B₄C -20-50 μm	10.00	5.00	7.77	0.63	0.15	0.16	0.09	0.08	0.10	Balance

Table 2.Machining parameter

Cutting Speed	75,120, and 180 m/min
Feed Rate	0.1,0.2, and 0.3 mm/rev
Depth of Cut	0.5, 0.75 and 1.0 mm
Tool Holder	PCLNR 25*25 M 12
Tool Insert	CNMA 120408 (1600) Grade)

Table - 3 Experimental Layout using an L27 orthogonal array and corresponding response values for PCD 1600 grade

	Machining Parameter			Response(s)	
Group No.	A Cutting Speed	B Feed	C Depth of cut	power consumed in Kw *	Surface Roughness (Ra) in Micron *
1	1	1	1	0.34	1.60
2	1	1	2	0.41	1.79
3	1	1	3	0.30	1.83
4	1	2	1	0.47	3.77
5	1	2	2	0.57	3.95
6	1	2	3	0.42	3.37
7	1	3	1	0.54	6.53
8	1	3	2	0.64	6.74
9	1	3	3	0.50	6.76
10	2	1	1	0.49	1.4
11	2	1	2	0.53	1.65
12	2	1	3	0.39	1.86
13	2	2	1	0.52	1.91
14	2	2	2	0.77	3.77
15	2	2	3	0.59	2.67
16	2	3	1	0.73	7.17
17	2	3	2	0.9	6.75
18	2	3	3	0.66	6.9
19	3	1	1	0.70	1.6
20	3	1	2	0.76	1.89
21	3	1	3	0.57	1.98
22	3	2	1	0.79	3.53
23	3	2	2	1.30	3.78
24	3	2	3	1.13	3.36
25	3	3	1	1.12	6.8
26	3	3	2	1.34	7.12
27	3	3	3	1.0	7.06

* Average value of three trails

RESULTS AND DISCUSSIONS

Power Consumed

Figs 3 and 4 show the plot between cutting speed and power consumed for depth of cut 0.5 mm and 1.0 mm respectively. It was observed that power consumed increased as cutting speed increased for all combinations of machining conditions. Power consumed was greater at higher cutting speed and it decreased at the lower cutting speed. It was observed that more power was required to pull than to cut the particles [4]. Power consumed increased as the content of reinforcing particles increased. At higher cutting speeds removal of hard silicon particles and boron carbide particles from aluminum matrix becomes easier [7],[8]. It was also observed that(fig.3), with a feed rate of 0.3 mm/rev, machining with PCD 1600 grade insert is more effective than with a feed rate of 0.1 and 0. 2 mm/rev at 0.5 mm depth of cut. This trend was changed when machining the work piece at 1.0 mm depth of cut, where power consumed was more at lower cutting speed for feed rate 0.3 mm/rev compared to 0.2 mm/rev. This is true at higher depth of cut where power required is high irrespective of feed rate. At 0.5 mm depth of cut, power consumed was less at higher cutting speed at 0.3 mm feed rate compared to other feed rates.

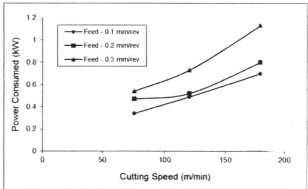

Fig 3 Cutting speed Versus Power consumed (Depth of cut 0.50 mm)

When machining the work piece at 1.0 mm depth of cut, power consumed was slightly greater than machining the work piece at 0.5 mm depth of cut. It was observed that, to decrease the power consumed and to improve the tool life, it is necessary to machine the work piece at lower depth of cut with low feed rate.

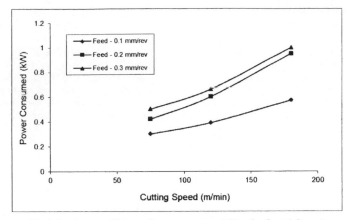

Fig 4 Cutting speed Versus Power consumed (Depth of cut 1.0 mm)

Surface Roughness

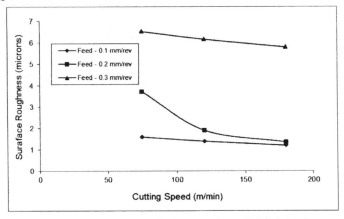

Fig 5.Cutting speed Versus Surface Roughness (DOC=0.5mm)

The turning operation was performed at feed rates 0.1, 0.2 and 0.32 mm/rev with depth of cut 0.5, 0.75 and 1.0 mm respectively. The influence of surface finish on cutting speed is represented in figs 5 and 6. The figures show the relationship pertaining to depth of cut 0.5 mm and 1.0 mm respectively for machining the samples. General trend of the graphs show that average surface roughness value Ra decreased as the cutting speed increased. Similar trend exists for other depth of cut also. In both the figures machining with low feed rates shows good surface finish irrespective of depth of cut. It is believed that depth of cut has less influence on surface roughness. Machining with depth of cut 0.5 mm, and feed 0.1 mm/rev shows better surface roughness at higher cutting speed. This is due to the fact that at higher cutting speed removal of hard silicon particles from the matrix becomes easier [5].

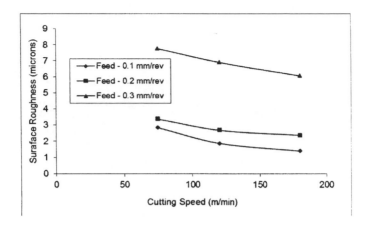

Fig 6.Cutting speed Versus Surface Roughness (DOC= 1.0mm)

Tool Wear:

From the above observations best machining parameter was determined at cutting speed of 180 m/min, feed rate of 0.1mm/rev and depth of cut of 0.5 mm.

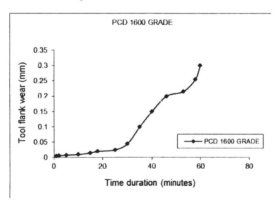

Fig – 7 Time duration versus tool flankwear

Setting this cutting condition as a constant parameter and the samples machined for a time duration of 60 minutes a tool flank wear study was carried out. The tools were monitored for normal types of wear, namely flank wear, crater wear and nose wear using a tool maker's microscope. Tool flank wear was caused by abrasive nature of the hard particles present in the work piece. At low cutting speed a worn flank encouraged the adhesion of work piece material on the tool insert and formed a built-up-edge feature [1,12,14].

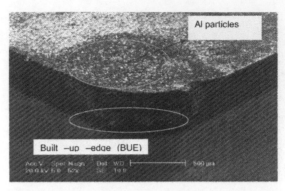

Fig – 8 SEM image of PCD 1600 grade insert after machining the workpiece for 60 minute

duration

It is proved that silicon and boron carbide particles, which have high hardness abrade the cutting tool. It is observed that the tool life of PCD 1600 grade is performing well in the chosen cutting condition.

CONCLUSIONS
1. Power consumed is less at lower cutting speeds because of less friction between tool and work piece interface.
2. Primary wear mechanism is believed to be abrasion between reinforcing particles and cutting tool material.
3. Tool wear is more in flank portion of the PCD tool.
4. Surface finish improves at higher cutting speeds. Surface finish is good at higher cutting speeds with low depth of cut at lower feed rate
5. Tool wear strongly depends on cutting speed followed by depth of cut.
6. Surface roughness strongly depends on feed rate. As feed rate increases surface roughness also increases.
7. It is concluded that PCD 1600 grade is preferred for chosen cutting parameters.

REFERENCES
1. Tomac N and Tonnessen K (1992),"Machinability of particulate Aluminum Metal Matrix Composites", . Annals of the CIRP, Vol. 41, pp. 55-58.

2. Weinert. K (1993), "A consideration of tool wear mechanism when machining metal matrix composites (MMC)", CIRP Ann, Vol. 42, pp. 95-98.

3. D. Rouby and P. Reynaud, Fatigue behaviour related to interface modification during load cycling in ceramic-matrix fibre composites, *Compos Sci Technol* **48** (1993), pp. 109–118.

4. Lane G, (1992) , "The effect of different reinforcement on PCD tool life for aluminium composites", Proceedings of the Machining of Composites Materials Symposium, ASM Materials Week, Chicago, IL, pp. 3-15.

5. Manna A and Bhattacharyya B (2003), "Study on Different Tooling Systems during Turning for Effective Machining of Al-SiC MMC", Journal of Production Engineering Institution of Engineers, Vol. 83, pp. 46-50.

6. Muthukrishnan N and Paulo Davim J (2009), "Optimization of machining parameters of Al-SiC –MMC with ANOVA and ANN analysis", J. Mater. Process Techno, Vol. 209. pp. 225-232.

7. Muthukrishnan N,Murugan.M, and Prahlada Rao K (2008), "Machinability issues in turning of Al-SiC (10p) metal matrix composites", Int J Adv Manuf. Technol, Vol. 38. pp. 21- 218.

8. Paulo Davim. J, and Montiro Baptista A (2000), "Relationship between cutting force and PCD cutting tool wear in machining silicon carbide reinforced aluminum", J. Mater. Process Technol, Vol. 103, pp. 417-423.

9. Pramanik A , Zhang LC and Arsecularatne JA (2006), "Prediction of cutting forces in machining of metal matrix composites", International Journal of Machine Tools & Manufacture, Vol. 46 pp. 1795 – 1803.

10. Heat P (1991), "Cutting composites with PCD", Comp. Manuf. Vol. 7, pp 1-3.

11. Andrewes C, Feng H, and Lau W (2005), "Machining of an aluminum/SiC composite using diamond inserts", J.Mater. Process. Techno, Vol. 102, pp. 25-29.

12. Caroline J.E, Andrews, His-yung Feng, and Lau W.M (2000), "Machining of an aluminum - SiC composite using diamond inserts", Journal of Materials Processing Technology, Vol. 102, pp. 25-29.

13. G.A. Chadwick G.A and Heat P (1990), "Machining metal matrix composites", Met. Mater. Vol. 6, pp 73-76.

14. T. Clyne T and Withers P (1995), "An introduction to metal matrix composites", Camb. Solid State Sci.SER. pp 1-10.

15. F. Christin, Design, fabrication and application of C/C, C/SiC and SiC/SiC composites. In: W. Krenkel, R. Naslain and R. Schneider, Editors, High temperature ceramic matrix composites **vol. 4**, Wiley-VCH Press, Weinheim (2001), pp. 31–43.

16. G.N. Morscher, G. Ojard, R. Miller, Y. Gowayed, U. Santhosh and J. Ahmad et al., Tensile creep and fatigue of Sylramic-iBN melt-infiltrated SiC matrix composites: retained properties, damage development, and failure mechanisms, Compos Sci Technol **68** (2008), pp. 3305–3313.

FRETTING FATIGUE FAILURE OF ENGINEERING CERAMICS

C. Wörner *
K.-H. Lang *

* Karlsruhe Institute of Technology, Institute for Applied Materials (IAM-WK), Karlsruhe, Germany

ABSTRACT

Rollers made of engineering ceramic are very well suitable for hot rolling of high-strength steels. During the shaping process the ceramic rollers are highly loaded. The appearing complex loading is caused by a cyclic loading from the rotation of the rollers and high Hertzian stresses connected with friction, caused by the contact between the rollers and the rolling stock during the shaping process.

To investigate the lifetime behavior of potential roller materials, fretting fatigue tests are carried out in a modified four-point-bending unit, where a fretting pad is pressed onto the tensile loaded side of the bending bar and oscillated. In the test series, carried out on silicon-nitride and alumina, different types of the fretting loading were realized.

In all test series the ceramics are damaged by the fretting loading by the initiation and/or activation of cracks. In static tests a fretting fatigue loading leads to lower residual strengths.
The strong tensile field in the specimen caused by the oscillating fretting pad leads to an activation of the surface defects, but doesn't influence the R-curves of the materials.

In cyclic tests the additional cyclic four-point-bending-loading causes a strong degradation of the reinforcement elements, which leads to a more lower R-curve than expected from the pure four point bending and the lifetime gets significant lower.

The variation of the fretting loading helps to separate and understand the damage mechanisms.

INTRODUCTION

Recently, components made from engineering ceramics are increasingly applied for tasks, where they replace high strength steels or other hard metals. Advantages are the lower density and the higher specific strength, in comparison to metals or cemented carbides. One example is the rolling of high strength steel wires, where silicon-nitride rollers are used instead of a roller made from hard metal. This causes an increase of 100 % of service time [1]. The service time of the ceramic rollers is limited by the deterioration of the roll pass. Fretting fatigue causes spalling of material from the rollers after a certain number of revolutions.

In principle, fretting fatigue can arise whenever in the contact zone between two loaded bodies cyclic relative movements appear. Even very small relative movements of only a few μm suffice to induce fretting fatigue. During the rolling process of steel wire such a load situation occurs. The wire entering the process zone possesses a lower speed than the roller surface. With the onset of the shaping process the Hertzian stresses increase and the rolling stock is accelerated. Crossing the neutral point where rolling stock and the surface of the rollers posses the same speed, the wire is further accelerated and ejected. The different speeds of the rolling stock and the surface of the rollers cause friction in the contact zone. In addition the rollers experience a cyclic load due to the rotation. Hence, according to [2], all preconditions for fretting fatigue are fulfilled. In the present work this kind of complex loading is simulated in laboratory tests by performing standard four-point-bending tests with superposed fretting load on the tensile loaded side of the sample.

TESTED MATERIAL, EXPERIMENTAL METHODS AND ANALYSIS

Tested material

The fretting fatigue tests were performed on two different types of engineering ceramics: specimens produced from alumina from FriaTec (identification F99.7) and specimens produced from silicon nitride from CeramTec (identification SL200BG).

F99.7 was extensively tested under different load conditions in [3] [4] [5]. Data for these materials can be found in [4] and [5]. A selection of physical properties of F99.7 and SL200BG are listed in Table I.

Table I: charact. data of the ceramic materials [4] [5] [6]

		Alumina	Silicon Nitride
Density	[g/cm³]	3.9-3.95	3.19
Compression strength	[MPa]	3500	2500
Bending strength (4-Point)	[MPa]	384	1044
Young's modulus	[10³ MPa]	380	310
Weibull modulus (m)		22	14±2
Poisson's ratio (v)		0.22	0.27
max. working temperature	[°C]	1950	1300

F99.7 is a commercially available engineering ceramic material, purchased from FriaTec as blanks. The grains are globular with a median grain diameter of 7 μm. The portion of the glassy phase with 0.3 % is very low and the material is not completely free of pores. A micrograph of the microstructure is shown in Figure 1 a).

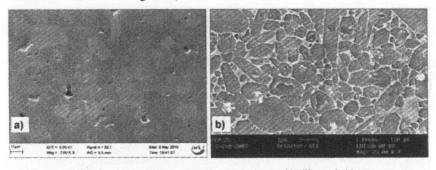

a) alumina   b) silicon nitride

Figure 1: micrographs of the tested materials

SL200BG specimens were produced using powder obtained from CeramTec. The blanks were isostatically cold-pressed at IAM-KM of KIT and sintered by CeramTec. SLB200BG has elongated grains with a median grain diameter of 0.7 μm and an aspect ratio of three. The portion of the glassy phase is relatively high with 14 %. The glassy phase can be seen in the micrograph section in Figure 1 b) as the light grey phase between the elongated grains.

From the blanks of the two materials four-point-bending bars according to DIN 843-1 [7] (equal to ASTM C 1161-20c [8]) with dimensions of 4x3x50 mm³ were produced. The edges are chamfered to eliminate sharp-edge-effects. The tensile sides of the specimens was fine grinded to minimize defects caused by the production process.

Experimental methods

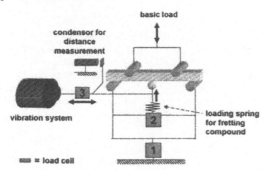

Figure 2: scheme of the experimental setup

All tests are carried out using electro-mechanical testing machines from the Instron E series. For testing purposes, an existing experimental setup [3] was modified according to [9]. Figure 2 shows a scheme of the experimental setup. Four-point bending loading is realized according to DIN 843-1 [7].

The basic load is induced by a servo-electric linear actuator with a maximum force of 6 kN from above via the inner supporting spans. The fretting load can be applied using different fretting pads: balls with a diameter of 5 mm or 15 mm or a pin with a diameter of 2 mm. The fretting pad is pressed against the tensile side of the specimen centered between the two supports. The force acting against the specimen is adjusted by a spring and a lever system. It is held constant during the whole test. The fretting pad is fixated in a retention system and can precisely be moved across the surface of the four-point-bending specimen by a permanent-magnet excited oscillator of the type TIRA540. The appearant friction force is registered by a load cell. A radiation furnace allows tests to be carried out at elevated temperatures up to 1200° C.

Experimental procedure

Two types of experimental procedure were applied: the static tests and the cyclic test.
For static testing the specimen is put into the four-point-bending test rig and fastened with a force of 150 N, to avoid movement of the specimen. The fretting pad is pressed against the tensile side with a force of 10 N and oscillated at a afrequency of 40 Hz with defined sliding amplitude for a given number of cycles. At alumina, fretting fatigue tests were made with a ball of 5 mm diameter at a sliding amplitude of $s_a = 300$ µm, at silicon nitride with a pin at 2 mm diameter at an amplitude of $s_a = 100$ µm. The fastening force of the test rig is held constant during the procedure at a defined low level to avoid effects of sub critical crack growth [10]. After the predefined number of cycles is reached, the fretting pad is removed from the specimen and the fracture strength is determined by loading the specimen via the four-point-bending rig at a loading rate of $\dot{\sigma} = 1000$ MPa/s until breakage. A test series consists of at least five specimens. To determine the initial strength of the material at least 15 undamaged specimens are tested.

For cyclic testing the specimen is fitted into the testing rig and loaded with the mean load. After the mean load is reached, the fretting pad is pressed, centered between the supports, against the tensile loaded side of the specimen. The dynamic four-point-bending loading and the fretting loading start cycling at the same time with the same frequency of 40 Hz if not declared differently. The stress ratio of the cyclic basic load is $R = 0.5$, the sliding amplitudes are the same as in the static testing. The experiment ends with the breakage of the specimen or reaching the ultimate number of cycles of $N = 10^7$ cycles. For every load case at least ten specimens are tested.

Analytical evaluation

Cyclic loading may result in an alteration of R-curves. Hereby the loading causes a degradation of the effectiveness of reinforcement elements up to the complete loss of effectiveness. The reduction of R-curves can be determined from cyclic load-reduction-experiments on notched specimens [11].

Schwind describes in [5] a further method for estimating sections of altered R-curves. Here, the initial crack length in strength tests, resp. residual strength tests from the static testing, is estimated by the tangent-method [12]. According to equation (eq. 1), with "i" being the value of an individual specimen and "n" being the total number of specimens investigated a probability of failure F can be assigned:

$$F = \frac{i}{n+1}$$ (eq. 1)

With this, every specimen i can be correlated with an initial crack length a_i, known from the strength measurement, calculated with the tangent method. Under the assumption that the starting-size of the damage inducing defects is always equally distributed, every number of cycles to failure in cyclic testing at a certain loading condition can be assigned to a definite initial crack length.

If failure appear on at least two specimens with different loading conditions, the stress intensity factor K can be determined according to equation (eq. 2), in dependency of the load amplitude and the initial crack length. In equation (eq. 2) "σ" describes the stress, "a" the crack length and "Y" a geometrical factor, depending to the crack form and size. The stress intensity factor K can now be plotted.

$$K = \sigma\sqrt{a\pi}Y$$ (eq. 2)

The resulting K-characteristics depict an envelope of the reduced R-curves. This allows the R-curves to be estimated qualitatively in a section. It only allows calculating a part in the transitional area. K_0 and an eventually appearance of a plateau (often described as K_{IC}) can't be described with this method.

For the determination of the alteration of R-curves due to fretting fatigue loading lifetime, data points with approximately the same number of cycles have to be in an area where residual strength measurements, resp. inert strength measurements for the pure cyclic loading, has been carried out.

It is possible to calculate the crack length even at fretted specimen with the tangent method and the monotonic R-curve. To be sure, there is no damage and/or reduction of the reinforcement elements by the fretting loading, that affect the crack length, several fracture surfaces of fretted specimen where investigated with SEM and the total crack length was measured. This measured crack length was compared to the calculated crack length.

RESULTS

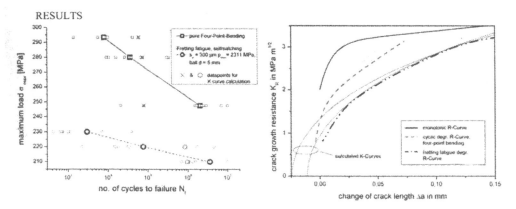

Figure 3: SN curves Alumina   Figure 4: K- and R-curves Alumina

Figure 3 shows different SN-curves of tests from alumina under cyclic and fretting fatigue loading. The SN-curve of pure cyclic four-point-bending shows a clear reduction of the lifetime with increasing maximum tensile stresses. The fretting fatigue SN-curve was determined with a ball as the fretting pad and a friction stroke of $s_a = 300$ μm. It is clearly visible, that the additional fretting loading reduces the endurable four-point-bending loading significantly. Obviously the additional fretting loading damages the material severely and reduces the lifetime.

Figure 4 shows different K-curves and R-curves. The continuous line depicts the monotonic R-curve of the tested material. The dashed line shows an altered R-curve which was estimated after a cyclic loading with 10^5 cycles. The enveloping K-curves were determined from the three data points marked by X in Figure 3. It can be seen that alumina, even though it does not possess a pronounced R-curve behavior, but instead a large transition region, experiences a degradation of reinforcement elements through pure cyclic loading. This degradation alters the characteristic of the R-curve. It leads to the conclusion, that the reduction of the maximal endurable load with increasing lifetime is not induced by pure sub critical crack growth.

The dash-dotted line shows the degraded R-curve after 10^6 cycles of fretting fatigue tests, calculated from the circled data points in Figure 3. Here the transition region is even wider and the whole R-curve is shifted to smaller K-values compared to the R-curve after pure cyclic loading. This pronounced alteration of the R-curve is caused by the superimposed fretting loading which leads to a degradation of the effectiveness of reinforcement elements.

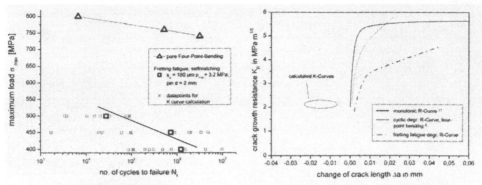

Figure 5: SN-Curves Silicon-nitride   Figure 6: K- and R-Curves Silicon-nitride

Figure 5 shows results from cyclic fretting fatigue tests on silicon-nitride with a silicon-nitride pin as a fretting pad. It also shows a life-time curve of pure cyclic loading from [5]. It can be seen, that the additional fretting loading, even with the low contact pressure, has a great influence on the endurable cyclic four-point bending loading. This leads to the expectation, that additional fretting loading causes accessory damage to the material, which exceeds the pure erosion of surface material. The determination of K-curves in Figure 6 after approximately 10^5 cycles shows a clear degradation of the R-curve compared to the monotonic R-curve from [13]. Compared to the cyclic R-curves from [5], which were obtained after approximately $2.3 \cdot 10^5$ cycles, the R-curve after fretting fatigue loading show a distinct reduction of the transition region and a shift to longer crack lengths. This leads to the conclusion, that the additional fretting load severely reduces the effect of reinforcement elements, even at only the half of the loading time.

SUMMARY, CONCLUSION AND OUTLOOK

By pure fretting loading, the initial defects in the material are enlarged and/or new cracks are generated and/or activated in the contact zone of the fretting loading. These enlarged and/or new cracks reduce the residual strength of the material according to the number of applied fretting cycles. But the local stresses caused by the fretting loading are not high enough to damage or destroy the reinforcement elements significantly. Is a cyclic four point bending loading superimposed the SN-curves are massively shifted to lower endurable maximum stresses compared to the SN-curves of pure four-point-bending loading. This lifetime reduction is a direct result of the reduction of the effectiveness of the reinforcement elements. This behavior can be described with the degraded R-curve. The degradation after superposed fretting loading is much higher then after pure four-point-bending. However because there could not be observed any degradation of the reinforcement elements after pure fretting loading the fatigue effects can not be separated.

In continuative studies the fretting fatigue behavior of ceramics should be investigated in tests with high strength metallic fretting pads. Also the length of the starting cracks will be determined using materialographic methods. For this, a dye impregnation method proposed in [14] and [15] for ceramic materials should be used. In this method Palladium is impregnated under high pressure into the cracks, which can be detected on the fractured surface with back scatter electron emission after testing. This gives the possibility to see the starting crack length and as well as the length of the stable grown crack. So a more detailed description of the K- and R-curves will be possible.

ACKNOWLEDGMENT

This work was funded by the German Research Foundation (DFG) within the Center of Excellence SFB 483 in subprojects C6.

REFERENCES

[1]A. Kailer and T. Hollstein. Walzen mit Keramik. Fraunhofer IRB Verlag, 2004.
[2]D.A. Hills and D. Nowell. Mechanics of fretting fatigue. Kluwer Academic Publishers, 1994.
[3]Rachid Nejma. Verformungs- und Versagensverhalten von Aluminiumoxidkeramik unter isothermer und thermisch-mechanischer Ermüdungsbeanspruchung. Shaker, Aachen, 2007.
[4]Thomas Schalk. Reibermüdungsverhalten ingenieurkeramischer Werkstoffe. Shaker, Aachen, 2010.
[5]Thomas Schwind. Untersuchungen zum zyklischen Ermüdungsverhalten von Si_3N_4 und Al_2O_3 ausgehend von natürlichen Fehlern. Shaker, Aachen, 2010.
[6]Hermann Salmang and Horst Scholze. Keramik. Springer, Berlin; Heidelberg [u.a.], 2006.
[7]Deutsches Institut für Normung e.V. DIN EN 843-1 Hochleistungskeramik - Mechanische Eigenschaften monolithischer Keramik bei Raumtemperatur - Teil 1: Bestimmung der Biegefestigkeit. Berlin, 2008.
[8]ASTM International. ASTM C 1161-02c - Standard Test Method for Flexural Strength of Advanced Ceramics at Ambient Temperature. ASTM International, 2008.
[9]M. Okane, Y. Mutoh, Y. Kishi, and S. Suzuki. Static and Cyclic Fretting Fatigue Behaviour of Silicon Nitride. Fatigue Fract. Engng. Mater. Struct. Vol. 19, No. 12, 1996. p. 1493-1504.
[10]Dietrich Munz and Theo Fett. Ceramics. Springer, Berlin; Heidelberg [u.a.], corr. 2. print. edition edition, 2001.
[11]J.J. Kruzic, R.M. Cannon, J.W. Ager III, and R.O. Ritchie. Fatigue threshold R-Curves fpr predicting reliability of ceramics under cyclic loading. Acta Materialia 53, 2005. p. 2595-2605.
[12]Dietrich Munz. What Can We Learn from R-Curve Measurements? Journal of the American Ceramic Society, 90, 2007. p. 1-15.
[13]Stefan Fünfschilling. Mikrostrukturelle Einflüsse auf das R-Kurvenverhalten bei Siliciumnitridkeramiken. Schriftenreihe des Instituts für Keramik im Maschinenbau ; 53. KIT Scientific Publishing, Karlsruhe, 2010.
[14]Wataru Kanematsu, Mutusuo Sando, Lewis K. Ives, Ryan Marinenko, and George D. Quinn. Dye Impregnation Method for Revealing Machining Crack Geometry. Journal of the American Ceramic Society, 84, 2001. p. 795-800.
[15]Wataru Kanematsu and Lewis K. Ives. Propagation Behavior of Machining Cracks in Delayed Fracture. Journal of the American Ceramic Society, 87, 2004. p. 500-503.

Microstructure and Mechanical Properties of Monolithic and Composite Systems

LOW CTE & HIGH STIFFNESS DIAMOND REINFORCED SiC BASED COMPOSITES WITH MACHINEABLE SURFACES FOR MIRRORS & STRUCTURES

M.A. Akbas, D. Mastrobattisto, B. Vance, P. Jurgaitis, S. So, P. Chhillar and M.K. Aghajanian
M Cubed Technologies, Inc.
921 Main Street
Monroe, CT 06468

ABSTRACT

Diamond reinforced reaction bonded silicon carbide composites have unique properties such as very high stiffness, low density, low thermal expansion coefficient and high thermal conductivity making them attractive materials for high precision optical and structural components. However, their use in high precision equipments was limited due to significant difficulties in high tolerance machining of these super hard composites. In this present work, machineable diamond reinforced SiC composites were fabricated through forming hybrid monolithic microstructures with diamond free machineable surfaces. This is achieved by coating diamond reinforced SiC green performs with SiC slurries. Coated preforms were then infiltrated with molten silicon. Resulting fully dense monolithic structures have surfaces free of diamond particles thus they can be ground, lapped and polished using conventional technologies.

INTRODUCTION

Diamond is a unique particle reinforcement material in ceramic matrix composites thanks to its extraordinary thermal and mechanical properties such as ultra high Young's modulus and high thermal conductivity together with low thermal expansion coefficient [1-2]. Due to these unique combination of properties, diamond has been previously used as a particulate reinforcement in many ceramic matrix composites such a silicon carbide-diamond [3], alumina-diamond [4], ZnS-diamond [5], Si_3N_4-diamond [6] and cordierite-diamond [7]. Among these many possible combinations, SiC diamond composites are uniquely suited for high precision applications such as wafer chucks for semiconductor capital equipments, ultra stable optical substrates for high speed laser scanning mirrors and high thermal conductivity substrates for high power laser optics. This is thanks to its unique combination of physical properties such as high specific stiffness (>500 GPa with a density below 3.3 g/cc), low thermal expansion coefficient (< 2ppm/°C) and high thermal conductivity (>250 W/m. K). Figure 1 compares common engineering materials used in industry with respect to their mechanical and thermal stability. Mechanical and thermal stability is defined by Young's modulus divided by density and thermal conductivity divided by thermal expansion coefficient respectively. As it can be seen from this chart, diamond reinforced SiC based composites located at the upper right quadrant of this chart show superior thermal and mechanical stability significantly better than many of their counterparts. Moreover, these composites can be processed in the absence of high pressure to form rather large and complex shapes using reaction bonding technology. Details of processing of diamond reinforced reaction bonded silicon carbide composites can be found elsewhere [8-10]. However, it should be noted that, diamond reinforced reaction bonded silicon carbide composites can be machined to fabricate rather complex structures in green form and resulting structures can then be infiltrated by molten silicon to produce fully dense structures larger than

135

300 mm with dimensional tolerances well below 500 micron. However, once densified, final machining of these super hard composites is very difficult. Thus achieving demanding dimensional tolerance requirements of semiconductor or optical capital equipment components requires development of new innovative technologies.

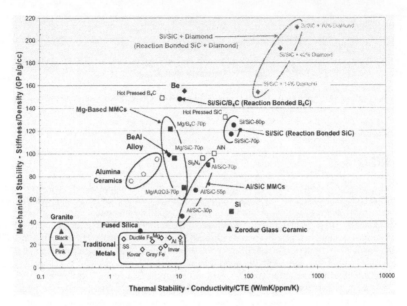

Figure 1. Mechanical and thermal stability chart for common engineering materials. Thermal stability is defined by thermal conductivity divided by thermal expansion coefficient. Mechanical stability is defined by Young's modulus divided by density.

In the present work, green diamond reinforced silicon carbide preforms were machined to form into complex shapes. Functional surfaces of these structures, that require higher degree of flatness or dimensional accuracy, were coated with diamond free SiC slurries. Coated performs were then infiltrated with molten silicon to form monolithic hybrid reaction bonded ceramic composites with machineable diamond free surfaces. Thus, the resulting hybrid monolithic composites can be ground, lapped or polished on fully densified state to achieve demanding dimensional tolerances without the need of additional coating step.

EXPERIMENTAL PROCEDURES

Starting materials to fabricate green preforms were commercially available green silicon carbide particles with purity better than 99.5% and an average particle size of 50 micron. The SiC particles were dispersed in an aqueous slurry together with commercially available diamond particles with an average particle size of 22 micron. Diamond reinforced green ceramic preforms with 80% theoretical density were formed by casting of SiC/diamond slurries into molds. These preforms contained 40 vol. % diamond particles and were strong enough to be machined in the green state. Details of composite processing can be found elsewhere [8-10]. Generic scanning mirror substrate shapes with 100 mm in diameter with back side rib structure were machined using standard CNC machines (Acer VMC-2040L). Figure 2 is a secondary electron SEM micrograph showing the typical microstructure of the preform. It shows a nominally 20 to 120 micron distribution of blocky-shaped SiC particles together with homogeneous distribution of diamond particles in matrix.

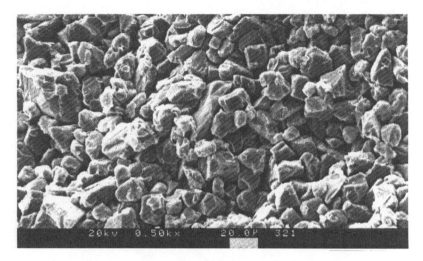

Figure 2. A secondary electron micrograph of diamond reinforced silicon carbide performs.

Machineable diamond free surface layers were formed on machined preform by spray coating SiC slurries directly onto preform surfaces. The coating slurry was formulated using SiC powders with mean particle size of approximately 12 micron. SiC powders were dispersed in aqueous slurry using a dispersant, high molecular weight organic binder, wetting and antifoam agents. This slurry was then sprayed directly over mirror surfaces to form 100 to 300 micron thick coatings. Following drying and initial heat treatments, Si infiltration of the coated perform was done at a peak temperature of nominally 1500°C in vacuum. Infiltrated composites were grit blasted to remove any residual excess silicon pearls on the surfaces.

The density of resulting infiltrated composites were measured using water immersion (ASTM B 311). Young's modulus was determined by ultrasonic pulse echo (ASTM E 494) technique. Cross-sectional samples for microstructural analysis were prepared using standard metallographic techniques and microstructures were characterized using scanning electron microscopy (ISI-SR-50) equipped with back scattered electron detector and EDAX spectroscopy.

Diamond free surfaces of fully dense infiltrated composites were ground using standard ceramic machining techniques followed by lapping and polishing. Surface topography of lapped and polished samples was characterized using 300 mm Zygo ™ phase shift laser interferometer (model number MK 11-01).

RESULTS AND DISCUSSION

Infiltrated density and Young's modulus of 40 vol. % diamond reinforced silicon carbide composites were measured to be 3.27 g/cc and 570 GPa respectively. These results indicate full densification with negligible residual porosity. No apparent cracks on coating were observed after infiltration. This can be attributed to good match of thermal expansion coefficient of the coating and diamond reinforced matrix. Figure 3 is the picture of the infiltrated diamond reinforced reaction bonded silicon carbide generic scanning laser mirror substrate. It shows defect free optical surface at top after lapping and rough polishing.

Figure 3.. A photograph of 100 mm generic laser scanning mirror after infiltration, machining and lapping.

Figure 4 provides a secondary electron SEM micrograph of a typical fracture surface of Si infiltrated diamond reinforced silicon carbide matrix. It indicates a fully infiltrated dense microstructure consisting of SiC grains and a homogeneous distribution of diamond particles well bonded into ceramic matrix. The bonding is evident by transgranular fracture mode observed in the diamond particles. Microstructure contains small amount of residual silicon phase much less than initial 20 vol. % porosity. This is due to the reaction between molten silicon and diamond to form secondary silicon carbide. Figure 5 is a polished cross-sectional view showing the diamond free surface coating and transition to diamond reinforced matrix. The coating layer can easily be identified by the lack of diamond particles within the coating area as marked on micrograph. While diamond free surfaces can be precision machined to achieve dimensional tolerances below 10 micron, diamond reinforced substrate provides high mechanical and thermal stability. Moreover, there is no apparent interface between surface coating layer and the composite matrix. Atomic level continuity is maintained during the homogeneous solidification of molten Si. The SEM micrograph shows a gradual transition from diamond reinforced matrix to diamond free surface layer. It is believed that this gradual transition of the microstructure is responsible from successful accommodation of stresses due to CTE mismatch.

Figure 4. A Secondary electron micrograph of the fracture surface of diamond reinforced reaction bonded silicon carbide composite matrix.

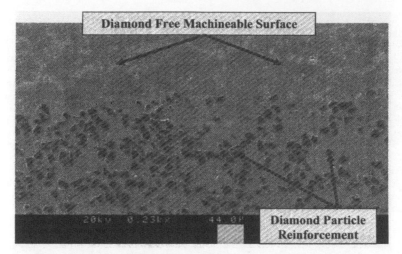

Figure 5. A SEM micrograph of infiltrated composites showing the diamond reinforced matrix and transition to diamond free surface layer creating a stock for the final machining.

Diamond free functional optical surface of the mirror was ground using plated diamond wheels and lapped using traditional techniques (see figure 3). Figure 6 is phase shift laser histogram of the lapped surface showing a global surface flatness of a peak to valley (PV) value better than 160 nm (λ/4).

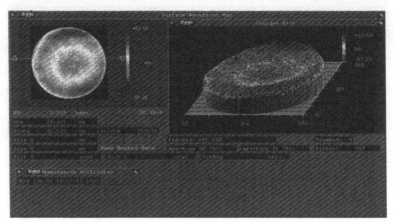

Figure 6. Phase shift laser interferometer histogram of the lapped and rough polished surface of the generic diamond reinforced SiC laser scanning mirror showing PV value of approximately λ/4.

SUMMARY

Monolithic diamond reinforced reaction bonded silicon carbide composites with diamond free, machineable surfaces were fabricated by infiltration of coated performs. Resulting ceramics were fully dense and crack free with excellent mechanical and thermal stability. Diamond free functional surfaces were successfully machined and lapped to demonstrate high precision tolerances.

REFERENCES
1. R.M. Cherenko and H.M. Strong, "Physical Properties of Diamond", General Electric CRD Rept. No: 75CRD089, General Electric Co., Schenectady, NY, Oct. (1975).
2. K.E. Spear, "Diamond – Ceramic Coating of the Future", J. Am. Ceram. Soc., 72[2], 171-91 (1989).
3. M. Shimano and S. Kume,"HIP Sintered Composites of C (Diamond)/SiC", J. Am. Ceram. Soc. 87[4] 752-55 (2004).
4. J. Liu and D. Ownby, "Normal Pressure Hot Pressing of α- Alumina-Diamond Composites", J. Am. Ceram. Soc., 74 [10], 2666-68 (1991).
5. D.S. Farquhar, R. Raj and L. Phoenix, "Fracture and Stiffness Characteristics of Particulate Composites of Diamond in Zinc Sulfide", J. Am. Ceram. Soc., 73[103], 3074-80 (1990).
6. T. Noma and A. Sawoka, "Fracture Toughness of High Pressure Sintered Diamond/Silicon Nitride Composites", J. Am. Ceram. Soc., 68[10], C-271-C-273 (1985).
7. D.P.H. Hasselman, K.Y. Donaldson, J. Liu, L. Gouckler and P.D, Ownby, "Thermal Conductivity of Particulate Diamond Reinforced Cordierite Matrix Composites", J. Am. Ceram. Soc. 77[7], 1757-60 (1994).
8. M.A. Akbas, D. Mastrobattisto and W. Vance, "Methods for Forming Machineable Surfaces on Diamond Reinforced Composite Bodies and Articles Formed Thereby", United States Patent Application, Filed on March 23, 2009 and 61/217,912 Filed on June 6, (2009).
9. P. G. Karandikar and S. Wong, "Microstructural Design of Si-B₄C-Diamond System", CESP Vol. 32 [5] (2011) 61-70.
10. S. Salamone and O. Spriggs, "Effect of Reaction Time on Composition and Properties of SiC-Diamond Ceramic Composites", CESP Vol. 32 [2] (2011) 213-222.

TAILORING MICROSTRUCTURES IN MULLITE FOR TOUGHNESS ENHANCEMENT

D. Glymond*, M. Vick[o,+], M.-J. Pan[o], F. Giuliani[*,+], L.J. Vandeperre*
*Centre of Advanced Structural Ceramics & Department of Materials, Imperial College London, South Kensington Campus, London SW7 2AZ, UK
[+]Department of Mechanical Engineering, Imperial College London, South Kensington Campus, London SW7 2AZ, UK
[o] Naval Research Laboratory, Washington, DC 20375

ABSTRACT
    Mullite is considered a promising candidate for future structural applications such as ceramic recuperators in turbo propelled engines, due to its highly favourable properties at high temperatures. In order for it to be viable for structural applications its relatively weak fracture toughness needs to be improved. A reliable way of improving fracture toughness in a range of materials is to tailor the microstructure to contain elongated grains capable of bridging cracks. In this paper, the tailoring of mullite microstructures using a range of processing methods are reported: reactive sintering of mixtures of alumina and silica, sol-gel synthesis of mullite and the use of sol-gel derived additives to enhance the sintering of commercial mullite powders. The difference in morphologies produced as well as the influence on indentation fracture toughness is described.

INTRODUCTION
    Mullite is becoming an important material in many high temperature structural applications due to its low coefficient of thermal expansion, good creep resistance and strength at these temperatures.[1 2] The main drawback of mullite with regards to its structural feasibility is its fracture toughness of only 1.8-2.8 MPa m$^{1/2}$ [3-4]. There are a variety of methods to increase the toughness of a ceramic, one of which is the elongation of the grains to create crack bridging, and even R curve effects. It has been shown in silicon nitride[5] and silicon carbide[6] that sintering using a liquid phase can lead to elongated grains due to dissolution and precipitation. In mullite grain elongation has previously been shown to occur during reactive sintering of alumina and silica[7] therefore reactive sintering was included as one of the methods for producing mullite with elongated grains in this study. However, the microstructures with elongated grains previously reported also contained excess glass, which might be detrimental to the creep resistance. The aim therefore was to investigate whether elongated grains could be produced with minimal residual glass.
    Moreover, mullite produced via reactive sintering requires very high sintering temperatures, whereas sol-gel derived mullite has been shown to have a lower temperature of formation[8-9]. Therefore a second aim is to clarify whether elongated grains could be obtained from sol-gel precursors for mullite.
    Finally, a novel route for processing mullite at lower temperatures is also proposed: the use of sol-gel derived sintering additives for mullite. Two variants have been investigated: sol-gel precursors with a composition in the mullite region (30 wt% SiO$_2$) to take advantage of the lower processing temperatures of sol-gel derived mullite and sol-gel precursors with a composition close to the binary eutectic in the Al$_2$O$_3$-SiO$_2$ phase diagram (92 wt% SiO$_2$), to promote the presence of a limited amount of liquid phase and hence elongated grain formation.
    The effect of these processing methods on both the microstructure and indentation fracture toughness is reported.

EXPERIMENTAL

Processing

Sol-gel synthesis of mullite and additive pre-cursors started by preparing a boehmite sol from aluminium nitrate nonahydrate (Fluka, UK) and ammonium hydroxide via a process reported elsewhere[9].Tetraethyl orthosilicate (TEOS, Sigma Aldrich, UK) was added to the boehmite sol dropwise as the silicon donor to form the precursors. The ratio of boehmite sol to TEOS was set to yield pre-cursors with 30 wt% SiO_2 to produce a pre-cursor for mullite or with 92 wt% SiO_2 to produce a low melting point (~1580 °C) sintering additive. Precipitation from the sol was achieved by addition of ammonium hydroxide to bring the pH to 8.0. The solution was stirred for a further 24 hours, concentrated at 50 °C to a quarter of its original volume, dried at 70 °C, and then calcined at 600 °C to remove excess organics.

The pre-cursors were added as a sintering additive at a level of 10 wt% to a commercial fine mullite powder (KM 101 from KCM corporation, USA). For the purpose of comparison, samples consisting of the pre-cursor with the mullite composition and of the commercial mullite powder alone were produced also.

For reactive sintering following the method of Sacks and Pask[7], mixtures of commercial alumina (CT3000SG, Almatis, Germany) and silica (Microsilica 95, Elkem, UK) in various compositions were prepared. The silica powder was calcined at 600 °C to remove its more volatile impurities before use.

All mixtures were prepared by ball milling in acetone for 5 hours. Poly-ethylene glycol (PEG 400, Alfa Asaer, UK) was added as a binder and pellets (8 mm) were produced by uni-axial compaction at 200 MPa before calcination at 1500 °C for 10 hours, then sintering at 1550 °C for 4 hours or at 1700 °C for 1 hour.

Characterisation

The density of the sintered pellets was determined using Archimedes' principle, with % density values from a comparison to the pure mullite density value of 3.16 g cm^3. The microstructure was characterised by observing fracture surfaces in a scanning electron microscope (JEOL-5610LV,Jeol, Japan) and the crystalline phases were determined by X-ray diffraction using a copper Kα-source (PW 1700, Philips, The Netherlands). The indentation toughness, K_{ind}, was determined from the crack lengths emanating from 5 kg Vickers indents (Indentec, Zwick-Roel, Germany). K_{ind} was calculated using the method developed by Anstis et al.[10]

$$K_{ind} = \xi \sqrt{\frac{E}{H} \frac{F}{c^{3/2}}}$$

where ξ is a material independent constant (0.016), E is the Young's modulus, H is the

Hardness, F is the force and c is the crack length. The results reported are the average of five indents on each sample. Although indentation toughness is deemed to give optimistic results by some, this is not always the case[11]. Indentation toughness is a good tool for characterisation when sample size is too small for standard testing, and a large number of samples are tested[12]. The equation by Anstis et al. was used as it generally gives a less optimistic result than other indentation equation[10,12-13].

RESULTS

Figure 1a shows the microstructure obtained with commercial mullite powder without additives after sintering at 1700 °C. There is no elongated grain formation and as shown in Table 1, the material does not fully densify even when sintered at 1700 °C. In sharp contrast, the sol-gel derived mullite precursor reaches full density after sintering at 1550 °C, see Figure 1b. However, in the sol-gel derived material fracture is transgranular.

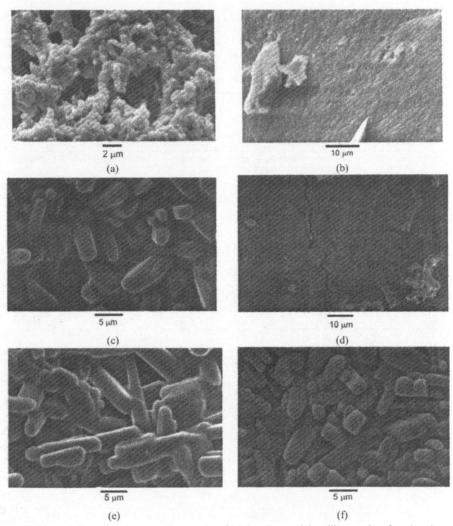

Figure 1. Scanning electron micrographs of (a) the commercial mullite powder after sintering at 1700°C for 1 hour, (b) the sol-gel derived mullite precursor after sintering at 1550 °C for 4h, (c)

reactive sintering of a mixture with 25 wt% SiO$_2$ for 4 h at 1550 °C and (d) the same mixture after 1 h at 1700 °C, (e) 29 wt% SiO$_2$ after 1 h at 1700 °C and (f) 27 wt% SiO$_2$ after sintering at 1700 °C for 1h.

The results in Table 1 also show that reactive sintering at 1550 °C does not lead to full densification, and the microstructure shown in Figure 1c illustrates that even when the SiO$_2$ content is as low as 25 wt%, a glassy phase remains between the grains when sintered at 1550 °C.

However, reactive sintering at 1700 °C leads to high densities and most of the intergranular phase disappears, see Figure 1d-1f. For all reactive sintering routes, fracture clearly occurred inter-granular, but only compositions with at least 27 wt% SiO$_2$ retained elongated grains.

The commercial mullite powder with 10 wt% additives only shows limited improvement in densification at 1550 °C, which is below the expected melting point of the low melting additive but at the temperature where the sol-gel derived mullite densifies fully. As shown in Figure 2, the grains also remain small and equi-axed. At 1700 °C, both types of additives improve densification to 95-96%. Large grains have now formed in both, but only the low melting point additive instigates elongated grain formation, see Figure 3. Fracture appears intergranular for the additive of mullite composition whereas in the sample with the low melting point additive, mixed mode fracture occurs.

Table 1. Density and indentation toughness of the different compositions and sintering methods

Sample	Sintering at 1550 °C (4hours)		Sintering at 1700 °C (1 hour)	
	Density	K$_{ind}$ (MPa m$^{1/2}$)	Density	K$_{ind}$ (MPa m$^{1/2}$)
KM 101 mullite	61.2%	-	80.4%	1.71±0.2
Sol-gel mullite	99.8%	-	-	-
Reactive sintering				
71wt% Al$_2$O$_3$– 29 wt% SiO$_2$	81.5%	1.84±0.2	99.5%	2.03±0.2
72 wt% Al$_2$O$_3$ – 28 wt% SiO$_2$	85.3%	1.75±0.2	97.2%	1.93±0.2
73 wt% Al$_2$O$_3$ – 27 wt% SiO$_2$	83.8%	1.66±0.2	99.1%	1.98±0.2
74 wt% Al$_2$O$_3$ – 26 wt% SiO$_2$	82.8%	1.84±0.2	98.4%	2.15±0.2
75 wt% Al$_2$O$_3$– 25 wt% SiO$_2$	88.5%	1.83±0.2	98.5%	2.08±0.2
90 wt% KM 101 mullite + 10 wt% sol gel additive				
70 wt% Al$_2$O$_3$ – 30 wt% SiO$_2$	74.1%	2.10±0.2	95.5%	2.57.±0.2
8 wt% Al$_2$O$_3$ – 92 wt% SiO$_2$	78.6%	1.64±0.2	96.3%	2.25±0.2

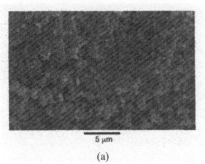 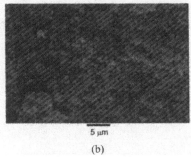

5 µm   5 µm

(a)   (b)

Figure 2. Scanning electron micrographs of the commercial powder with 10 wt% sol-gel derived sintering additive after sintering for 4 h at 1550 °C. (a) sol-gel additive of mullite composition (30 wt% SiO$_2$) and (b) low melting point sol-gel additive (92 wt% SiO$_2$).

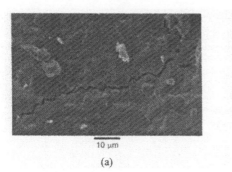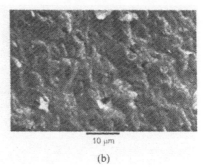

(a)   (b)

Figure 3. Scanning electron micrographs of the commercial powder with 10 wt% sol-gel derived sintering additive after sintering for 1 h at 1700 °C. (a) sol-gel additive of mullite composition (30 wt% SiO_2) and (b) low melting point sol-gel additive (92 wt% SiO_2).

Examples of X-ray diffraction patterns are shown in Figure 4. All routes lead to the formation of mullite as the major crystalline phase with some evidence for other phases such as alumina, cristobalite, quartz and sillimanite. Some mullite even formed when the low melting point additive (SG8/92) was heated to 1550 °C on its own. Cristabolite was observed when the sol-gel derived mullite precursor was used as a sintering additive. The formation of cristoballite has been seen previously in sol gel derived products and it formation depends on the silica particle size in the sol gel precursor[14].

Table 1 illustrates that the relative density has a major influence on the indentation toughness, which improves from around 1.6-2 MPa m$^{1/2}$ after sintering at 1550 °C to 2-2.6 MPa m$^{1/2}$ after sintering at 1700 °C. The general appearance of the cracks surrounding the Vickers indents is illustrated in Figure 5a, and larger magnification micrographs for the different processing routes are shown in Figure 5b-f. Given the large extent of intergranular fracture observed on the fracture surfaces of so many samples, the cracks are surprisingly straight. Nevertheless there is some evidence of crack bridging: in Figure 5b and d, small grains sticking out of the fracture surfaces underneath the surface can be seen. Since these materials did not contain long grains; see Figure 1a and 1d respectively, the potential for a strong crack deflection and bridging contribution is limited. The reactive sintering mixtures with more than 28 wt% SiO_2, which also showed the clearest elongation of the grains, show evidence of more extensive bridging, see Figure 5c. Surprisingly the indentation toughness of KM 101 mullite sintered with 10 wt% of a sol gel additive of the mullite composition is the highest value recorded despite the more equiaxed grains, see Figure 3a, and only limited evidence of crack deflection or bridging, see Figure 5e. However, also clear from Figure 3a is that in this material failure is clearly intergranular while for the slightly more elongated grains obtained by adding a low melting point sol gel additive, there is some evidence of mixed-mode fracture; see Figure 3b, leading to fairly straight crack paths around indents as well, Figure 5f.

DISCUSSION

The results show that if dense mullite is to be obtained at 1550 °C, i.e. below the eutectic melting point, the only feasibly route appears to be to use sol-gel derived mullite pre-cursors. Although the reaction mechanism for mullite formation in these pre-cursors is probably very similar to what happens during reactive sintering of alumina and silica mixtures, the much more intimate mixing of the silica and alumina leads to improved reaction kinetics – an effect also observed elswhere[9].

Addition of 10 wt% sol gel derived formulations to commercial KM mullite enhances the densification but reaching high densities still requires high sintering temperatures. For the liquid phase forming additive this is perhaps not surprising as melt formation is only expected at 1580 °C. For the samples where the additive had the mullite composition and which does reach full density on its own, 10 wt% is probably not sufficient to overcome the limited spontaneous sintering of the commercial powder.

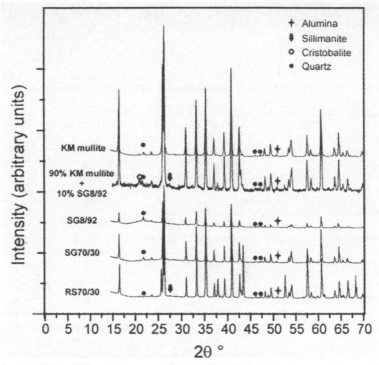

Figure 4. X-Ray Diffraction patterns of the mullite obtained via a range of routes. (SG = Sol Gel, RS = Reactive sintering. Unmarked peaks are attributed to mullite)

The reactive sintering route allows forming elongated grains at 1550 °C, and the growth ledges visible on the large grains at very high magnification, see Figure 6, together with the presence of a glassy phase in between the grains, see Figure1c, suggest dissolution-reprecipitation is indeed the mechanism which drives the elongation of the grains. However, here also full reaction requires higher processing temperatures. The range of compositions studied suggests that elongated grains are only obtained if there is a minimum of 28 wt% SiO_2 present. For lower silica contents, large grains form but these are more equiaxed in nature.

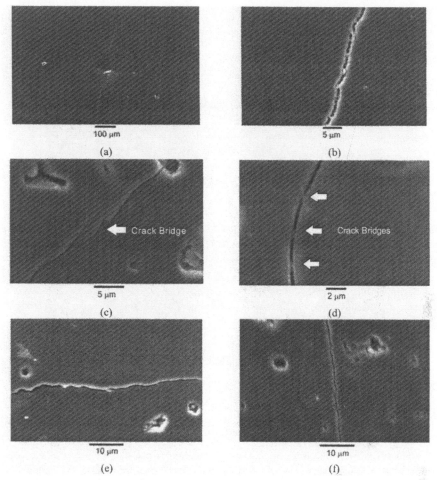

Figure 5. Scanning electron micrographs of cracks near Vickers indents for the different processing routes after sintering for 1 h at 1700 °C. (a-b) KM 101 mullite, (c) reactive sintering with 28 wt% SiO$_2$, (d) reactive sintering with 26 wt% SiO$_2$, (e)KM 101 mullite with 10wt% sol gel-additive (30 wt% SiO$_2$), (f) KM 101 mullite with 10wt% sol-gel additive (92 wt% SiO$_2$).

As could be expected, the toughness increases with density. Although the clearest examples of crack bridging have been observed in the reactively sintered mullites, the highest indentation toughness of 2.6 MPa m$^{1/2}$ was observed for the KM 101 mullite powder processed with 10 wt% sol-gel derived sintering additive of mullite composition after sintering at 1700 °C. This sample does contain large grains and shows crack deflection at the grain boundaries, but surprisingly the highest toughness is in a sample where the grains are equiaxed. Since large, equiaxed, grains were also obtained during reactive sintering (25 wt% SiO_2) while the toughness was lower, the difference in toughness might be due to differences in the fracture resistance of the grain boundary phases rather than the grain structure alone. Overall, however, the toughness values are lower than the values reported for the toughness of mullite-SiC-ZrO_2 composites[15-16], which range from 3.5-4.6 MPa m$^{1/2}$. However, optimisation of the microstructures in terms of grain boundary phases, grain shapes and grain size distributions might lead to comparable toughness values.

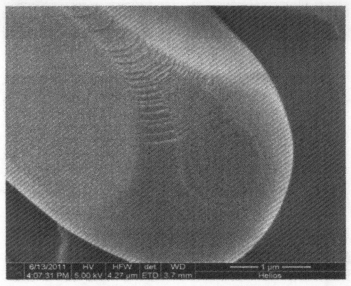

Figure 6. Detail of an elongated grain after reactive sintering a mixture containing 25 wt% SiO_2 for 4 h at 1550 °C

CONCLUSIONS

Direct sintering of sol-gel derived mullite precursors allows the formation of dense mullite at the lowest processing temperatures.

In line with previous reports, reactive sintering of alumina and silica is a very effective method for producing elongated grains. These form at relatively moderate temperatures but do require high temperatures for full densification. Despite evidence of extensive intergranular failure, the toughness remains moderate at 2 MPa m$^{1/2}$.

Addition of 10 wt% sol-gel derived mullite pre-cursors to mullite powders improves the densification of the latter. This additive promotes the growth of the grains but does not lead to high aspect ratio grains. However, the indentation toughness of this material at 2.6 MPa m$^{1/2}$ was higher than that of samples with elongated grains produced by reactive sintering. Addition of 10 wt% of a sol-gel derived low melting temperature additive (92 wt% SiO_2 / 8 wt% Al_2O_3) also enhances the densification of the mullite powder and stimulates the formation of high aspect ratio grains above its melting point. However, mixed mode fracture leads to a more limited toughness.

Compared to the mullite-ZrO_2 composite approach, the improvements in toughness through self-reinforcement appear moderate. It is suggested that the optimisation of the grain boundary phases could improve this.

ACKNOWLEDGEMENTS
DG, FG and LV thank the US Office of Naval Research and the Office of Naval Research Global for funding this work through grant N62909-10-1-7083.

REFERENCES

1.   Lessing, P. A.; Gordon, R. S.; Mazdiyasni, K. S., Creep of Polycrystalline Mullite. *J Am Ceram Soc* **1975,** *58* (3-4), 149-149.
2.   Dokko, P. C.; Pask, J. A.; Mazdiyasni, K. S., High-Temperature Mechanical-Properties of Mullite under Compression. *J Am Ceram Soc* **1977,** *60* (3-4), 150-155.
3.   Kanzaki, S.; Tabata, H.; Kumazawa, T.; Ohta, S., Sintering and Mechanical-Properties of Stoichiometric Mullite. *J Am Ceram Soc* **1985,** *68* (1), C6-C7.
4.   Mah, T. I.; Mazdiyasni, K. S., Mechanical-Properties of Mullite. *J Am Ceram Soc* **1983,** *66* (10), 699-703.
5.   Li, C.-W.; Yamanis, J., *Super-Tough Silicon Nitride with R-Curve Behavior.* John Wiley & Sons, Inc.: 2008; p 632-645.
6.   Sciti, D.; Guicciardi, S.; Bellosi, A., Effect of annealing treatments on microstructure and mechanical properties of liquid-phase-sintered silicon carbide. *J Eur Ceram Soc* **2001,** *21* (5), 621-632.
7.   Sacks, M. D.; Pask, J. A., Sintering of Mullite-Containing Materials .1. Effect of Composition. *J Am Ceram Soc* **1982,** *65* (2), 65-70.
8.   Varma, H. K.; Mani, T. V.; Damodaran, A. D.; Warrier, K. G.; Balachandran, U., Characteristics of Alumina Powders Prepared by Spray-Drying of Boehmite Sol. *J Am Ceram Soc* **1994,** *77* (6), 1597-1600.
9.   Cividanes, L. S.; Campos, T. M. B.; Rodrigues, L. A.; Brunelli, D. D.; Thim, G. P., Review of mullite synthesis routes by sol-gel method. *J Sol-Gel Sci Techn* **2010,** *55* (1), 111-125.
10.   Anstis, G. R.; Chantikul, P.; Lawn, B. R.; Marshall, D. B., A Critical Evaluation of Indentation Techniques for Measuring Fracture Toughness: 1, Direct Crack Measurements. *J Am Ceram Soc* **1981,** *64* (9), 533-538.
11.   Orange, G.; Fantozzi, G.; Cambier, F.; Leblud, C.; Anseau, M. R.; Leriche, A., High-Temperature Mechanical-Properties of Reaction-Sintered Mullite Zirconia and Mullite Alumina Zirconia Composites. *J Mater Sci* **1985,** *20* (7), 2533-2540.
12.   Ponton, C. B.; Rawlings, R. D., Vickers indentation fracture toughness test Part 2 Application and critical evaluation of standardised indentation toughness equations. *Mater Sci Tech-Lond* **1989,** *5* (10), 961-976.
13.   Osendi, M. I.; Baudin, C., Mechanical properties of mullite materials. *J Eur Ceram Soc* **1996,** *16* (2), 217-224.
14.   Fahrenholtz, W. G.; Smith, D. M.; Cesarano, J., Effect of Precursor Particle-Size on the Densification and Crystallization Behavior of Mullite. *J Am Ceram Soc* **1993,** *76* (2), 433-437.
15.   Bender, B. A.; Michael, V.; Ming-Jen, P., Pressureless Sintering of Mullite-Ceria-Doped Zirconia-Silicon Carbide Composites. In *Mechanical Properties and Performance of Engineering Ceramics and Composites V,* John Wiley & Sons, Inc.: 2010; pp 15-25.
16.   Garrido, L. B.; Aglietti, E. F.; Martorello, L.; Camerucci, M. A.; Cavalieri, A. L., Hardness and fracture toughness of mullite-zirconia composites obtained by slip casting. *Mat Sci Eng a-Struct* **2006,** *419* (1-2), 290-296.

MICROSTRUCTURES OF La-DOPED LOW THERMAL EXPANSION CORDIERITE CERAMICS

Hiroto Unno[1], Shoichi Toh[2], Jun Sugawara[1], Kensaku Hattori[1], Seiichiro Uehara[3] and Syo Matsumura[2,4]

[1]Ceramics Division, Krosaki Harima Corporation, Fukuoka 806-8586, Japan, [2]HVEM Laboratory, Kyushu University, Fukuoka 819-0395, Japan, [3]Department of Earth and Planetary Sciences, Kyushu University, Fukuoka 812-8581, Japan, [4]Department of Applied Quantum Physics and Nuclear Engineering, Kyushu University, Fukuoka 819-0395, Japan

ABSTRACT

Microstructures of a La_2O_3-doped cordierite ceramic used as a low thermal expansion material were investigated by transmission electron microscopy (TEM) including energy-dispersive spectroscopy (EDS) and X-ray diffraction (XRD) analyses in connection with the thermal expansion properties. TEM and scanning transmission electron microscopy (STEM)-EDS revealed that cordierite grains possess accumulated dense strain, and the grain boundaries are enriched with La and Si, forming an intergranular glassy phase. In-situ TEM observation heated up to 800 °C deduced that the accumulated lattice strain and the lattice parameters change with increase of temperature, although the microstructure remains almost unchanged. It was shown by conventional and interferometric dilatometry that the volume of sintered material contracts with increase of temperature, but it expands with further rise in temperature, drawing a minimum thermal expansion coefficient at around room temperature. XRD analyses revealed that the volume thermal expansion of cordierite grains is in reasonable agreement with that of the intergranular phase at a temperature of 800 °C or lower, while there is a significant difference in both the expansions at higher temperature. Therefore, the unchanged grain volume around room temperature, which is caused by the contraction along the a-axis as revealed by XRD analyses, is the reason for the low thermal expansion coefficient of the sintered material. We considered that residual stresses resulting from the thermal expansion mismatches at higher temperature has little effect on the dimensional stability of the sintered material, because the presence of glass would enable stress-free conditions at temperatures above the glass transition.

INTRODUCTION

Recently, there have been strong demands in the field of precision engineering towards thermally and mechanically stable dimension. To meet the requirement, low thermal expansion materials such as SiO_2-Al_2O_3-P_2O_5-Li_2O based glass ceramics[1] and cordierite, $Mg_2Al_4Si_5O_{18}$ based polycrystalline ceramics[2-4] are being used as mirror substrates, primary standards and other precision components. Particularly, the glass ceramic has an extremely low thermal expansion coefficient of < 0.03×10^{-6}/K at room temperature[1]. However, early studies have found that the glass ceramic exhibits pronounced shrinkage during a short span of several years even at room temperature, depending on prior thermal histories[5,6]. It has also been discovered that the glass ceramic shows a length hysteresis upon cycling a large temperature interval[6].

On the other hand, cordierite originally has a relatively low thermal expansion coefficient, and numerous studies have been carried out in order to elucidate the mechanism of the thermal expansion in natural and synthetic cordierites, either pure or substituted[7-13]. Cordierite occurs in two polymorphic forms: the high-temperature disordered hexagonal form (space group P6/mmc) stable between 1450 °C and its melting point (around 1460 °C), and the low-temperature ordered orthorhombic form (space group Cccm) stable below 1450 °C. Cordierite undergoes slight thermal expansion parallel to the plane of the characteristic six-membered rings (along the a and b axes) and slight contraction perpendicular to the rings (along the c axis). Recent studies have found that a cordierite based polycrystalline ceramic doped with La_2O_3 has an extremely low thermal expansion coefficient of < 0.03×10^{-6}/K at room temperature and high dimensional stability against changes in time[2,3]. Table 1 summarizes the known physical properties of the La-doped cordierite ceramic. However, to the best of our knowledge, no microscopical study of the La-doped cordierite has been performed in connection with the thermal expansion properties as yet. The fundamental mechanism to control the thermal expansion properties including the dimensional stability against changes in time is still an open question.

The present work has been conducted to study the microstructures of the La-doped cordierite ceramic by transmission electron microscopy (TEM) and X-ray diffraction (XRD) analyses. Special emphasis has been placed on the thermal expansion mismatches between cordierite and other phases constituting this multi-phase ceramic, taking advantage of analytical Rietveld method of XRD data, to get insights into the mechanism of the low thermal expansion and the high dimensional stability.

Table 1. Physical properties of the La-doped low thermal expansion cordierite ceramic.

Bulk density	g/cm^3	2.55
Young's modulus	GPa	140
Poisson's ratio	-	0.31
Flexural strength	MPa	230
Fracture toughness	$MPa^{1/2}$	1.2
Hardness HV	GPa	8.1
Thermal expansion coefficient	$10^{-6}/K$ (23 °C)	< 0.03
Thermal conductivity	W/m K (23 °C)	4.2
Specific heat	J/g K	0.78

EXPERIMENTAL

The material used in this study was cordierite with 2.9 wt% of La_2O_3. This study dealt with pure cordierite crystallized from molten glass as a starting raw material, but commonly other minor and/or secondary phases such as mullite, corundum, spinel, forsterite, clinoenstatite, phosterite and cristobalite were observed together with the cordierite phase. Mixed powders of the commercial-grade synthetic cordierite (99.4 % purity, average particle size of 2.0 μm) and reagent-grade $La(OH)_3$ (99.9 % purity, average particle size of 1.4 μm) were wet milled together with organic binder and deionized water. The slurry thus obtained was dried using a spray dryer and then isostatically pressed at room temperature into green compacts of 100 mm × 100 mm × 35 mm. The green compacts were annealed once at 350 °C for 5 h in air to burn the organic additives out and then sintered at 1360 °C for 4 h in an Ar-gas flow atmosphere.

Discs of 3 mm in diameter and 100 μm in thickness were cut out from the sintered compacts, followed by mechanical polishing, dimpling and Ar ion thinning in order to be electron transparent for TEM observation. Microstructural characterization was performed at accelerating voltage of 200 kV using an analytical TEM, JEM-ARM200F, incorporating a spherical aberration corrector for electron optic system. The microscope was equipped with a silicon-drift-type energy-dispersive X-ray detector, EX-24220M6G5T that collects X-rays from TEM samples at a large solid angle of up to 0.8 steradians from a detection area of 100 mm^2 for chemical analysis. The electron beam probe used in the energy-dispersive spectroscopy (EDS) measurement was < 0.5 nm in diameter. Microstructures were also examined with a scanning electron microscopy (SEM; Ultra55, Carl Zeiss AG), equipped with a Shottky thermal field emission gun and four electron detectors. In this work, backscattered electron (BSE) images were observed with an energy-selective backscattered (EsB) detector in the lens column. In-situ TEM observation heated from room temperature up to 800 °C was carried out with JEM-2010EFF accelerated at 200 kV with an omega-type energy filter taken with an electron probe size of around 2 nm to characterize the high temperature behavior of the La-doped cordierite.

X-ray powder diffraction was performed with Cu-K_α radiation (Ultima IV, Rigaku, Japan) to identify the phases formed in the sintered material and determine the lattice parameters. A fine powder particle size of less than 5 μm in diameter was investigated to allow higher measuring accuracy. The X-ray diffraction data were collected between the 2θ range of 4 to 80 ° with a step size of 0.02 ° and a step time of 2 s in static air. The sample was heated from -150 to 1200 °C at a rate of 10 °C min⁻¹ and a soaking time of 5 min was employed at each temperature step before the scan. The quantitative determination of the lattice parameters was performed from X-ray diffraction data by using Rietveld refinement[14]. Full-pattern Rietveld refinement was performed with RIETAN-FP, available in the software package REITAN-VENUS[14]. The refinement involved the following parameters: a scale factor, zero displacement correction, unit cell parameter, peak profile parameters and overall temperature factor.

The linear thermal expansion was measured between -160 and 140 °C by a double-passed Michelson interferometric dilatometer (LIX-2, ULVAC-RIKO Inc., Japan) at a heating rate of 2 °C min⁻¹. Classical dilatometry was also carried out on the long rod (20 mm size) using a thermomechanical analyzer (TMA8310, Rigaku, Japan) in the temperature range of 30 to 1200 °C. A ramp of 10 °C min⁻¹ was used during heating.

RESULTS AND DISCUSSION

Macroscopic thermal expansion of the sintered material around room temperature

Cordierite containing 2.9 wt% La_2O_3 was sintered at 1360 °C for 4 h which measuring the macroscopic expansion by interferometric dilatometry, as illustrated in Fig 1. The thermal expansion of the sintered specimen decreases with increasing temperature, reaching a minimum at around room temperature. After that the thermal expansion begins to increase with increasing temperature. In previous works regarding synthetic cordierites, temperatures giving a minimum thermal expansion were around 200 °C or more, that is, the cordierites exhibited negative thermal expansion around room tempearture[10,15]. Thus, it is obvious that the thermal expansion behavior is changed with addition of La_2O_3. The coefficient of thermal expansion of the sintered material, which can be obtained by differentiating the thermal expansion curve, has been 0.03 × 10⁻⁶/°C at 23 °C. This value is comparable to that of low thermal expansion glass ceramics[1].

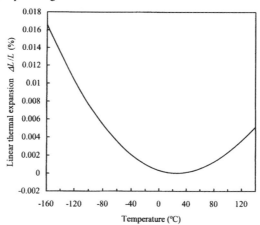

Figure 1. Macroscopic expansion of cordierite containing 2.9 wt% La_2O_3 sintered at 1360 °C for 4 h, as measured by interferometric dilatometry.

Microstructures of the sintered material

Figure 2 shows the TEM bright-field image of the low thermal expansion cordierite. The cordierite grains after sintering for 4h exhibit curved or faceted boundaries with grain sizes of 0.5 to 2.0 μm. Some grains possess a closely spaced pattern of cross-hatched microstructure in some areas and lack contrast in others. In earlier works[16,17], it has been established that the cordierite initially crystallized from molten glass is well crystalline, defect free, hexagonal and homogeneous. With continued annealing, the hexagonal cordierite first develops a tweed microstructure indicating orthogonal modulations in the degree of order and lattice strain. With further annealing, this microstructure progressively coarsens, eventually developing cross-hatched twins within the orthorhombic phase. In the present work, the similar microstructure is recognized in the TEM image as shown in Fig 2. The contrast observed in the grain interiors is, therefore, thought to be due to accumulated lattice strain which is correlated with transformation from metastable hexagonal to orthorhombic cordierite. As indicated by arrows in Fig 2, some inclusions are also recognized in the grain interior. The inclusion has been identified as the La-enriched particle, having a reticulated structure by EDS analyses.

Figure 2. TEM bright-field image showing overall grain morphology. Arrows indicate the La-enriched particle embedded in the cordierite grain interior.

Figure 3 illustrates the scanning transmission electron microscopy (STEM) micrograph of the typical microstructure. The corresponding EDS composition maps of La, Al and Mg show almost reversed contrast with each other. The doped lanthanum ions are segregated at boundaries between cordierite grains, and aluminum and magnesium ions are uniformly distributed in the grain interior. Silicon and oxygen ions are present along the grain boundaries similarly to the grain interior. It has been deduced together with selected-area diffraction that the sintered cordierite ceramic is mainly composed of cordierite grains and intergranular glassy phase enriched with La and Si. Moreover, Al-rich spherical inclusions are found in the cordierite grains in both the STEM micrograph and EDS map for Al, as indicated by arrows in Fig 3. The STEM image also shows the presence of white spots in the grain interior, which can be identified as a slightly Si-enriched amorphous phase, as revealed by both the Fourier transformed diffraction pattern from high-resolution TEM image and EDS analysis.

The BSE image of the sintered cordierite, taken with an EsB detector, shows the overall grain morphology and distribution and is illustrated in Fig 4. Like the results of STEM-EDS analysis shown in Fig 3, doped lanthanum ions are segregated at boundaries between cordierite grains, which appear with light grey in Fig4. The volume fraction of amorphous phase along the grain boundary has been estimated to be 9.47 % by using the binary images of BSE micrographs. Further, it is also found that the microstructure consists of large grains in a matrix of small equiaxed grains. Overall, these microscope observations clearly show that the low thermal expansion ceramic consists of homogeneously distributed cordierite grains with accumulated lattice strain and intergranular glassy phase enriched with La and Si, except for a small volume fraction of some inclusions.

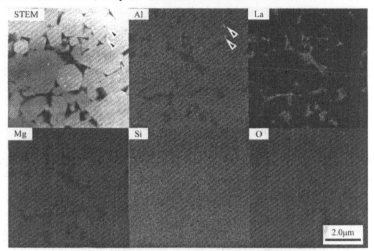

Figure 3. STEM bright-field image and EDS composition maps of Al, La, Mg, Si and O. Arrows indicate the Al-rich particles embedded in the cordierite grain interior.

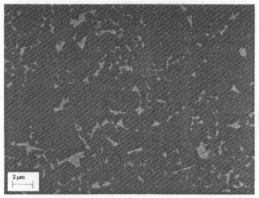

Figure 4. BSE micrograph of the typical microstructure.

In-situ TEM bright-field images which show the variation of the microstructure heated from room temperature up to 800 °C, are illustrated in Fig 5. The cordierite grain occupies almost all of the observed area, as shown in Fig5. The La-rich inclusion and intergranular glassy phase are also recognized at the upper left and the lower right of the images, respectively. No significant change with temperature can be seen in terms of the shape of the grain boundary and the inclusion, therefore, the microstructure almost keeps unchanged with increase of temperature. With elevating temperature, however, the spaced cross-hatched microstructure in the grain interior gradually become vague, showing clear contrast at 500 °C, and finally becoming devoid of any contrast over large crystal volumes after heating at 700 °C or higher. The contrast variation suggests that the accumulated lattice strain and the lattice parameters of the cordierite grain change with increase of temperature.

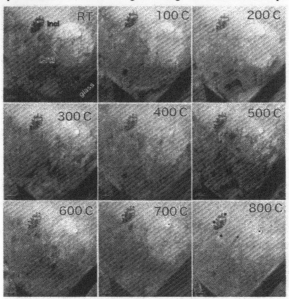

Figure 5. In-situ TEM bright-field images heated from room temperature up to 800 °C. The designations Crd, Incl and glass refer to cordierite, inclusion and glass phase, respectively.

Thermal expansion of cordierite grains and the sintered material

The variation in the lattice parameters of cordierite grains possessing dense lattice strains, which was suggested in a set of in-situ TEM images, was characterized using X-ray powder diffraction analyses. It is well known that the anisotropic coefficient of thermal expansion of ceramics creates grain-level stresses upon cooling from the firing temperature. It is also possible that expansion mismatches between cordierite grains and other phases constituting multi-phase ceramics are reason for the residual micro-stresses. Previous works[15,18] have investigated whether the micro-stresses could be released after powdering and to what extent the powder particle size could influence those stresses, by comparison between compact and powder data. It was found that a fine particle size powder had less stress locked in the microstructure than a compact sample, but it was not completely stress-free. However, the accumulated lattice strain in the grain interior is believed to be stable after powdering. Thus, the present work using a particle size of < 5 μm allows observation of lattice expansion of cordierite grains possessing dense lattice strains, not taking into account the micro-stresses. Qualitative analyses were conducted by comparing the diffraction lines with the standard PCPDF files using a search-match method. According to the diffraction patterns, no hexagonal polymorph of cordierite (called indialite, PDF-48-1600) was detected, although it is well known that these materials may contain a few weight percent of it. In the following, the material will be considered single orthorhombic cordierite (PDF-12-0303) as a crystalline phase. The variations of lattice thermal expansion along the a, b and c-axes with temperature are given in Fig 6, assuming that the lattice expansion is zero at room temperature. With heating, the length along the b-axis increases, and that along the c-axis slightly decreases; that is, the cordierite undergoes thermal expansion along the b-axis and slight contraction along the c-axis. These tendencies are in relative agreement with previous studies[10,15], where cordierites monotonically expand by around 0.5 % along the b-axis and contract by around 0.1 % along the c-axis up to 1200 °C, although the absolute values of thermal expansion coefficients are different from each other. On the contrary, there is a noteworthy difference in thermal expansion along the a-axis: the length first decreases with increase of temperature to -50 °C, after which it remains almost unchanged to 400 °C. It, then gradually increases with further rise in temperature. Previous studies have suggested that the lattice parameter along the a-axis monotonically increases with temperature similarly to that of the b-axis[10,15]. This is a distinct difference between the cordierite in this study and that previously reported on synthetic cordierites with stoichiometric composition[10,15]. The most obvious explanation for the difference in thermal expansion along the a-axis is the dopant effect of La. This result raises the possibility that incorporating La ions into structural channels causes the distortion of the characteristic six-membered rings in such a direction as to near six-fold rotational symmetry, which is consistent with the contraction or the unchanged length with temperature along the a-axis.

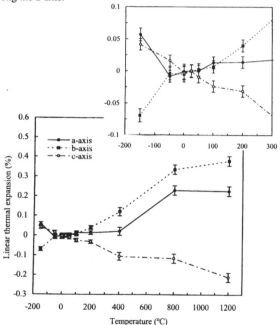

Figure 6. The linear thermal expansion along the a-, b- and c-axes as a function of temperature. Inset shows an enlarged view of the low-temperature region.

The volume expansion of cordierite grains around room temperature as a function of temperature from -160 to 140 °C, which is calculated from the lattice parameters, is shown in Fig 7. The macroscopic expansion of the sintered material has been included for comparison; it was measured by interferometric dilatometry and it was assumed that the expansion is zero at room temperature. Both the volume variations with temperature are in reasonable agreement with each other, although the cordierite grain expands slightly less than the sintered specimen. Therefore, the data indicates that the unchanged grain volume around room temperature, which is caused by the contraction along a-axis, is reason for the low thermal expansion coefficient of the sintered material. The slight difference in both the volume variations is due to the thermal expansion of the intergranular glassy phase. Thus, it should also be noted that the intergranular glassy phase expands slightly more than the cordierite grain around room temperature.

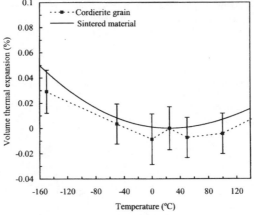

Figure 7. The volume variation in cordierite grains as a function of temperature compared with that of the sintered material, as measured by interferometric dilatometry.

Figure 8 compares the lattice and macroscopic volume expansions in the temperature range of -160 to 1200 °C. The grain volume remains almost unchanged until 400 °C, where it begins to increase with further rise in temperature. Eventually it is almost saturated at 800 °C or higher, while that of the sintered material monotonically increases with temperature. Therefore, these results indicate that there is a significant difference in thermal expansion between cordierite grains and the intergranular phase at higher temperature, whereas both the volume variations are in relative agreement at a temperature of 800 °C or lower. The expansion mismatches of this sort could generate residual micro-stresses. However, it is quite possible that the residual stresses at higher temperature are relatively small, because the presence of glass would enable stress-free conditions at temperatures above the glass transition (~900 °C for lanthanum aluminosilicate glass[19]). Therefore, it is considered that the small residual stresses result in the high dimensional stability of the La-doped cordierite. However, there is a need for further research to better understand how the anisotropic coefficient of thermal expansion affects the residual stresses and the resulting dimensional stability of the sintered material, because the present work only assumes average volume expansion of the cordierite grains. It is interesting to note that with increasing of temperature up to 400 °C, the intergranular glassy phase expands greatly as compared to the almost unchanged volume for cordierite grains. Therefore, it is quite possible that decreasing the volume fraction of the glassy phase provides a low thermal expansion coefficient over a wide temperature range.

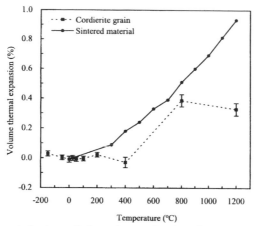

Figure 8. The volume variation in cordierite grains as a function of temperature compared with that of sintered material, as measured by conventional dilatometry.

CONCLUSION

In the present study, microstructures of a La_2O_3-doped cordierite ceramic were investigated by TEM including EDS and XRD analyses in connection with the thermal expansion properties. After sintering the La_2O_3-doped cordierite ceramic, the grains possessed accumulated dense strain, and the grain boundaries were significantly enriched with La and Si, forming an intergranular glassy phase. It has been revealed by in-situ TEM observations that the accumulated lattice strain and the lattice parameters change with an increase of temperature, although the microstructure remains almost unchanged. Conventional and interferometric dilatometry deduced that the volume of sintered cordierite contracts with increase of temperature, but it expands with further rise in temperature, drawing a minimum thermal expansion coefficient at around room temperature. By comparison of the volume variations between the sintered material and cordierite grains, it was revealed that the volume thermal expansion of cordierite grains was in reasonable agreement with that of the intergranular phase at temperature of 800 °C or lower, while there is a significant difference in both the expansions at higher temperature. Therefore, the unchanged grain volume around room temperature, which is caused by the contraction along the a-axis, is the reason for the low thermal expansion coefficient of the sintered material. It was considered that the thermal expansion mismatches at higher temperature has little effect on the dimensional stability of the sintered material because of the stress relaxation resulting from the presence of the intergranular glassy phase. However, there is a need for further research to better understand how the anisotropic coefficient of thermal expansion affects the residual stresses and the resulting dimensional stability of the sintered material.

ACKNOWLEDGEMENT

Financial support from an innovation promoting program of New Energy and Industrial Technology Development Organization in Japan is gratefully acknowledged. We wish to thank the staff of the fine ceramics plant of Krosaki Harima Corporation for their support.

REFERENCES
[1]H. Bach and D. Krause, Low Thermal Expansion Glass Ceramics, Second Edition, *Springer*, 2005.
[2]H. Unno, J. Sugawara, N. Kosugi, Y. Kuma, K. Abe and T. Aoki, NEXCERA, Zero Thermal Expansion Ceramics for Ultra Precision Applications, *Proc. of the 10th Euspen International Conference 2010*, **1**, 160-163 (2010).
[3]J. Sugawara, H. Unno, N. Kosugi, Y. Kuma, K. Abe, NEXCERA, Zero Thermal Expansion Ceramics for Optical Applications, *Proc. of ASPE 2010 Annual Meeting*, **50**, 404-407 (2010).
[4]Y. Kondo, K. Sasajima, S. Osawa, O. Sato and M. Komori, Traceability Strategu for Gear-pitch-measuring Instruments: Development and Calibration of a Multiball Artifact, *Meas. Sci. Tech.*, **20**, 1-8 (2009).
[5]F. B. Helms, H. Darnedde and G. Exner, Längenstabilität bei Raumtemperatur von Proben der Glaskeramik Zerodur, *Metrologia*, **21**, 49-57 (1985).
[6]R. Schödel and G. Bönsch, Interferometric Measurements of Thermal Expansion, Length Stability and Compressibility of Glass Ceramics, *Proc. of the 3th Euspen International Conference*, 691-694 (2002).
[7]M. F. Hochella Jr., G. E. Brown Jr., F. Ross and G. V. Gibbs, High-Temperature Crystal Chemistry of Hydrous Mg- and Fe-Cordierite, *Am. Mineral.*, **64**, 337-351 (1979).
[8]D. L. Evans, G. R. Fischer, J. E. Geiger and F. W. Martin, Thermal Expansion and Chemical Modifications of Cordierite, *J. Am. Ceram. Soc.*, **63**, 629-634 (1980).
[9]H. Ikawa, T. Otagiri, O. Imai, K. Urabe and S. Udagawa, Thermal Expansion of High Cordierite and Its Solid Solutions, *J. Ceram. Soc. Jpn.*, **94**, 344-350 (1986).
[10]H. Ikawa, T. Otagiri, O. Imai, M. Suzuki, K. Urabe and S. Udagawa, Crystal Structures and Mechanism of Thermal Expansion of the High Cordierite and Its Solid Solutions, *J. Am. Ceram. Soc.*, **69**, 492-498 (1986).
[11]D. Mercurio, P. Thomas, J. P. Mercurio, B. Frit, Y. H. Kim and G. Roult, Powder Neutron Diffraction Study of the Thermal Expansion of a K-Substituted Cordierite, *J. Mater. Sci.*, **24**, 3976-3983 (1989).
[12]S. S. Vepa, and A. M. Umarji, Effect of Substitution of Ca on Thermal Expansion of Cordierite $(Mg_2Al_4Si_5O_{18})$, *J. Am. Ceram. Soc.*, **76**, 1873-1876 (1993).
[13]A. Miyake, Effect of Ionic Size on Thermal Expansion of Low Cordierite by Molecular Dynamics Simulation, J. Am. Ceram. Soc., **88**, 121-126 (2005).
[14]F. Izumi and K. Momma, Three-dimensional visualization in powder diffraction, *Solid State Phenom.*, **130**, 15-20 (2007).
[15]G. Bruno, A. M. Efremov and D. W. Brown, Evidence for and Caluculation of Micro-strain in Porous Synthetic Cordierite, *Scripta Materialia*, **63**, 285-288 (2010).
[16]A. Putnis, The Distortion Index in Anhydrous Mg-cordierite, *Contributions to Mineralogy and Petrology*, **74**, 135-141 (1980).
[17]A. Putnis, E. Salje, S. A. T. Redfern, C. A. Fyfe and H. Strobl, Structural State of Mg-cordierite 1: Order Parameters from Synchrotron X-ray and NMR Data, *Physics and Chemistry of Minerals*, **14**, 446-454 (1987).
[18]G. Bruno, A. M. Efremov, B. Clausen, A. M. Balagurov, V. Simkin, B. R. Wheaton, et al., Acta Mater., **58**, 1994 (2009).
[19]J. E. Shelby and J. T. Kohli, Rare-Earth Aluminosilicate Glasses, J. Am. Ceram. Soc., **73**, 39-42 (1990).

STRATEGIES TO OPTIMIZE THE STRENGTH AND FRACTURE RESISTANCE OF
CERAMIC LAMINATES

Raul Bermejo[a], Zdenek Chlup[b], Lucie Sestakova[c], Oldrich Sevecek[c], Robert Danzer[a]

[a] Montanuniversität Leoben, Institut für Struktur- und Funktionskeramik, Leoben, Austria
[b] Academy of Sciences of the Czech Republic, Institute of Physics of Materials, Brno, Czech Republic
[c] Materials Center Leoben Forschung GmbH, Leoben, Austria

ABSTRACT
    Layered ceramics, compared to conventional monolithic ceramics, are good choices for highly-loaded structural applications because they exhibit greater fracture resistance, higher strengths and better mechanical reliability (*i.e.* they are flaw tolerant materials). The use of tailored residual compressive stresses in the layers is a key parameter to adjust these properties. In this work ceramic laminates are analyzed which have internal residual compressive stresses. The most important factors having influence on the strength and fracture resistance of these laminates are discussed. A fracture mechanics analysis is employed to estimate the crack growth resistance of the material as a function of the crack length. It is found that the proper selection of a suitable strain mismatch (responsible for the generation of residual stresses), volume ratio of the layer materials and thickness and distribution of individual layers are crucial to achieve a high fracture resistance and/or a high lower limit (threshold) for strength in ceramic laminates. Design guidelines to avoid cracking of layers associated with high residual stresses are also provided. Design criteria to optimize strength and fracture resistance in advanced ceramics to be used in engineering applications are established.

INTRODUCTION
    For many decades nature has inspired research to develop high performance materials by combining ceramics with other ceramics, metals or polymers [1-3]. Examples of advanced biological structures include the extraordinarily high toughness and strength of mollusk shells, which are related to the fine-scale of their microstructures; a laminate of thin calcium carbonate crystallite layers consisting of 99% calcium carbonate ($CaCO_3$) and tough biopolymers, arranged in an energy-absorbing hierarchical microstructure [4]. The strength and toughness of such layered structures are significantly higher than those of their constituents, for instance yielding an increase in toughness values of one order of magnitude [5, 6]. Nevertheless, the replication of architectural features found in nature (at the micro and nano scales) into macro scale structural engineering materials at a reasonable cost is still not possible. But it would be very welcome to build these toughness raising architectural features into durable materials which possess also additional functional and/or structural properties such as, for example, high temperature stability and specific electric functionality. The concepts of layered ceramics, functionally graded materials and coatings could allow tailoring of the surface and bulk properties of advanced engineering components with the purpose of enhancing their structural integrity as well as adding multi-functionality, which would translate into higher efficiency and better performance of these components.
    The use of colloidal processing has to some extent enabled the reduction of the critical size of the flaw causing the failure of the material, thus yielding relative high strength ceramics [7]. However, in the last decades a "flaw-tolerant" design approach in layered ceramic–ceramic composites has also been attempted to improve fracture resistance, strength and mechanical reliability of the ceramic materials [3; 8-14].
    Ceramic laminate systems are in general produced via the powder route, which also involves a high temperature sintering step. At the sintering temperature significant diffusion occurs, and any stress at the micro as well as at the macro scale will be relaxed within a relatively short time span.

But at room temperature significant diffusion in ceramic systems does not exist and stresses cannot relax. Therefore the differential thermal shrinking of the constituents during the cooling from sintering to room temperature causes significant mismatch strains which translate into residual stresses in the layers [15; 16]. Mismatch strains can also result from other reasons, e.g. phase transformations of constituents of the layer materials [17-19]. The residual stress field developed in the layers can be used as key element to improve the fracture behavior of the system. In this regard, various toughening mechanisms can take place in layered ceramic structures. The most important ones, among others, are crack deflection, crack bifurcation and interface delamination as well as crack shielding, which are all triggered by compressive residual stresses and/or elastic mismatch between layers [20-30]. Efforts to combine both approaches have recently been attempted by some authors, where, under some particular conditions, interface delamination can even be produced in laminates with strong interfaces, thus taking advantage of various toughening mechanisms in a unique design [31; 32].

Crack deflection and interface delamination exist for a special group of laminates with "weak interfaces" [2; 33-37], i.e. in this kind of multilayer the interface between the adjacent layers has a very low fracture resistance. An approaching crack deflects into this interface which protects the structure against catastrophic failure. Although the strength of these materials may not be in general significantly improved, since it is given by the strength of the constituent layer, the work of fracture of the system can be enhanced by about one order of magnitude compared to monolithic ceramics.

Crack shielding is the dominant mechanism in laminates with strong interfaces between the layers [3; 8-14; 38]. It should be noted that, in nature, crack shielding is in general related to gradients in the elastic properties. In the mollusk shells, to give an example, the differences in the elastic moduli between the ceramic particles and the biopolymer exceed a factor of several hundred, which causes significant elastic shielding, if the crack propagates from the ceramic to the polymer [4]. But in the case of ceramic–ceramic composites differences of the elastic constants of the constituents are much smaller, i.e. smaller than a factor of two or even much less, and elastic shielding is much less pronounced. In this case significant shielding is caused by compressive residual stresses. Depending on the relative coefficients of thermal expansion of the layers (and/or of the volume changes of constituents) the residual stresses in the external layer can be compressive or tensile. This has a strong influence on the fracture behavior. Laminates with compressive-stressed external layers usually have high strength and excellent wear resistance. Therefore, they can be used, for instance, in cutting tools (see for example [39; 40]). Laminates with tensile-stressed external layers (internally balanced by compressive-stressed layers) ensure especially high strength and mechanical reliability [3; 12-14; 41].

Generally, ceramic laminates with residual stresses can have an arbitrary sequence of layers (architecture), but they are mostly fabricated as symmetric plates (with respect to the middle plane) to avoid curving of the laminate. Much research has been conducted on alumina/zirconia laminates, because simultaneous sintering of both materials is easily possible and laminates made of those materials have strong interfaces. Most of the works consider relatively simple laminates, which are composed of two types of layers (A and B). Each layer type has the same thickness and the layers are built up in a periodic sequence (A, B, A, …, A, B, A). Thus, only one parameter, the so-called "layer thickness ratio", has to be optimized in order to find the highest possible fracture resistance [19; 25; 42]. However, little effort has been devoted to searching for an optimal design of multilayer ceramic laminates, which also allows a non-periodic sequence of layers.

In this paper we review concepts and introduce new recommendations to the optimization of fracture resistance, strength and mechanical reliability of ceramic laminates designed with residual stresses and strong interfaces. The compressive stresses are considered to be the driving force for the improvement in the mechanical behavior. We analyze the propagation of a surface crack oriented perpendicularly to the layer plane under the influence of the residual stress field. The selective location of the compressive stresses is investigated. Laminates having internal layers under compression are considered. The concepts are demonstrated for laminates with a periodic sequence of layers and then extended to symmetric laminates having a non-periodic layer sequence.

RESIDUAL STRESSES IN LAYERED CERAMICS

We consider layered ceramics where different materials are sealed together at high temperatures and which are subsequently cooled down to room temperature. The layer materials undergo differential dimensional changes during cooling (caused for example by differential thermal shrinking, chemical reactions and/or volume changes originated from phase transformations), which cause a mismatch strain $\Delta\varepsilon$ between adjacent layers. This develops the residual stress state of the laminate. If the interfaces are strong (i.e. the laminate interface does not spontaneously crack) an alternating compressive–tensile residual stress state is generated in the layers.

For ideal elastic materials, neglecting the influence of the external surfaces (where stresses may relax) and considering the laminate as an infinite plate, the stress field has been determined analytically and experimentally [16; 43-48]. In the present work the analysis has been made for thermal mismatch strains (where the mismatch only results from different thermal strains of the different layers) but the results remain valid for mismatch strains between layers which result from any other reason. In each layer a homogeneous and biaxial residual stress state exists. The stress magnitude $\sigma_{res,i}$ can be defined as:

$$\sigma_{res,i} = \frac{E_i}{1-v_i}(\bar{\alpha} - \alpha_i)\Delta T = \frac{E_i}{1-v_i}\Delta\varepsilon_i \tag{1}$$

where E_i, v_i and α_i are material properties of the ith layer (Young's modulus, Poisson's ratio and coefficient of thermal expansion). $(\bar{\alpha} - \alpha_i)\Delta T = \Delta\varepsilon_i$ is the mismatch strain of the ith layer. The temperature difference is: $\Delta T = T_0 - T_{Ref}$. The reference temperature T_{Ref} refers to the temperature at which the laminate is considered to be stress free. T_0 is the room temperature. The coefficient $\bar{\alpha}$ is given as an average expansion coefficient of the laminate:

$$\bar{\alpha} = \sum_{i=1}^{N} \frac{E_i t_i \alpha_i}{1-v_i} \Big/ \sum_{i=1}^{N} \frac{E_i t_i}{1-v_i} \tag{2}$$

with t_i being the thickness of the ith layer and N the number of layers.

Note that the reference temperature T_{Ref} is, in practice, not easy to determine. It may be slightly lower than the sintering temperature, since – if temperatures are reduced after sintering – the diffusion does not stop abruptly but it becomes slower and slower. In practice normalization based on Eq. (1) and additional residual stress measurements have to be performed to determine T_{Ref}.

If only two types of layer materials (A and B) are represented in the laminate (this is the case we consider in the following), Eq. (2) can be written in the form:

$$\bar{\alpha} = \left(\frac{E_A \alpha_A}{1-v_A} \cdot \sum_{i=1}^{n_A} t_{A,i} + \frac{E_B \alpha_B}{1-v_B} \cdot \sum_{i=1}^{n_B} t_{B,i}\right) \Big/ \left(\frac{E_A}{1-v_A} \cdot \sum_{i=1}^{n_A} t_{A,i} + \frac{E_B}{1-v_B} \cdot \sum_{i=1}^{n_B} t_{B,i}\right) \tag{3}$$

The magnitude of the residual stresses depends on the properties of A and B and of the ratio between the total thickness of the layers of type A ($T_A = \sum t_{A,i}$) and B ($T_B = \sum t_{B,i}$). The ratio of total thickness of the layer materials equals to their volume ratio: $T_B / T_A = V_B / V_A$. It is interesting to note that the magnitude of the residual stresses only depends on this volume ratio and not on the thickness of the individual layers i:

$$\sigma_{res,i} = f_i\left(\sum_{j=1}^{n_A} t_{A,j} \Big/ \sum_{j=1}^{n_B} t_{B,j}\right) = f_i\left(T_B/T_A\right) = f_i\left(V_B/V_A\right) \tag{4}$$

This is a very important aspect, which has not adequately been addressed in the past. It can provide the designer with more flexibility in order to tailor the disposition and thickness of layers when searching for an optimal design for a given level of residual stresses. Further consequences are explained as follows.

Although residual stresses are the key feature to enhance the mechanical properties of many layered systems, some negative effects of high residual stresses should be considered (see Fig. 1). For instance, while compressive stresses are beneficial in acting as a "shielding" mechanism against crack propagation, tensile stresses can cause cracking of the layer, if its strength is overcome. A typical example of that is "tunneling cracks", which may appear at the surface of the tensile layers [24; 49-51] and which can affect the structural integrity of the laminate. They can be avoided by lowering the tensile stresses in the corresponding layers.

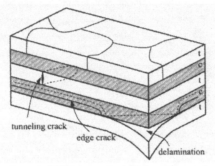

Figure 1. Different cracks in a layered ceramic due to residual stresses.

Another important aspect is the free surface of the material. It is well known that stresses at the free surface of layered materials are different from those within the bulk. In the region far from the free edges (i.e. in a finite plate), biaxial residual stresses parallel to the layer plane exist, and the stresses perpendicular to the layer plane are negligible [52]. Near the free edges, however, the residual stress state is no longer biaxial since the edge surface must be traction-free. As a result, a stress component perpendicular to the layer plane appears at the free edge [52-55]. This stress has a sign opposite to that of the biaxial stresses in the interior. Hence, for a compressive layer sandwiched between two tensile layers, a tensile residual stress perpendicular to the layer plane exists near the free surface of the compressive layer. The maximum tensile residual stress at the free surface equals the biaxial compressive residual stress in the interior. Although such tensile residual stress decreases rapidly from the edge surface to become negligible at a distance of the order of the compressive layer thickness, defects at the surface may be activated and cracks may initiate. An example is the so-called "edge cracks", initiating from pre-existing flaws, encountered at the free edges of the compressive layers [56]. Hence, preventing the initiation of edge cracking is important for the structural integrity of the component and should be considered in the multilayer design [57].

APPROACH TO DETERMINE FRACTURE RESISTANCE AND STRENGTH IN CERAMIC LAMINATES

Assumptions and Simplifications

In order to study the influence of the architecture of multilayer ceramics on the crack shielding some assumptions regarding geometry and material properties are beneficial:

- The laminate is described as an infinite plate and the influence of free surfaces is neglected.
- We consider symmetrical laminates (to avoid some curving) made of two materials (A and B).
- The coefficients of thermal expansion of A and B are different, which induces residual stresses in the layers. The magnitude of residual stresses in each layer depends mainly on the volume fraction V_B/V_A of material A and B as defined in Eq. (3) and Eq. (4).

First we restrict the analysis to periodic structures, where all layers made of material A have the thickness t_A and all B-layers have the thickness t_B. Figure 2 shows a typical example. For such laminate made from N layers ($(N+1)/2$ A-layers and $(N-1)/2$ B-layers) it holds that:

$$\frac{T_B}{T_A} = \frac{N+1}{N-1} \cdot \frac{t_B}{t_A} = \frac{V_B}{V_A} \tag{5}$$

(Later we extend the analysis to architectures where the thickness of the layers made of the same material can be different, i.e. non-periodic laminates).

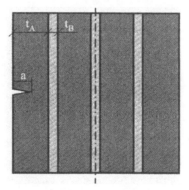

Figure 2. Symmetric, periodic multilayer architecture (ABABABA) with A and B layers of thickness t_A and t_B respectively. A through-thickness edge crack of length a propagates perpendicularly to the layers plane.

- We also restrict the analysis to laminates with high number of layers, i.e. $N \to \infty$. Then Eq. (5) can be approximated by: $t_B/t_A \approx T_B/T_A = V_B/V_A$. For the stresses this simplification yields:

$$\frac{\sigma_A}{\sigma_B} = -\frac{t_B}{t_A} \tag{6}$$

- For the description of the crack shielding caused by residual stresses the elastic properties of materials A and B are considered to be equal. Note that under that assumption it holds $E_A = E_B = E$; $v_A = v_B = v$. Also fracture resistance and strength of individual layers are considered equal for convenience, *i.e.* $K_{c,A} = K_{c,B} = K_c$ and $\sigma_{c,A} = \sigma_{c,B} = \sigma_c$.
- Fracture of a laminate always starts from the surface layer, as observed experimentally under flexural bending. We assume that the fracture initiating flaw can be described by a straight through-thickness edge crack of length a (see Fig. 2). Then a 2-D model can be used to describe the behavior of the crack.
- For convenience the external stress field is assumed to be uniaxial and homogeneous, having the magnitude σ_{appl}.
- The geometric factor of a straight through-thickness edge crack in a laminate with an infinite thickness in a homogeneous uniaxial stress field is defined constant as $Y = 1.12$.

Fracture Mechanics Analysis

The fracture criterion for brittle materials is described by the Griffith/Irwin equation [58]:

$$K(a) \geq K_c \tag{7}$$

where

$$K(a) = \sigma Y \sqrt{\pi a} \tag{8}$$

is the stress intensity factor, which describes the stress field at the tip of a crack of length a under the action of the stress σ. K_c is the fracture toughness of the material and Y is the geometric factor.

The shielding effect caused by the residual stresses can be described by treating the residual stresses to be an additional external stress, giving an additional term $K_{res}(a)$ to the stress intensity factor at the crack tip, K_{tip}, which now reads:

$$K_{tip}(a) = K_{appl}(a) + K_{res}(a) \tag{9}$$

with $K_{appl}(a) = Y\sigma_{appl}\sqrt{\pi a}$ being the stress intensity factor caused by the applied stress. Thus, solving Eq. (9) for K_{appl}, the Griffith/Irwin criterion (Eq. (7)) becomes:

$$K_{appl}(a) \geq K_c - K_{res}(a) = K_R(a) \tag{10}$$

For $K_{res} < 0$, as it holds for the action of compressive stresses, $K_R(a) \geq K_c$, what is called an increasing fracture resistance curve (R-curve). This describes the "shielding" effect associated with the compressive stresses. If tensile residual stresses are acting, "anti-shielding" occurs and K_R decreases with increasing crack extension.

In the following the fracture resistance curves of several ceramic laminates are determined using the weight function method. The procedure of calculation is described in detail elsewhere [42;57].

RESULTS AND DISCUSSION

Qualitative Behavior of Cracks in Laminates with Internal Compressive Stresses
For laminates with internal compressive stress (here named as ICS-Laminates) the fracture resistance has to be determined using approximation methods such as the weight function method [59; 60]. At first we again restrict the analysis to symmetric and periodic laminates having infinite thickness.

Due to the tensile residual stresses in the outer A-layer the R-curve decreases in this layer, but it increases again in the following B-layer, which contains residual compressive stresses (see Eq. (9)). Figure 3 shows a typical R-curve of an ICS-laminate as a function of the crack length parameter $1.12\sqrt{\pi a}$ in the region of maximum shielding. Indeed, the maximum shielding (peak fracture resistance) is achieved for a crack length $a = t_A + t_B$.

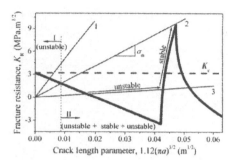

Figure 3. Fracture resistance in a typical ICS-laminate versus the crack size parameter $1.12\sqrt{\pi a}$.

It is assumed that no material flaws occur which are larger than the thickness of the first A-layer (i.e. $a \leq t_A$). For very small flaws having a size within region I unstable crack propagation occurs, which causes catastrophic failure. But that happens at very high stresses (see the slope of line 1 in Fig. 3). Larger cracks in region II first pop in to an even greater size if stress is applied (section of line 3 in Fig. 3 with the decreasing R-curve) and then stop (intersection point of line 3 with the increasing part of the R-curve in layer B in Fig. 3). This is caused by the shielding effect of the compressive layer. If the stress is further increased stable crack growth up to the crack length $a = t_A + t_B$ occurs. The slope of line 2 in Fig. 3 corresponds to the stress, where these cracks become unstable again and where catastrophic failure occurs. Thus, line 2 in Fig. 3 determines a threshold for the strength of the laminate, which is: $\sigma_{th} = K_{R,peak} / 1.12\sqrt{\pi(t_A + t_B)}$. Note that only the first two layers of ICS-laminates are important for the assessment of the fracture resistance.

Optimization of Periodic Laminates
In order to optimize the fracture resistance of the laminates, the volume ratio between materials A and B can be modified. For periodic laminates, this can be approximated to varying the layer thickness ratio, i.e. t_B/t_A. Since the maximal value of fracture resistance is achieved in the first two layers, considering $t_A + t_B = constant$ an optimal design for maximal shielding can be found for $t_B / t_A \approx 0.25$. Figure 4a shows the maximum shielding $K_{R,max}$ versus $t_A / (t_A + t_B)$ for laminates with different values of $t_A + t_B$. It can be seen that the maximum always occurs for $t_A / (t_A + t_B) \approx 0.8$

(which corresponds to $t_B / t_A \approx 0.25$) and that the shielding also increases with $t_A + t_B$. Similar behavior can be observed for the threshold strength of ICS-laminates. Since the threshold strength can be calculated from the maximum shielding as follows:

$$\sigma_{th} = \frac{K_{R,peak}}{1.12\sqrt{\pi(t_A + t_B)}} \quad (11)$$

and the denominator in Eq. (11) is constant, the threshold strength reaches its maximum for the ratio $t_B / t_A \approx 0.25$ or $t_A / (t_A + t_B) \approx 4/5$. It also increases with $t_A + t_B$, as seen in Fig. 4b.

We caution the reader, that the optimization process of ICS-laminates must consider the restriction of residual stresses to avoid cracking of the laminate due to tunneling and/or edge cracks (see more details in [57]).

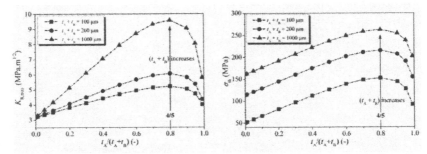

Figure 4. (a) Maximum fracture resistance and (b) threshold strength versus $t_A / (t_A + t_B)$ for ICS-laminates with different thickness of the first two layers $t_A + t_B$. Maximum shielding and strength are achieved for $t_B / t_A \approx 0.25$, increasing with $t_A + t_B$. (Broken lines are only guide to the eye).

Optimization of Non-Periodic Laminates

In this section we extend the analysis to non-periodic ICS laminates with finite thickness. We compare different laminates having the same tensile residual stress and the same compressive residual stress (i.e. the same V_B / V_A ratio) but a different thickness of the two outer layers, i.e. $t_A + t_B = constant$. Figure 5 shows R-curves for several non-periodic laminates (referred to as ICS2, ICS3 and ICS4). For comparison the R-curve of a periodic laminate (named ICS1) with 9 layers (taken from Fig. 3) is also plotted. It can be observed that when increasing the thickness of layer B (Laminate ICS2) the properties (e.g. σ_{th} and K_{Ic}) can be slightly increased. In the case that the thickness of layer A is reduced (ICS3), the threshold strength is significantly improved compared to ICS1 and ICS2. Moreover, if both layer A is reduced and layer B is increased in the same design (ICS4), both threshold strength and fracture resistance can be very much enhanced. Summarizing, the optimal design for ICS-laminates with non-periodic layers lies in the combination of both concepts: (i) first make the tensile layer A as thin as possible and (ii) the compressive layer B as thick as possible. This design ensures low decrease of K_R in the first thin tensile layer and high increase of K_R within the second thick compressive layer. When both steps are adopted in a unique non-periodic design, the mechanical properties of the system can be significantly improved with respect to monolithic materials.

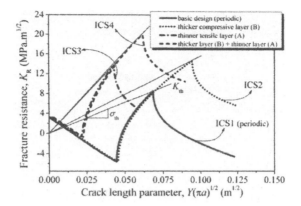

Figure 5. Fracture resistance (R-curves) for several non-periodic ICS laminates with different thickness of the first A and B layers. For comparison the R-curve of a periodic laminate with 9 layers ("basic design") is also plotted.

SUMMARY

Design guidelines for symmetric layered ceramics with strong interfaces and compressive residual stresses in the internal layers have been provided to optimize the strength and fracture resistance of the material. In laminates with a periodic layer distribution, the maximum fracture resistance can be obtained considering the combination of the first two layers. The first layer (with tensile residual stresses) should be as thin as possible, but thick enough to contain possible processing flaws. The second layer (compressive) must be as thick as possible. A minimum value for strength to guarantee mechanical reliability ("threshold strength") is found for these laminates, which can be defined independently of the nature of possible fracture origins, i.e processing flaws or external damage.

It has been demonstrated that laminates designed with a non-periodic distribution of the layers can yield significantly higher fracture resistance and threshold strength values than a periodic architecture. The idea is to make the first layer (tensile) very thin and the second layer (compressive) as thick as possible. The distribution of the inner layers can be used to tailor the residual stresses in an optimal way to fulfill the appropriate material volume ratio.

ACKNOWLEDGEMENT

Financial support by the Austrian Federal Government (in particular from the Bundesministerium für Verkehr, Innovation und Technologie and the Bundesministerium für Wirtschaft und Arbeit) and the Styrian Provincial Government, represented by Österreichische Forschungsförderungsgesellschaft mbH and by Steirische Wirtschaftsförderungsgesellschaft mbH, within the research activities of the K2 Competence Centre on "Integrated Research in Materials, Processing and Product Engineering", operated by the Materials Center Leoben Forschung GmbH in the framework of the Austrian COMET Competence Centre Programme, is gratefully acknowledged.

REFERENCES

[1] A. G. Evans, Perspective on the development of high-toughness ceramics, *J. Am. Ceram. Soc.,* **73**, 187-206 (1990).

[2] W. J. Clegg, K. Kendall, N. M. Alford, T. W. Button and J. D. Birchall, A simple way to make tough ceramics, *Nature,* **347**, 455-7 (1990).

[3] M. Rao, J. Sanchez-Herencia, G. Beltz, R. M. McMeeking and F. Lange, Laminar ceramics that exhibit a threshold strength, *Science,* **286**, 102-5 (1999).

[4] S. Deville, E. Saiz, R. K. Nalla and A. P. Tomsia, Freezing as a path to build complex composites, *Science,* **311**, 515-8 (2006).

[5] E. Munch, M. E. Launey, D. H. Alsem, E. Saiz, A. P. Tomsia and R. O. Ritchie, Tough, bio-inspired hybrid materials, *Science,* **322**, 1516-20 (2008).

[6] E. L. Maximilien and R. O. Ritchie, On the fracture toughness of advanced materials, *Advanced Materials,* **21**, 2103–10 (2009).

[7] F. F. Lange, Powder processing science and technology for increasing reliability, *J. Am. Ceram. Soc.,* **72**, 3-15 (1989).

[8] J. J. Hansen, R. A. Cutler, D. K. Shetty and A. V. Virkar, Indentation Fracture Response and Damage Resistance of Al_2O_3-ZrO_2 Composites Strengthened by Transformation-Induced Residual Stresses, *J. Am. Ceram. Soc.,* **71**, C501-C5 (1988).

[9] R. Lakshminarayanan, D. K. Shetty and R. A. Cutler, Toughening of layered ceramic composites with residual surface compression, *J. Am. Ceram. Soc.,* **79**, 79-87 (1996).

[10] J. Sanchez-Herencia, J. Moya and A. Tomsia, Microstructural design in alumina-alumina/zirconia layered composites, *Scripta Mater.,* **38**, 1-5 (1998).

[11] D. J. Green, R. Tandon and V. M. Sglavo, Crack arrest and multiple cracking in glass through the use of designed residual stress profiles, *Science,* **283**, 1295-7 (1999).

[12] M. Lugovy, V. Slyunyayev, V. Subbotin, N. Orlovskaya and G. Gogotsi, Crack arrest in Si_3N_4-based layered composites with residual stress, *Comp. Sci. Tech.,* **64**, 1947-57 (2004).

[13] V. M. Sglavo, M. Paternoster and M. Bertoldi, Tailored residual stresses in high reliability alumina-mullite ceramic laminates, *J. Am. Ceram. Soc.,* **88**, 2826-32 (2005).

[14] R. Bermejo, Y. Torres, C. Baudin, A. J. Sánchez-Herencia, J. Pascual, M. Anglada and L. Llanes, Threshold strength evaluation on an Al_2O_3–ZrO_2 multilayered system, *J. Eur. Ceram. Soc.,* **27**, 1443-8 (2007).

[15] H. Tomaszewski, Residual Stresses in Layered Ceramic Composites, *J. Eur. Ceram. Soc.,* **19**, 1329-31 (1999).

[16] D. J. Green, P. Cai and G. L. Messing, Residual stresses in alumina-zirconia laminates, *J. Eur. Ceram. Soc.,* **19**, 2511-1517 (1999).

[17] A. J. Sanchez-Herencia, C. Pascual, J. He and F. Lange, ZrO_2/ZrO_2 layered composites for crack bifurcation, *J. Am. Ceram. Soc.,* **82**, 1512-8 (1999).

[18] M. G. Pontin, M. Rao, J. Sanchez-Herencia and F. Lange, Laminar ceramics utilizing the zirconia tetragonal-to-monoclinic phase transformation to obtain a threshold strength, *J. Am. Ceram. Soc.,* **85**, 3041-8 (2002).

[19] R. Bermejo, Y. Torres, A. J. Sanchez-Herencia, C. Baudin, M. Anglada and L. Llanes, Residual stresses, strength and toughness of laminates with different layer thickness ratios, *Acta Mater.,* **54**, 4745-57 (2006).

[20] W. J. Clegg, The fabrication and failure of laminar ceramic composites, *Act. Metall. Mater.,* **40**, 3085-93 (1992).

[21] A. J. Phillipps, W. J. Clegg and T. W. Clyne, Fracture Behaviour of Ceramic Laminates in Bending - 1. Modelling of Crack Propagation, *Act. Metall. Mater.,* **41**, 805-17 (1993).

[22] M. Y. He, A. G. Evans and J. W. Hutchinson, Crack deflection at an interface between dissimilar elastic materials: Role of residual stresses, *Int. J. Solids and Struct.,* **31**, 3443-55 (1994).

[23] M. Oechsner, C. Hillman and F. Lange, Crack bifurcation in laminar ceramic composites, *J. Am. Ceram. Soc.,* **79**, 1834-8 (1996).

[24] A. J. Sanchez-Herencia, L. James and F. F. Lange, Bifurcation in alumina plates produced by a phase transformation in central, alumina/zirconia thin layers, *J. Eur. Ceram. Soc.*, **20**, 1297-300 (2000).

[25] M. Lugovy, N. Orlovskaya, V. Slyunyayev, G. Gogotsi, J. Kübler and A. J. Sanchez-Herencia, Crack bifurcation features in laminar specimens with fixed total thickness, *Comp. Sci. Tech.*, **62**, 819-30 (2002).

[26] M. G. Pontin and F. F. Lange, Crack bifurcation at the surface of laminar ceramics that exhibit a threshold strength, *J. Am. Ceram. Soc.*, **88**, 1315-7 (2005).

[27] K. Hbaieb, R. McMeeking and F. Lange, Crack bifurcation in laminar ceramics having large compressive stress, *Int. J. Solids and Struct.*, **44**, 3328-43 (2007).

[28] H. Bahr, V. Pham, H. Weiss, U. Bahr, M. Streubig, H. Balke and V. Ulbricht, Threshold strength prediction for laminar ceramics from bifurcated crack path simulation, *Int. J. Mat. Research*, **98**, 683-91 (2007).

[29] C. Chen, R. Bermejo and O. Kolednik, Numerical analysis on special cracking phenomena of residual compressive inter-layers in ceramic laminates *Engng. Fract. Mech.*, **77**, 2567-76 (2010).

[30] L. Náhlik, L. Šestáková, P. Hutař and R. Bermejo, Prediction of crack propagation in layered ceramics with strong interfaces, *Engng. Fract. Mech.*, **77**, 2192-9 (2010).

[31] R. Bermejo and R. Danzer, High failure resistance layered ceramics using crack bifurcation and interface delamination as reinforcement mechanisms, *Eng. Fract. Mech.*, **77**, 2126-35 (2010).

[32] R. Bermejo and R. Danzer, Failure resistance optimisation in layered ceramics designed with strong interfaces, *J. Ceram. Sci. Tech.*, **01**, 15-20 (2010).

[33] O. Prakash, P. Sarkar and P. S. Nicholson, Crack deflection in ceramic/ceramic laminates with strong interfaces, *J. Am. Ceram. Soc.*, **78**, 1125-7 (1995).

[34] D. B. Marshall, P. E. D. Morgan and R. M. Housley, Debonding in multilayered composites of zirconia and LaPO₄, *J. Am. Ceram. Soc.*, **80**, 1677-83 (1997).

[35] W. J. Clegg, Design of ceramic laminates for structural applications, *Mater. Sci. Tech.*, **14**, 483-95 (1998).

[36] A. Ceylan and P. A. Fuierer, Fracture toughness of alumina/lanthanum titanate laminate composites with weak interface, *Materials Letters*, **61**, 551-5 (2007).

[37] H. Tomaszéwski, H. Weglarz, A. Wajler, M. Boniecki and D. Kalinski, Multilayer ceramic composites with high failure resistance, *J. Eur. Ceram. Soc.*, **27**, 1373-7 (2007).

[38] R. Bermejo, Y. Torres, M. Anglada and L. Llanes, Fatigue behavior of alumina-zirconia multilayered ceramics, *J. Am. Ceram. Soc.*, **91**, 1618-25 (2008).

[39] D. Jianxin, D. Zhenxing, Y. Dongling, Z. Hui, A. Xing and Z. Jun, Fabrication and performance of Al2O3/(W,Ti)C + Al2O3/TiC multilayered ceramic cutting tools, *Materials Science and Engineering: A*, **527**, 1039-47 (2010).

[40] J. Gao, Y. He and D. Wang, Fabrication and high temperature oxidation resistance of ZrO2/Al2O3 micro-laminated coatings on stainless steel, *Materials Chemistry and Physics*, **123**, 731-6 (2010).

[41] N. Orlovskaya, M. Lugovy, V. Subbotin, O. Rachenko, J. Adams, M. Chheda, J. Shih, J. Sankar and S. Yarmolenko, Robust design and manufacturing of ceramic laminates with controlled thermal residual stresses for enhanced toughness, *J. Mater. Sci.*, **40**, 5483-90 (2005).

[42] R. Bermejo, J. Pascual, T. Lube and R. Danzer, Optimal strength and toughness of Al2O3-ZrO2 laminates designed with external or internal compressive layers, *J. Eur. Ceram. Soc.*, **28**, 1575-83 (2008).

[43] H. J. Oël and V. D. Fréchette, Stress distribution in multiphase systems: I, composites with planar interfaces, *J. Am. Ceram. Soc.*, **50**, 542-9 (1967).

[44] T. Chartier, D. Merle and J. L. Besson, Laminar ceramic composites, *J. Eur. Ceram. Soc.*, **15**, 101-7 (1995).

[45] V. M. Sglavo and M. Bertoldi, Design and production of ceramic laminates with high mechanical resistance and reliability, *Acta Mater.*, **54**, 4929-37 (2006).

[46] V. M. Sglavo and M. Bertoldi, Design and production of ceramic laminates with high mechanical reliability, *Comp B: Engineering.*, **37**, 481-9 (2006).

[47] A. Costabile and V. M. Sglavo, Influence of the Architecture on the Mechanical Performances of Alumina-Zirconia-Mullite Ceramic Laminates, *Adv. Sci. Tech.*, **45**, 1103-8 (2006).

[48] M. Leoni, M. Ortolani, M. Bertoldi, V. M. Sglavo and P. Scardi, Nondestructive Measurement of the Residual Stress Profile in Ceramic Laminates, *J. Am. Ceram. Soc.*, **91**, 1218-25 (2008).

[49] S. Ho and Z. Suo, Tunneling cracks in constrained layers, *J. Appl. Mech.*, **60**, 890-4 (1993).

[50] C. Hillman, Z. Suo and F. Lange, Cracking of Laminates Subjected to Biaxial Tensile Stresses, *J. Am. Ceram. Soc.*, **79**, 2127-33 (1996).

[51] R. Bermejo, A. J. Sanchez-Herencia, C. Baudín and L. Llanes, Tensiones residuales en cerámicas multicapa de Al$_2$O$_3$-ZrO$_2$: naturaleza, evaluación y consecuencias sobre la integridad estructural, *Bol. Soc. Esp. Ceram. V.*, **45**, 352-7 (2006).

[52] S. Ho, C. Hillman, F. F. Lange and Z. Suo, Surface cracking in layers under biaxial, residual compressive stress, *J. Am. Ceram. Soc.*, **78**, 2353-9 (1995).

[53] M. Rao and F. Lange, Factors affecting threshold strength in laminar ceramics containing thin compressive layers, *J. Am. Ceram. Soc.*, **85**, 1222-8 (2002).

[54] M. Y. He, A. G. Evans and A. Yehle, Criterion for the avoidance of edge cracking in layered systems, *J. Am. Ceram. Soc.*, **87**, 1418-23 (2004).

[55] A. J. Monkowski and G. E. Beltz, Suppression of edge cracking in layered ceramic composites by edge coating, *Int. J. Solids and Struct.*, **42**, 581-90 (2005).

[56] C. Chen, R. Bermejo and O. Kolednik, Numerical analysis on spceial cracking phenomena of residual compressive inter-layers in ceramic laminates, *Engng. Fract. Mech.*, **77**, 2567-76 (2010).

[57] L. Sestakova, R. Bermejo, Z. Chlup and R. Danzer, Strategies for fracture toughness, strength and reliability optimisation of ceramic-ceramic laminates, *Int. J. Mat. Research*, **102**, 613-26 (2011).

[58] A. A. Griffith, The phenomenon of rupture and flow in solids, *Phil. Trans. Roy. Soc. London*, **A221**, 163-98 (1921).

[59] T. Fett and D. Munz, Stress Intensity Factors and Weight Functions. Southampton, 1997.

[60] H. F. Bueckner, Weight functions for the notched bar, *Zeitschrift für Angewandte Mathematik und Mechanik*, **51**, 97-109 (1971).

INVESTIGATION OF CRITICAL FIBER LENGTH IN PHENOL MATRIX BASED SHORT FIBER CFRP BY DOUBLE OVERLAP JOINTS

Daniel Heim, Swen Zaremba, and Klaus Drechsler
Institute for Carbon Composites, Mechanical Engineering, Technical University of Munich
Garching, Bavaria, Germany

ABSTRACT

For short fiber CFRP based on a phenol matrix the influence of fiber length in terms of tensile strength and shear strength is investigated. The effect of fiber length within a representative phenol matrix structure (high porosity) is studied by a double lap joint principle. Therefore, a laminate with three layers is manufactured in such a way that each layer exhibits a cut. The two outer layers have the joint at the same position; the middle layer's joint has an offset which is adequate to the overlap length. Specimens with an overlap length ranging from 4 mm to 100 mm are mechanically tested. The CFRP material shows a nonlinear increase in tensile strength for short overlaps. As the overlap length increases, tensile strength reaches an upper limit asymptotically. This overlap length behavior is already known from adhesive double lap joints. The transition point between the rise and the asymptotic region is to determine the optimal overlap length. Due to contrary effects in short fiber composites (e.g. heterogeneity with increasing fiber length), the optimum length for the overlap is not the same as for the continuous fiber composites. If the model system is appropriate for short fiber composites, the optimal overlap length corresponds to an upper limit fiber length. The double lap joint principle could be an effective way of determining a maximum fiber length for a specific fiber matrix combination with a realistic matrix structure.

INTRODUCTION

The interaction of fibers and matrix is crucial for composite performance, especially in short fiber reinforced composites. This type of composites is apparent in everyone's daily life. For example cooking spoons, bath tubes or secondary structures of cars, trains or aircrafts are made out of these composites. The majority is made out of glass fiber with thermoplastics or thermosetting resins. With increasing requirements in terms of mechanical properties or light weight design, carbon fibers are getting more important. Besides fiber type, the fiber length plays a key role for mechanical properties of short fiber reinforced composites. A simple model is the critical fiber length.[1] Until the critical fiber length is not exceeded the mechanical properties increase. If the critical fiber length is exceeded, fiber breakage will occur and no further improvement, e.g. in tensile strength, is possible. In this study an investigation of the critical fiber length for a carbon fiber with a phenol matrix is performed by a double lap joint principle.

EXPERIMENTAL

The model system is a double lap joint, because compared with a single lap joint the normal stresses (peal stresses) can be neglected in this setup. The overlap length x (cf. Figure 1) is carried out by a cut in the outer layers (1st and 3rd layer) at the position x_0. The middle layer (2nd layer) is cut at position $x_0 + x$. The three layers are unidirectional carbon fiber prepreg with a phenol matrix. The layers were sliced by a cutting machine, followed by hand laminating and hot pressing. After having glued glass fiber tabs (50 mm length) onto the CFRP plates at both ends of each side, the CFRP plates were cut into specimens of 390 mm x 25 mm x 1 mm (L x W x T) with a water cooled diamond blade saw.

Mechanical tests were performed according to DIN EN 2561 on a Zwick Roell Z250 with a preforce of 200 N (v_{pre} = 1 mm/min). During testing the crosshead speed was 2 mm/min.

Tensile strength σ_f is calculated according to equation (1), where F is the measured force, w is the width, and t is the thickness of the specimen (cf. Figure 2). To calculate the average shear strength $\bar{\tau}$ equation (2) is used, where F is the measured force, w is the width of the specimen, and x is the overlap length (cf. Figure 3).

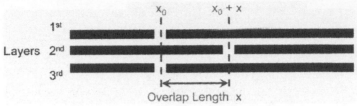

Figure 1. Schematic of the double lap joint laminates.

$$\sigma_f = \frac{F}{w \cdot t} \tag{1}$$

$$\bar{\tau} = \frac{F}{2 \cdot w \cdot x} \tag{2}$$

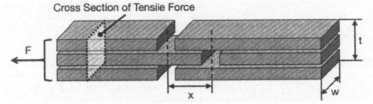

Figure 2. Schematic of tensile strength of UD laminates with overlap.

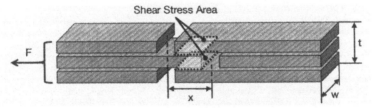

Figure 3. Schematic of average shear strength of UD laminates with overlap.

In total 20 different overlap lengths in the range of 4 mm to 100 mm were investigated. For each overlap length at least six specimens were tested. Additionally tensile test on 30 unidirectional specimens of the same geometry and layup without any cuts were completed. The aim is to measure the composite strength and compare these results with the tensile strength of overlap specimens. All tensile specimens were produced in four batches. Microsections were prepared to determine the porosity and the fiber volume fraction (cf. Figure 4).

Figure 4. Microsections of tensile specimen a) low porosity b) high porosity.

RESULTS AND DISCUSSION

Unidirectional Specimens without Cuts

In the case of tensile specimens without cuts the load is equally distributed among the three layers within the area of interest. All specimens of this type showed a linear behavior till ultimate failure. Fiber volume fraction of these samples was 53 ± 5 %. Average ultimate tensile strength was 1100 MPa. The first standard deviation was 107 MPa and therefore about 10 % of the average value. Scattering within and between batches was randomly distributed (cf. Figure 5). The cause for the scattering is the different amount of porosities, which varies from 2 % to 18 % (cf. Figure 4). This high porosity arises from steam produced by polycondensation of the phenolic matrix. Moreover the specimens show some waviness, especially between areas with low and high porosity.

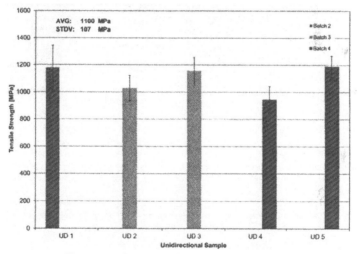

Figure 5. Tensile strength of UD laminates without cuts.

Unidirectional Specimens with an Overlap
Figure 6 shows photographs of a 20 mm overlap joint. In Figure 6 a) a gap is visible in the butt splice area, which consists out of pure matrix in the outer layers. Part b) of Figure 6 shows the two parts of a fractured specimen, which were pulled apart from each other during mechanical testing.

The load transfer for unidirectional sample with an overlap is not straight forward. Far away from the overlap area the load is equally distributed among the three layers. In the overlap region load transfer is as follows. Before the position x_0 the loads from the 1st and 3rd layer are transferred via shear forces to the 2nd layer (cf. Figure 1). The complete tensile force is conducted by the 2nd layer at point x_0. This point represents the first weak point of the set up. In the region from x_0 to $x_0 + x$ the tensile force is transferred to the 1st and 3rd layer by shear force, this region is the second weak point of the set up. As a consequence two failure modes are possible. In the first case the average shear strength of the matrix is higher than the shear stress in the overlap area. For this situation tensile fiber fracture of the 2nd layer will occur at position x_0. In the second case shear stress in the overlap area is higher than the average shear strength of the matrix, so the interface between the middle and the outer layer will be destroyed. The result will be a fiber pull out in the region x_0 to x_0+x.

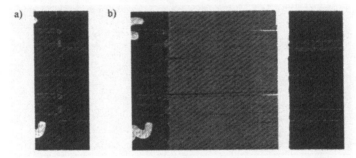

Figure 6. Photographs of 20 mm overlap joint: a) butt splice with a gap in the outer layers b) tested sample with the two pieces pulled apart from each other.

Test results show a x^{-1} decrease of the average shear strength with increasing overlap length x (cf. Figure 7). In Figure 9 tensile strength over overlap length is plotted. The dotted line is the tensile strength of one fiber layer (about 367 MPa), which is one third of the average tensile strength determined by the unidirectional specimens without cuts. All tested specimens have lower tensile strength than the tensile strength of one fiber layer. Furthermore all samples exhibit the failure of the interface between the middle and the outer layers (cf. Figure 6 b). This indicates that the interface is the weakest link and that the average shear strength of the matrix is lower than the occurring shear stresses (cf. Figure 11). The low average shear strength is mainly caused by the high porosity inside the matrix.

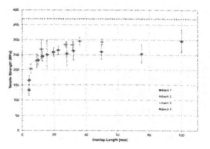

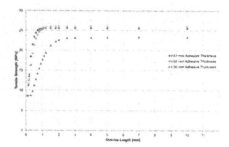

Figure 7. Experimental tensile strength of UD laminates with overlap.

Figure 8. Calculated (ESAComp 4.1) tensile strength of exemplary adhesive double lap joints with different overlap lengths.

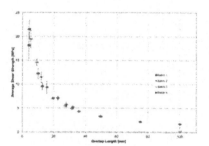

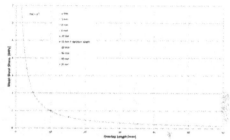

Figure 9. Experimental average shear strength of UD laminates with overlap.

Figure 10. Calculated (ESAComp 4.1) average shear strength of exemplary adhesive double lap joints with different overlap lengths.

Nevertheless the tensile strength shows an interesting behavior. In Figure 8 and Figure 10 some exemplary calculations for double lap joint behavior were performed with the software ESAComp 4.1. Due to the lack of material properties for the used phenol matrix the in ESAComp 4.1 included material data for UD T300 epoxy matrix (FVF 60%) and adhesive HYSOL EA 9321 were used for the exemplary simulations instead. In this model the phenol matrix is an equivalent of the Hysol EA 9321 adhesive and used carbon fibers are the equivalent of the UD T300 epoxy composite. It is know from adhesive theory, that there is an additional excessive increase of the shear stress, if the overlap length is too short (cf. Figure 11).[2,3] This causes a lowering in tensile strength of double lap joints. For longer overlap lengths the excessive increase vanishes and the tensile strength is constant. The same behavior is observed for the tensile strength of the unidirectional samples with an overlap. As the overlap length increases, tensile strength reaches an upper limit asymptotically. The transition point between the rise and the asymptotic region can be used to determine the optimal overlap length.

Due to contrary effects in short fiber composites (e.g. heterogeneity with increasing fiber length [4-6]) the optimum length for the overlap is not the same as the optimum fiber length. But if the model system is appropriate the optimal overlap length corresponds to an upper limit for the fiber length.

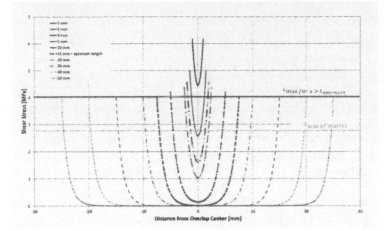

Figure 11. Calculated (ESAComp 4.1) shear stress distribution of exemplary adhesive double lap joints with different overlap lengths.

CONCLUSION

Tensile tests were performed on unidirectional carbon fiber reinforced phenol matrix specimens without any cuts and with an overlap length ranging from 4 mm to 100 mm. The uncut specimens had a tensile strength of 1100 ± 107 MPa. The overlap specimens always failed because of shear stresses and not of fiber fracture. Hence no critical fiber length could be determined, but a novel approach to evaluate the upper limit for an optimal fiber length in short carbon fiber reinforced phenol matrices was implemented. If the mechanical properties for a certain application are sufficient, fibers with less tensile strength could be used, because of the matrix is the weakest link. If it is necessary to further increase the mechanical performance of the short fiber composite, the porosity have to minimized, e.g. lower heating rates, or a phenol matrix with higher shear strength. In principle the double lap joint could be an effective way of determining a maximum fiber length for a specific fiber matrix combination with a realistic matrix structure.

ACKNOWLEDGEMENT

The authors gratefully acknowledge the support by the Faculty Graduate Center of Mechanical Engineering of TUM Graduate School at Technical University of Munich, Germany.

REFERENCES

[1] W. D. Callister, Materials Science and Engineering: An Introduction, *John Wiley & Sons Inc.,* New York, 1994.
[2] O. Volkerson, Neue Untersuchungen zur Theorie der Klebeverbindungen *Jahrbuch 1963 WGLR* 1963, 299-306.
[3] M. Goland, E.Reissner, The stress in cemented joints *Journal of Applied Mechanics* 1944, 11, A17-A27.

[4] L. T. Harper, T. A. Turner, N. A. Warrior, J. S. Dahl, C. D. Rudd, Characterisation of random carbon fibre composites from a directed fibre preforming process: Analysis of microstructural parameters *Composites Part A: Applied Science and Manufacturing* 2006, 37, 2136–2147.

[5] L. T. Harper, T. A. Turner, N. A. Warrior, J. S. Dahl, C. D. Rudd, Characterisation of random carbon fibre composites from a directed fibre preforming process: effect of fibre length *Composites Part A: Applied Science and Manufacturing* 2006, 37, 1863-1878.

[6] D. A. Trapeznikov, A. I. Toropov, O. D. Loskutov, Modelling approach to optimization of mechanical properties of discontinuous fibre-reinforced C/C composites, *Composites* 1992, 23, 174–182.

MECHANICAL AND MICROSTRUCTURAL CHARACTERIZATION OF C/C-SIC MANUFACTURED VIA TRIAXIAL AND BIAXIAL BRAIDED FIBER PREFORMS

Fabian Breede, Martin Frieß, Raouf Jemmali, Dietmar Koch, Heinz Voggenreiter
Department of Ceramic Composites and Structures
Institute of Structures and Design
German Aerospace Center (DLR)
Stuttgart, Germany

Virginia Frenzel, Klaus Drechsler
Institute of Aircraft Design, University of Stuttgart
Stuttgart, Germany

ABSTRACT

Carbon-carbon silicon carbide (C/C-SiC) composite plates were manufactured based on triaxial and biaxial braided fiber preforms. The liquid silicon infiltration (LSI) method was used to achieve the SiC matrix. In a first processing step intermediate modulus carbon fiber (T800 12k) preforms were manufactured via braiding technique in ±45° and ±75° fiber orientation and an optional 0° orientation. In the next step the fiber preforms were processed to carbon fiber reinforced plastics (CFRP) by warm pressing (using phenolic resin) and by resin transfer molding (RTM, using aromatic resin), respectively. The CFRP have a fiber volume content of about 60 percent. The main objective of this work was to investigate the mechanical properties under tensile and bending mode at room temperature depending on fiber orientation and processing route. The change in fiber orientation as well as the influence of triaxial fiber preforms in contrast to biaxial fiber performs will be discussed. The use of different CFRP process techniques resulted in different final matrix structures and therefore in different microstructures of the resulting C/C-SiC composites which will be compared based on CT and SEM analysis.

INTRODUCTION

Ceramic matrix composites (CMC) are favorable as structural materials in aerospace engineering where high specific strengths and resistance to elevated temperatures are desirable[1-2]. Carbon-carbon silicon carbide (C/C-SiC) CMC are promising candidates for thermo structural components, since they retain the advantage of a dense silicon carbide matrix while providing an enhanced degree of damage tolerance[3]. Liquid silicon infiltration (LSI) provides a simple processing route which generates CMC with a dense silicon carbide matrix with lower open porosities (below 5%) as compared to chemical vapor infiltration (CVI) and polymer impregnation and pyrolysis (PIP). Different fabric reinforcement based bi-directional and stitched C/SiC composites using LSI have been investigated and published in literature[4-7]. Mechanical properties, e.g. flexural strength, tensile strength and fracture toughness are important design parameters. These properties depend strongly on composition, microstructure and fiber orientation, which on the other hand depend on physical properties like porosity, size, shape and distribution of those pores and siliconization parameters[1-8].

In contrast to bi-directional or filament winding based composites here the potential of braided based CMC was evaluated. The primary advantage of braided fiber preforms is the ability to adapt the fiber orientation to specific structural needs as well as adding a third fiber orientation to improve in-plane quasi-isotropic material behavior.

The aim of the present work is first to characterize the microstructure with emphasis on the different fiber architectures and with respect to the different CFRP processing techniques using two types of precursor systems. The second is to investigate the mechanical behavior of the different composites at room temperatures and its relation to the microstructures.

EXPERIMENTAL PROCEDURE

Composite preparation

Preforming and CFRP: Biaxial and triaxial braided fibrous preform layers in the size range of 300mm x 300mm using high strength T800 12k carbon fiber were manufactured with a fiber orientation of ±45° and ±75° at the Institute of Aircraft Design at the University of Stuttgart, see Figure 1 and 2. First tubular preforms were braided and afterwards cut to layers. Six to eight layers were stacked and then CFRP green body plates were fabricated by means of resin transfer molding (RTM) and warm pressing with a fiber volume content of about 60%. An overview on materials and fabrication parameters used is given in Table I.

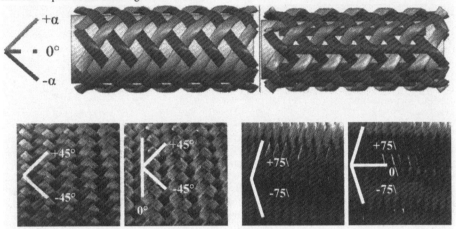

Figure 1. Schematic sketch of biaxial (left) and triaxial (right) braiding.

Figure 2. Different braided preform types. Biaxial ±45°, triaxial ±45°, biaxial ±75° and triaxial ±75°, respectively.

Table I. Overview of raw materials and fabrication parameters.

CFRP process route	RTM (I-type)				Warm pressing (HP-type)			
Precursor	Aromatic resin (XP60)				Phenolic resin (JK60)			
Fiber preform type	Biaxial		Triaxial		Biaxial		Triaxial	
Braiding angle	±45°	±75°	±45°	±75°	±45°	±75°	±45°	±75°
Plate-ID	I703	I705	I704	I711	HP641	HP642	HP640	HP643
Fiber volume content (FVC)	62%	60%	61%	65%	62%	58%	54%	70%
FVC w/o 3. fiber orientation	-	-	46%	55%	-	-	41%	62%
Fiber type	T800 12k carbon fiber							

C/C preforming (pyrolysis): The CFRP processing was then followed by carbonization at 900°C in nitrogen atmosphere and 1650°C in vacuum conditions. During the carbonization the polymer matrix is converted to amorphous carbon which goes along with a mass and volume reduction. The main goal during pyrolysis is the formation of a defined micro crack system with dense C/C-bundles. The advantage of the formation of dense C/C-bundles is that most of the carbon fibers are protected from degradation during the next processing step which is siliconization.

Siliconization: The porous C/C-preform plates were infiltrated by molten silicon at temperatures above 1450°C under vacuum conditions by means of capillary action. Liquid silicon reacts with the carbon at the surface of the micro crack system, consisting of pores and channels, forming the SiC matrix. Due to the creation of dense C/C-bundles most of the carbon fibers remain undamaged and provide damage tolerance without complex and expensive fiber coating. The resulting CMCs typically exhibit an open porosity lower than 5% and no further matrix densification steps are necessary resulting in a short and cost efficient fabrication route.

Material characterization and mechanical testing

Density, open porosity and geometrical shrinkage

The evolution of density and open porosity were determined by means of Archimedes' method according to DIN EN 1389 and are presented in Table III. The geometrical shrinkage during carbonization was conducted by digital linear measurement and is presented in Table IV.

Non-destructive characterization and microstructure observation

Non-destructive testing was carried out in order to assure material quality after each processing step. A homogenous distribution of pores and micro cracks during carbonization are very essential for a successful siliconization. Therefore CT-analysis [Nanotom, Phoenix x-ray Systems] and mercury porosimetry method [Pascal 140/240, CE Instruments, DIN 66133] were used to characterize the developed pore systems after carbonization.

Cross-sections of the fabricated C/C-SiC composites were ground and polished. The microstructural observations of the fiber architecture and the composition of the different braided composites were conducted using a scanning electron microscope (SEM) [Zeiss Ultra 55].

Flexural and tensile tests

All mechanical investigations were conducted at room temperature. Four to six specimens for each tested fiber orientation were cut. Tested fiber orientations were 0/90°, ±45° and ±15° with respect to the loading direction. The 0/90° test specimens were cut from plates with a fiber orientation of ±45°. The flexural material parameters were measured by three point-bending tests on a universal testing machine [Zwick Roell 1494] according to DIN EN 658-3 with a crosshead speed of 1mm/min. The flexural specimen size was found to be 3mm x 10mm x 90mm. Tensile properties were received from testing using a universal testing machine [Zwick Roell 1475] according to DIN EN 658-1 with a crosshead speed of 3mm/min. All tensile test specimens with a size of 3mm x 10mm x 150mm were provided with two strain gauges. Tests were carried out to determine maximal fracture strength, Young's modulus and elongation at fracture. The three point bending test also provided the minimal interlaminar shear strength. Tensile tests with a fiber orientation of ±45° provided the shear modulus according to ASTM D 3518. The fracture surfaces of tested specimens were examined by SEM.

RESULTS AND DISCUSSION

Preform characterization

The fabricated C/C-SiC composites are a unique kind of multilayer braided fabric based CMC. As stated in Table I eight composite panels were fabricated. Six to eight preform layers w. r. t to the desired global fiber volume fraction were used. The resulted areal weights of the braided preforms are presented in Table II. The preform layup was set to be the same orientation except for the ±45° triaxial based composites. Here, the preform layer orientation was changed for consecutive layers by 90° to achieve a quasi-isotropic in-plane material behavior.

Table II. Areal weight (g/m²) for single preform layers.

±45°		±75°	
Biaxial	Triaxial	Biaxial	Triaxial
370	490	350	410

Density and open porosity

The density and open porosity of the as-received braided C/C-SiC composites are presented in Table III showing average values for I- and HP-type, respectively. The RTM infiltration results in a lower open porosity in the CFRP state compared to warm pressing. The increase of the open porosity during carbonization was found to be in the range of 7.1% to 17.8% for the RTM based composites. However, due to the lower carbon yield, after carbonization a significantly higher open porosity is measured for the RTM based material which remains even after liquid siliconization (2.7% vs. 1.5%). The open porosity of warm pressed based composites was in the range of 9.0% to 12.2%.

At ±45° the RTM based composite showed significantly higher open porosity (16-18% vs. 10-12%) after carbonization and higher material density (1.91-1.98g/cm³ vs. 1.80g/cm³) in the C/C-SiC state in contrast to the warm pressed based composite. No major variations were observed comparing biaxial and triaxial architectures at ±45°. No significant differences were observed at ±75° comparing both manufacturing routes. Here open porosity and material density was found to be in the same range. Hence, biaxial composites with a fiber orientation of ±75° showed a considerably lower increase in open porosity (7-9% vs. 10-12%) due to higher shrinkage (4.2% vs. 0.3%) in the transverse direction compared to its triaxial counterpart, also see Table IV. In addition to a reduced increase in open porosity, the thickness reduction was found to be drastically lower within the biaxial ±75° composite. It decreased from typically 10% to about 1.5% as a result of the high transverse shrinkage. Average shrinkage values of both fabrication routes for all different braided composites are listed in Table IV. No significant differences were observed in the shrinkage behavior comparing both processing routes. Here only the fiber architecture influenced the shrinkage. No significant dimensional changes after siliconization were observed.

Table III. Density and open porosity of braided C/C-SiC plates for all process steps.

Braiding architecture and angle	Material state Plate-ID	CFRP		C/C		C/C-SiC	
		I	HP	I	HP	I	HP
Biaxial ±45°	Density / g/cm³	1.50	1.48	1.40	1.40	1.98	1.80
	Open porosity / %	1.2	2.1	15.8	10.5	2.7	1.3
Triaxial ±45°	Density / g/cm³	1.48	1.49	1.36	1.37	1.91	1.81
	Open porosity / %	1.2	2.7	17.8	12.2	3.4	1.4
Biaxial ±75°	Density / g/cm³	1.54	1.49	1.52	1.44	1.80	1.76
	Open porosity / %	1.6	1.0	7.1	9.0	2.6	2.0
Triaxial ±75°	Density / g/cm³	1.52	1.48	1.45	1.40	1.86	1.82
	Open porosity / %	0.4	1.8	11.6	9.9	2.1	1.3
Average values	Density / g/cm³	1.52 (0.02)	1.49 (0.01)	1.43 (0.06)	1.40 (0.02)	1.89 (0.08)	1.80 (0.02)
	Open porosity / %	1.1 (0.5)	1.9 (0.6)	13.1 (4.1)	10.6 (0.9)	2.7 (0.5)	1.5 (0.3)

(Values in brackets represent the standard deviation)

Table IV. Shrinkage behavior of different braided C/C-SiC composites during carbonization for both manufacturing routes.

Braiding angle	±45°		±75°	
Preform type	biaxial	triaxial	biaxial	triaxial
Transverse	0.2 %	0.1 %	4.2 %	0.3 %
Longitudinal	0.2 %	0.2 %	0.2 %	0.2 %
Through thickness direction	10.7 %	9.0 %	1.5 %	8.5 %

Mechanical behavior and microstructural characterization

Typical tensile stress-strain curves and flexural stress-strain curves are shown in Figure 3. representing the fabricated braided composite material. Generally speaking, comparing the two different fabrication routes (HP- and I-type, respectively) little differences were observed in terms of the resulting stress-strain characteristics for the same preform-type and fiber orientation. For this reason here only stress-strain curves of the composite material HP-type are shown.

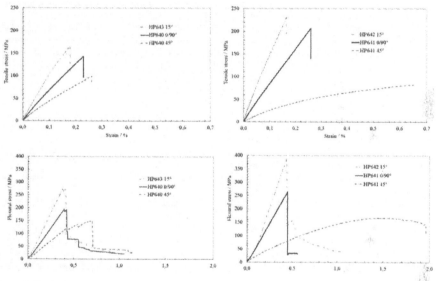

Figure 3. Typical stress-strain curves of braided C/C-SiC composites (HP-type). Triaxial (left), biaxial (right), tensile test (top) and flexural test (bottom).

The more or less nonlinear tensile stress–strain behavior of the braided C/C-SiC composite can be qualitatively understood by the process induced damage and damage accumulation occurring in the composite under increasing tensile stresses. The damage accumulation included micro-crack propagation, matrix cracking, debonding and sliding of the fiber, and fiber failure[8-9]. It is evident, that the slopes of the stress-strain curves, which represent the stiffness, show different characteristics. As mentioned earlier the modulus of elasticity and ultimate strength are strongly influenced by fiber orientation and fiber volume fraction. Maximum values are reached at a fiber orientation of 15°, followed by 0/90° and 45°. The modulus of elasticity increases with the amount of fibers orientated mainly in loading direction, see Figure 3. During tensile testing the stress drops to zero after failure. Hence, this was not the case during flexural testing. Here the composite material showed somehow a damage tolerant fracture behavior. The scatter in material properties observed for each fiber orientation may be explained by diverging fiber volume fractions (considering the global V_f and V_f in load direction) and by variation of the developed microstructures resulting from different precursor systems, see Table I. Both influence the resulting composition of the fabricated composite. Comparing the microstructures of HP-type with I-type the difference is clearly noticeable. The HP-type composites exhibited very dense C/C-bundles with minor or no silicon carbide within the C/C-bundles, whereas the I-type material showed a significantly higher content of silicon carbide resulting in increased fiber degradation during siliconization, see Figure 6 and 7. Figure 6 shows microstructures comparing warm pressed based and RTM based composites with biaxial fiber architecture and Figure 7 shows microstructures based on triaxial fiber architecture. Both, biaxial and triaxial panels from I-type (I703 and I704), showed the highest C/C-SiC density values in the range of 1.91-1.98 g/cm³. The increased open porosity after carbonization was an indicator for a higher degree of siliconization, also see Table III. All other C/C-SiC panels showed a material density in the range of 1.76-1.86 g/cm³.

Flexural strength and tensile strength data for different fiber orientations are presented in Figure 4 and 5 for a better qualitative comparison. For better understanding, the fiber orientation was visualized inside the bars within the graphs. Here, black lines represent the fiber orientation resulting from the braiding angle and red lines represent the third fiber orientation. At 0/90° the biaxial composite HP-type exhibited significantly higher strength values. This can be explained by a higher fiber volume fraction in loading direction due to eight biaxial preform layers, whereas the triaxial composite consisted only of six triaxial preform layers. The drastic discrepancy of HP to I-type considering tensile strength values at 0/90° was most likely a result of premature failure in the clamping section. The tensile strength values were expected to be in the range of the triaxial HP-type composite as they were during flexural testing. Nevertheless, I-type composites revealed a reduction in ultimate strength compared to HP-type at 0/90° due to a higher concentration of SiC going along with explicit fiber degradation. At a fiber orientation of ±45° all composites showed similar ultimate strength and no major difference due to the different microstructures was observed. At ±15° both composite types, HP and I, showed a dense formation of carbon fibers surrounded by carbon matrix with very thin inter- and translaminar silicon carbide bars. Both biaxial braided composite types exhibited the lowest material density at a fiber orientation of ±15° within a range of 1.76-1.80 g/cm³ due to the increased transverse shrinkage. With the latter leading to only a minor fiber degradation during siliconization and most fibers orientated in load direction, biaxial braided composites showed the highest ultimate strength values. Strength values of triaxial braided composites were smaller in contrast to biaxial composites. Here, the third fiber orientation at 90° impeded the shrinkage in transverse direction resulting in larger translaminar cracks causing a higher silicon carbide concentration, see Figure 8. Generally speaking the biaxial composites surpassed the triaxial composites in terms of mechanical strength for all tested fiber orientations (except for ±45°) as a result of lower degree of siliconization. At ±45° the third fiber orientation in every other layer eventually led to higher strength values. I-type composites at ±45° did not show that characteristic as a result of higher degree of siliconization

The shear modulus of elasticity was found to be in the range of 6-10 MPa for biaxial and 15-16 MPa for triaxial C/C-SiC composites. The shear modulus of elasticity was determined by tensile test at a fiber orientation of ±45°.

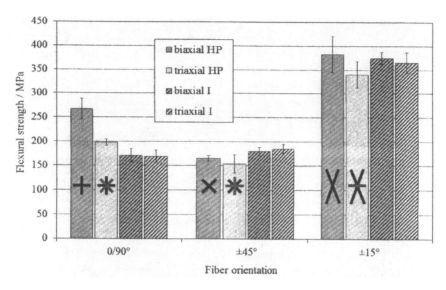

Figure 4. Flexural strength of braided C/C-SiC composites for different fiber orientations.

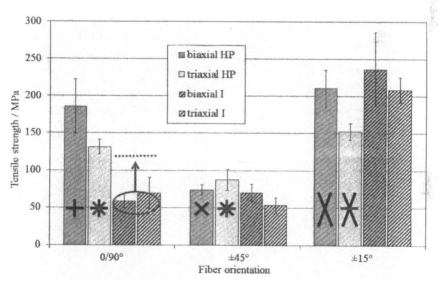

Figure 5. Tensile strength of braided C/C-SiC composites for different fiber orientations. Marked I-type specimens in 0/90° orientation failed at low stresses due to inaccurate specimen preparation.

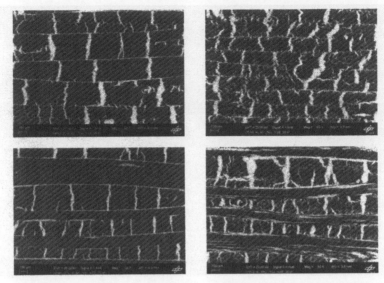

Figure 6. Microstructure comparison of biaxial braided C/C-SiC fabricated with warm pressed (left) and RTM processed (right) precursor. Fiber orientation: ±45° (top), 0/90° (bottom).

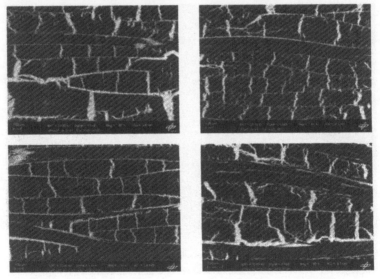

Figure 7. Microstructure comparison of triaxial braided C/C-SiC fabricated with warm pressed (left) and RTM processed (right) precursor. Fiber orientation: 45°/0°/-45° (top), 0/±45°/90° (bottom).

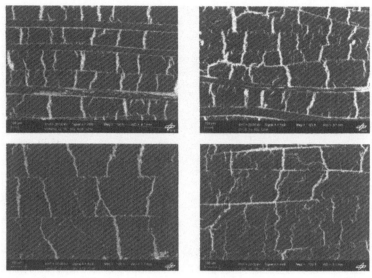

Figure 8. Microstructure comparison (polished surfaces) of biaxial (bottom) and triaxial (top) braided C/C-SiC fabricated with JK60 (HP-type, left) and XP60 (I-type, right) precursor. Fiber orientation: 15°.

To analyze the pore system in more detail, CT-imagining was performed on selected specimens after pyrolysis. This technique helps to further understand the differences in the developed micro crack systems with respect to changing fiber architectures and manufacturing parameters[10]. Here, the triaxial HP-type composite (0°/45°/90°) was selected as an example. Figure 9 shows a 3D-view of the pore system. Clearly visible are the main translaminar cracks in all three fiber orientations used in this composite. Crack width analysis and pore volume determination are important tools to help improving the manufacturing of C/C-SiC materials using LSI process.

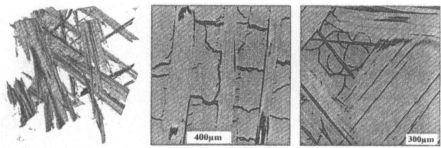

Figure 9. CT analysis of micro crack system of braided composites (HP-type, triaxial, 0/90°) after carbonization (C/C). 3D-view of micro crack system (left), cross section view (middle) and top view (right).

In addition to CT-analysis mercury porosimetry measurements were conducted to determine the pore size distribution. A typical size distribution of pores and pore channels w. r. t. the relative pore volume for braided composites after carbonization (C/C) is given in Figure 10. Here the pore diameter represents the crack width, in other words the distance between the crack surfaces based on a plate model theory. The results showed a very good agreement with CT-analysis. About 65% of the total open porosity was provided by cracks with a diameter of about 10-50µm.

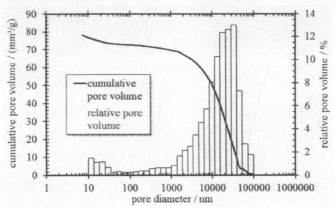

Figure 10. Typical pore size distribution of a braided composite after carbonization (C/C, HP-type, triaxial, 0/90°).

For further understanding of the fracture characteristic of the braided C/C-SiC material, the fracture surfaces of tested tensile specimens of both composite types were observed by SEM in this work. Figure 11 shows typical fracture surfaces of HP-type and I-type composites at 0/90° fiber orientation, respectively. Both fracture surfaces were ragged and exhibited holes and typical fiber pull out characteristics. At HP-type composites, entire C/C-segments were pulled out with no single fibers sticking out (fiber bundle pull-out), whereas the counterpart showed a more bushy fiber pull out characteristic (single fiber pull-out). The reason for this different fiber pull characteristic is attributed to the different microstructure composition explained early, also see Figure 6.

Figure 11. Typical fracture surfaces including fiber pull out of the braided C/C-SiC composite at a fiber orientation of 0/90°. HP-type (left) and I-type (right).

CONCLUSION

C/C-SiC composite panels based on biaxial and triaxial braided fiber preform were fabricated by LSI process for the first time at the DLR Stuttgart. Mechanical properties generally depend strongly on their microstructural composition, and therefore also on fiber orientation and fiber volume fraction, in combination with the respective manufacturing method and the precursor system. The following results summarize this study.

1. Using aromatic resin (RTM process) and phenolic resin (warm pressing) as carbon matrix precursor biaxial and triaxial based C/C-SiC composites can be fabricated using LSI route. The average material density was found to be 1.8-1.9 g/cm³ with an average open porosity below 3%.

2. Differences in microstructures were observed comparing the aromatic and phenolic precursor system. The phenolic precursor systems based C/C-SiC composite showed big, dense C/C-segments protecting most fibers from degradation by the highly reactive silicon. In contrast the aromatic resin system exhibited an affinity to higher silicon carbide concentrations within the C/C-bundles.

3. The mechanical behavior under tension and bending loads exhibited similar results for both precursor systems even though different microstructures had been observed. The scatter in the properties can be explained on the basis of different microstructure composition and fiber volume fraction in loading direction. The biaxial composite surpassed the triaxial composite in terms of mechanical strength for all tested fiber orientations (except for ±45°) as a result of lower degree of siliconization. At ±45° the third fiber orientation at 0° in every other layer eventually led to higher strength values within the HP-type. I-type composites at ±45° did not show that characteristic as a result of higher degree of siliconization. The multi-step fracture behavior during flexural testing and ragged fracture surfaces with fiber and fiber bundle pull-out indicate somewhat a non-brittle behavior of the fabricated C/C-SiC composites.

4. Pore size distribution and CT-analysis showed a very good agreement in terms of crack width and crack size distribution. Pore channels with a size of about 10-50μm accumulated to about 65% of the total pore volume.

ACKNOWLEDGEMENT

Financial support from the German Research Council (Deutsche Forschungsgemeinschaft DFG) in the framework of the *Sonderforschungsbereich Transregio 40* is gratefully acknowledged.

REFERENCES
[1] Jamet, J.F. and Lamicq, P.J. in: Naslain, R. (Eds.). High temperature Ceramic Matrix Composites. Woodhead, London, UK. p.735 (1993).
[2] Krenkel, W., Carbon fiber reinforced CMC for high performance structures. Int. J. Appl. Ceram. Technol, 1, 188-200 (2004).
[3] Naslain, R., Design, preparation and properties of non-oxide CMCs for application in engines and nuclear reactors: an overview. Composite Science and Technology, 64, 155–170 (2004).
[4] Krenkel, W. Applications of Fibre Reinforced C/C-SiC Ceramics. Ceramic Forum International , 80 (8), 31-38 (2003).
[5] Schulte-Fischedick, J et al. The morphology of silicon carbide in C/C–SiC. Materials Science and Engineering: A, Volume 332, Issues 1-2, Pages 146-152 (2002).
[6] Gern, F.H. et al. Liquid silicon infiltration: description of infiltration dynamics and silicon carbide formation. Composites Part A: Applied Science and Manufacturing, Volume 28, Issue 4, 355-364 (1997).

[7]Kumar, S. et al. Capillary infiltration studies of liquids into 3D-stitched C–C preforms: Part B: Kinetics of silicon infiltration. Journal of the European Ceramic Society, Volume 29, Issue 12, 2651-2657 (2009).

[8]Wang, M. et al. Characterization of microstructure and tensile behavior of a cross-woven C/SiC composite. Acta Materialia. Volume 44, Issue 4, 1371-1387 (1996).

[9]Baste, S. Inelastic behaviour of ceramic-matrix composites. Composites Science and Technology, Volume 61, Issue 15, 2285-2297 (2001)

[10]Schulte-Fischedick, J. The crack development on the micro- and mesoscopic scale during the pyrolysis of carbon fibre reinforced plastics to carbon/carbon composites. Composites Part A: Applied Science and Manufacturing. Volume 38, Issue 10, 2171-2181 (2007).

INFLUENCE OF FIBER FABRIC DENSITY AND MATRIX FILLERS AS WELL AS FIBER COATING ON THE PROPERTIES OF OXIPOL MATERIALS

Sandrine Hoenig, Enrico Klatt, Martin Frieß, Cedric Martin, Ibrahim Naji, and Dietmar Koch
Department of Ceramic Composites and Structures, Institute of Structures and Design, German
Aerospace Center (DLR)
Stuttgart, Germany

ABSTRACT
    Higher efficiency of jet engine and environmental friendly use are promised via an increase of
the combustion temperature. Oxide ceramic matrix composites (CMC) are reliable candidates for this
purpose. Within this scope, the German funded project HiPOC (High Performance Oxide Ceramics)
supports the further development of OXIPOL (Oxide CMC based on Polymers) materials at the
German Aerospace Center (DLR) in Stuttgart.
    The OXIPOL materials studied in this work are manufactured via the polymer infiltration and
pyrolysis process (PIP) and reinforced with alumina fibres (Nextel 610, 3M). A variation of the fibre
fabric density (i.e. filament number) and the fibre volume content is firstly investigated. Moreover, the
influence of ceramic fillers on the interlaminar shear strength (ILSS) as well as the influence of
different types of fibre coating (fugitive and lanthanum phosphate coating) on mechanical behaviour
are discussed.
    This work showed that OXIPOL materials with lower filament number present a more
homogeneous fibre coating in the bundle and a better matrix infiltration, leading to an improvement in
mechanical properties.

INTRODUCTION
    Efficiency of jet engines can be improved via an increase of inlet temperature combined with
lean combustion process. Current metallic alloys are limited by degradation in high temperature
environment while ceramic matrix composites (CMC) are reliable candidates. In particular, oxide
CMCs can be used in high temperature oxidizing environment, for example in the combustor liner.
Within this scope, the German funded project HIPOC [1] (High Performance Oxide Ceramics) supports
the further development of the OXIPOL (Oxide CMC based on Polymers) composites at the German
Aerospace Center (DLR) in Stuttgart.
    OXIPOL is built up via several polymer infiltration and pyrolysis (PIP) process steps [2,3]. The
PIP process presents the advantage of being faster and cheaper than other CMC manufacturing
processes. In this work, dried alumina weave cloths were stacked up and infiltrated with polysiloxane
precursors yielding an amorphous silicon oxycarbide SiOC matrix after pyrolysis. Oxide ceramic
alumina fibres (Nextel 610, 3M) were preferred to mullite fibres as they present higher tensile strength.
Their melting point of 2000 °C, their low density of 3.9 g/cm³, their small shrinkage after exposure
(less than 1% after 15h exposure at 1300°C), and their strength retention of about 70 % after 1000 h
exposure at 1200 °C make them relevant for the use in gas turbine applications. Awaited local
temperatures in a coated CMC combustor liner approaches 1100 °C up to 1300 °C. Amorphous SiOC
matrices are stable at temperature up to 1300 °C [4], thus this matrix is relevant for use in combustion
chambers. Crack deflections under loading inside OXIPOL are guaranteed via a weak fibre-matrix
interface. Fugitive coating or more stable coating in oxidizing environments (monazite coatings) can
deflect matrix cracks near fibres and allow a higher damage tolerance of the CMC [2] in jet engines. The
SiOC matrix is weakened during pyrolysis due to the building of matrix cracks. Dependant of the
ceramic yield of the polymer and the infiltration process, different crack forms appear during the
ceramisation whereas the polymer matrix converts to ceramic matrix and shrinks. Those cracks imply a
release of residual stresses within the matrix and a weak fibre-matrix interface. Using ceramic fillers

during the first polymer infiltration might be a way to limit the crack development during pyrolysis [5]. These fillers should also bring an increase of the flexural strength and of the interlaminar shear strength of the final CMC.

In this work, a variation of the fibre fabric density (i.e. filament number) is firstly investigated. The influence of the fibre volume content and ceramic fillers on the flexural strength and the interlaminar shear strength (ILSS) are then studied. Finally different kinds of fibre coating (fugitive and lanthanum phosphate coating) are studied before and after thermal exposure. Microstructure of the different plate configurations are analysed via scanning electron microscopy (SEM).

EXPERIMENTAL WORK

Matrix Processing via PIP

Damage tolerance of the OXIPOL materials are guaranteed by a weak fibre-matrix interface and a matrix as dense as possible. In this work, the OXIPOL materials are reinforced with alumina fibres Nextel 610 (3M). These oxide fibres proved to exhibit the best tensile properties compared to mullite fibres [2]. The weak interface is provided either by a carbon coating or a monazite coating firstly applied to the fabrics with the help of a Foulard facility. The amorphous silicon oxycarbide (SiOC) matrix of OXIPOL is then built up via several polymer infiltration and pyrolysis (PIP) loops. Due to a ceramic yield of the precursor polymer smaller than 100%, shrinkage of the matrix appears during pyrolysis and several PIP processes are then required for densifying the matrix. The first polymer infiltration by warm pressing of dried fabrics and MK powder (Wacker Chemie) were investigated in a precedent work [2]. Another infiltration method by resin transfer moulding (RTM) is also possible [3,6] ; in this case, the polymer is a liquid polysiloxane precursor composed of 50 mass-% MSE100 and 50 mass-% MK (Wacker Chemie). The warm pressing technology compared to the RTM route has the advantage to be a more cost and time effective method, moreover the MK powder used in the first infiltration has as higher ceramic yield (82 %) than the MSE100 precursor (15 %) [6]. In order to achieve a final open porosity of between 10-15 %, four to five reinfiltration and pyrolysis additional loops are required. OXIPOL parts are reinfiltrated either by plunging them in a liquid polysiloxane precursor bath (50 mass-% MSE100 and 50 mass-% MK) assisted by vacuum or by RTM method with the liquid precursor. Finally if a fugitive coating is applied, an oxidation step is performed with 20 h at 700 °C for burning the carbon fibre coating. Figure 1 gives an overview of the OXIPOL manufacturing processes via warm press technique.

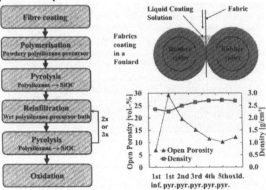

Figure 1. OXIPOL manufacturing process (warm pressing) and typical porosity/density evolution

Fabrics Density and Fibre Volume Content

The company 3M manufactured two kinds of 8-harness satin weave cloths with the alumina fibres Nextel 610. Typical properties given by 3M can be found in Table 1.

Table 1. Woven Fabrics Nextel 610, Typical Properties

Style	Weight (sized)	Thickness (sized)	Nominal Filament Count	Air Permeability w/o Sizing	Breaking Strength without Sizing	
					Warp	Fill
	$[g/m^2]$	[mm]	[-]	$[l/min/dm^2]$	[kg/cm]	[kg/cm]
DF-11	373	0.28	400	61	41	46
DF-19	654	0.48	750	40	54	46

These two types of fabrics (DF-11 and DF-19)were compared in this publication. The fabrics were coated in a Foulard facility with a 5 mass-% phenolic resin (JK60) diluted in 95 mass-% ethanol and then cured 2 h at 175 °C. The coated cloths were then pyrolysed under nitrogen atmosphere and a second coating loop in the Foulard facility was achieved [2]. This coating is further called coating A.

In the first part, the influence of the fabric density on tensile properties of OXIPOL plates will be quantified. For this purpose, CMC plates with a wished thickness of 3mm were produced. 14 plies of fabrics DF-11 were stacked up whereas eight plies of fabrics DF-19 were stacked up.

Secondly, the influence of fibre volume content on flexural properties and interlaminar shear strength (ILSS) will be discussed. CMC plates with fabrics DF-19 were manufactured with fibre volume content variations between 42.2 % and 57.7 %.

Ceramic Fillers

The presence of ceramic fillers during the first polymer infiltration should limit the crack development during pyrolysis, with expected increase of flexural properties and ILSS. Generally, active fillers are particles with composition that differ from those of the matrix [5]. A new kind of fillers was investigated here. After polymerisation of pure MK powder under pressure loading and vacuum, the polymer was pyrolysed up to 1000 °C, pure SiOC matrix was then produced. The obtained SiOC blocks were then grinded to powder with a milling facility; the coarse particles were filtered apart. These milled and filtered SiOC particles were used as fillers in different OXIPOL plates.

DF-11 and DF-19 fabrics were coated with the fugitive coating A. The first infiltration of CMC plates reinforced with fabrics DF-11 or DF-19 occured via warm press technique where MK powder precursor and SiOC fillers are mixed in different compositions. The percentage of SiOC fillers (x) compared to the initial matrix mass was fixed to 30, 40 and 50 mass-% respectively. After polymerisation, the plates were pyrolysed under nitrogen atmosphere. Four PIP processes with reinfiltrations in a resin bath under near-vacuum follow this step, finally the plates were oxidised 20 h at 700 °C.

Fibre Coating

Improvements on tensile and flexural properties due to fugitive coating in OXIPOL components were shown in precedent works [2,3]. An optimised fugitive coating was defined with two coating loops of the fabrics in a phenolic resin solution (5 mass-% JK60 diluted in 95 mass-% ethanol) and an intermediate pyrolysis of the coated fabrics (coating A). However, this coating proved to become closed at high temperature under oxidising environment: air reacts with SiOC matrix and SiO_2 layers can be observed with SEM as shown on Figure 2. The gaps between fibres and matrix tend to close, which lowers the mechanical properties of OXIPOL at high temperatures. A new thinner fugitive coating with a lower concentrated phenolic resin solution (2 mass-% JK60 diluted in

98 mass-% ethanol) was applied on fabrics with a Foulard facility. After each curing of the coated fabrics (2 h at 175 °C), the fabrics were pyrolysed at 1000 °C under nitrogen atmosphere. Five such coating loops were applied on the fabrics. This coating is further called coating B.

The tensile properties of OXIPOL plates with coating A and B (DF-19) were compared to those of an OXIPOL plate reinforced with eight sized plies of DF-19 cloths without fibre coating.

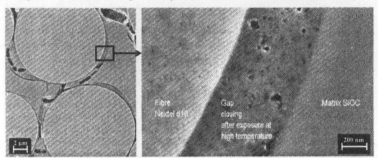

Figure 2. Microstructure of OXIPOL with fugitive coating after 20 h exposure at 1200 °C in air

LaPO$_4$ (monazite) coating is a relevant coating for oxide CMC based on Nextel 610 fibres. It is lightweight with a density of 5.13g/cm [7], its coefficient of thermal expansion is about 25 % higher than that of the alumina fibres [7], it is denser than other monazite and xenotimes [8] and monazite is well known and a commercially important mineral. Keller et al. demonstrated that LaPO$_4$ was an oxidation-resistant crack-deflecting fibre coating for CMCs [9]. Methods for applying LaPO$_4$ coatings on fibres rovings and fabrics were described and characterised by Fair et al. [10,11]. Based on these methods combined with processing in a Foulard facility, fabrics with alumina fibres Nextel 610 were coated. For testing this method, more permeable cloths were weaved: fibres Nextel 610 (3000 den, 3M) were plain weaved to a fabric weight of 340 g/m^2. This more permeable plain weave cloth and the fabrics DF-11 were coated with LaPO$_4$.

The alumina plain weave cloth and DF-11 fabrics were firstly desized at 700 °C for 10 min in air. A precursor solution of lanthanum citrate was prepared by dissolving lanthanum nitrate and citric acid in deionized water to a P:citrate ratio 1:2. This solution was chilled to about 5 °C and then mixed with a phosphoric acid just previous to the coating process to a La:P ratio from 1:1. The obtained final La:citrate:P ratio was then 1:2.5:1. The LaPO$_4$ precursor solution was then spilling in a Foulard facility (see figure 1). A pressure of the rubber roll of 3 bars and a rotation speed of 3 rpm were set. The desized fabrics were passed two times through the Foulard facility. The coated fabric were subsequently rinsed 5 min in a warm water bath at 90 °C, then 30 s in a room temperature bath of ammonia and deionized water and finally 30 s in a room temperature bath of deionised water. The cloths were dried in air at 110 °C for 30 min. The fabrics were coated a second time in the Foulard facility, rinsed and dried again. Finally the coated fabrics were burned 5 min in air at 900 °C. One coating loop includes the steps from first Foulard coating to burning at 900 °C. The plain weave cloth was coated with one LaPO$_4$-coating loop (coating C), whereas the DF11 cloth was coated with four LaPO$_4$-coating loops (coating D).

All coating variations used in this work are summarised in Table 2.

Table 2. Art of coatings in the different OXIPOL configurations

Appellation	w/o Coating	Coating A	Coating B	Coating C	Coating D
Description of the Coating	Sized and uncoated	Sized, 2 x 5 mass-% phenolic resin solution with intermediate pyrolysis	Sized, 5 x 2 mass-% phenolic resin solution with intermediate pyrolysis	1 x LaPO$_4$-coating loop	4 x LaPO$_4$-coating loops
Fibre-matrix Interface	-	Fugitive	Fugitive	Weak	Weak

Characterisation and Testing

The different OXIPOL variations were characterised by porosity measurements, calculation of the final volume portions and mechanical testing. The microstructure of the different OXIPOL plates were analysed via SEM.

The open porosity and density were determined by Archimedes method (DIN EN 993-1) after the first polymer infiltration, after each pyrolysis and after the optional oxidation of the OXIPOL plates. The final volume content of a plate was calculated via equation 1:

$$V_f = \frac{\text{number of cloths plies * fabric weight}}{\text{plate thickness * fibre density}} \qquad (1)$$

Firstly, the influence of the fabric density on tensile properties of OXIPOL plates was quantified. Tensile samples were water jet cut in OXIPOL plates (fabrics DF-11 or DF-19, coating A, V_f = 41.8-44.5%). Four to five tensile samples were tested acc. to DIN EN 658-1 at room temperature before and after exposure at 1200 °C in air for 20h.

Secondly, the influence of fibre volume content on flexural properties and interlaminar shear strength (ILSS) was studied. Four to five four-point and short three-point bending samples were water cut in OXIPOL plates (fabrics DF-19, coating A, V_f = 42.2-57.7 %) and tested at room temperature acc. to DIN EN 658-3.

Thirdly, the influence of the SiOC fillers on flexural properties and interlaminar shear strength (ILSS) was investigated at room temperature. Five to six four-point and short three-point bending samples were water cut from OXIPOL plates (fabrics DF-11 or DF-19, coating A, V_f = 33.6-42.2 %) and tested acc. to DIN EN 658-3.

Finally, different fugitive and LaPO$_4$ fibre coatings were characterised. The influence of the fugitive coatings A and B on the tensile properties were described. Four to five tensile samples were water jet cut in OXIPOL plates without fibre coating (DF19, V_f=34%) and with fugitive fibre coating (DF19, coating A or B, V_f = 42.2-50.1 %) . The samples were tested acc. to DIN EN 658-1 before and after thermal exposure at 1200 °C in air for 20 h. LaPO$_4$ fibre coatings (plain weave or DF-11, coating C or D, V_f = 42.4-44.8 %) were also characterised, via three-point bending tests acc. to DIN EN 658-3. Samples were tested at room temperature before and after thermal exposure in air at 1100 °C for 20 h. These properties were compared to those of the two following OXIPOL variations: "DF-19, uncoated, V_f= 40.5 %" and "DF-19, coating A, V_f= 50.5 %". The samples were water cut.

RESULT AND DISCUSSION

Variation of the Fabric Density

The following table lists the tensile properties of the two different OXIPOL configurations with cloths DF-19 (654 g/m^2) and DF-11 (373 g/m^2). Similar stiffness can be observed, even the tensile strength and strain at failure are comparable within experimental error (Table 3, Figure 3). Although the cloths DF-11 have a reduced aerial weight, in order to keep a similar fibre volume content more plies are required. The porosity is more homogeneously distributed than in a plate made of DF-19 as can be seen via SEM analyses on Figure 4, and the matrix is also more homogeneously distributed within the fibre bundles.

Table 3. Tensile properties of OXIPOL configurations with different fabric densities (coating A)

Exposure	Fabrics	E modulus [GPa]		Tensile Strength [MPa]		Strain at Failure [‰]	
		Average	Std. Dev.	Average	Std. Dev.	Average	Std. Dev.
RT	DF-19	89.9	14.2	124.0	25.5	1.51	0.54
RT	DF-11	95.0	7.5	99.8	31.6	1.02	0.35
20h 1200 °C	DF-19	98.5	8.4	57.5	11.1	0.46	0.20
20h 1200 °C	DF-11	97.9	3.4	50.3	12.0	0.29	0.24

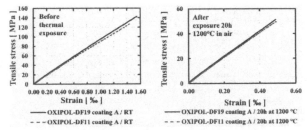

Figure 3. Typical stress-strain tensile curves before and after thermal exposure of OXIPOL with different fabric densities

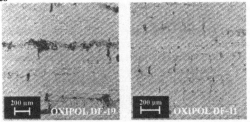

Figure 4. SEM analyses of OXIPOL plates (coating A) reinforced with DF-19 (left) or DF-11 (right)

Variation of the fibre volume content

With higher fibre volume content, higher density of OXIPOL can be reached with lower open porosity as shown on Figure 5. As the open porosity and geometric density of final OXIPOL plates were measured via the Archimedes method, the closed porosity was not considered separately. The results prove that higher fibre volume content of OXIPOL than the 42 % used in [2] can be reached.

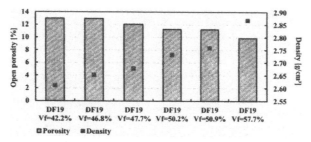

Figure 5. Open porosities and densities via Archimedes method (OXIPOL with coating A)

Typical four-point bending stress-displacement curves are represented in Figure 6. It can be observed that the flexural strength decreases with higher fibre volume content. Same phenomenon can be observed on the ILSS (within experimental error) which also decreases with higher fibre volume content. This can be relied to insufficient polymer infiltrations between some fabric plies, which lead to delaminations. The small ILSS of the OXIPOL materials investigated in this work are also due to this lack of matrix. However, a higher damage tolerance can be observed on plates with V_f over 50 % (Figure 6): a residual stress of 60 MPa can be observed at a deflection of 1.5 mm for OXIPOL with $V_f = 42.2$ % whereas this deflections reaches more than 2 mm for OXIPOL with $V_f = 50.2$ %. Indeed, with higher fibre volume content, crack deflections is more difficult due to more and thinner matrix regions between higher numbers of fabric plies.

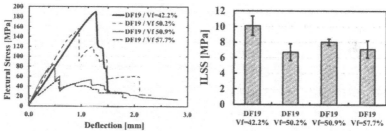

Figure 6. Typical four-point bending stress-displacement curves (left), and average ILSS (right)

Influence of the Ceramic Fillers
Ceramic SiOC fillers were mixed with the initial polymer matrix of the OXIPOL plates (coating A) reinforced with either DF-19 or DF-11. The results were compared to a reference plate without fillers (DF-19, coating A). Typical four-point bending stress-displacement curves are shown on Figure 7. For both, DF-19 and DF-11 the SiOC fillers cause a decrease of the flexural strength: within experimental error, this weakening is higher with increasing percentage of SiOC fillers, this correlates with the weakening of the ILSS (Figure 8). Indeed, the SiOC particles were too large for the required interstices between fabric plies: under pressure conditions of the warm press, this involved partial breaking of the fibre bundles (as shown on Figure 9). With higher percentage of SiOC fillers, more rovings were broken.

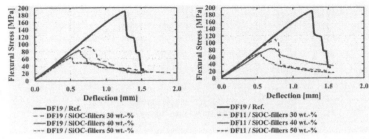

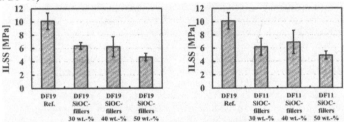

Figure 7. Typical four-point bending stress-displacement curves of OXIPOL with SiOC fillers (left: DF-19, right: DF-11)

Figure 8. ILSS of OXIPOL with SiOC fillers (left: DF-19, right: DF-11)

Though, an advantage of these SiOC fillers can be identified via the SEM pictures. Although they are identical to the final ceramic matrix composition (SiOC), it seems that the particles differ themselves from the "new" matrix. The SiOC particles inserted during the first infiltration kept almost their shape after reinfiltration and pyrolysis (e.g. no further shrinkage). This proves that the SiOC fillers can be used for minimizing the matrix shrinkage due to pyrolysis, but they have to be small enough for not damaging the rovings. Via the measurement of matrix layers in an OXIPOL plate with (DF-11 or DF-19, coating A), it is recommended to use SiOC-particles smaller than 50 μm in OXIPOL plates reinforced with DF-11 cloths and smaller than 80 μm with DF-19 cloths.

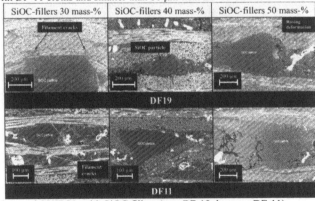

Figure 9. SEM analyses of OXIPOL with SiOC fillers (top: DF-19, bottom: DF-11)

Influence of the Fibre coating

Tensile properties of OXIPOL plates without coating, with the fugitive coating A found by Denis et al. in [2] and the thinner fugitive coating B were compared. Typical tensile stress-strain curves before and after thermal exposure at 1200 °C in air for 20 h are represented in Figure 10. Compared to the uncoated CMCs the coatings A and B present higher stiffness, higher tensile strength and higher strain at failure, this is also confirmed by Figure 11. So both fugitive coatings bring an improvement of tensile properties. Microstructural analyses show that the coating B is thin: it either is not to be detected around fibres via SEM analyses, or it adhered strongly to the matrix during pyrolysis, and was detached from the fibres.

Degradation of all the tensile properties can be observed after thermal exposure. The coating A presents always the best stiffness, tensile strength and strain at failure of all the samples. During exposure in air, oxidation occurs in pores of the matrix and in the interfacial gap between fibre and matrix: SiOC reacts with the air and builds SiO_2. This reaction is confirmed by SEM analyses (Figure 2). This leads to a brittle behaviour similar to monolithic ceramic and tensile properties near those of the uncoated samples. The fibre-matrix gap closing is faster if the fibre-matrix gap is thinner, which is the case for the coating B, and it is also facilitated with a porous matrix and higher concentration of SiOC. The coating A seems then to be the best coating for this test series, gap closing is not completed after 20 h of thermal exposure at 1200 °C but it would get closed for a longer exposure time. Thus, other kinds of fibre coatings for long time exposure under oxidizing environment like in combustion chambers have to be investigated.

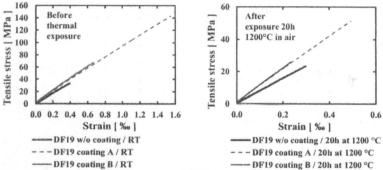

Figure 10. Typical tensile stress-strain curves of OXIPOL without and with different fugitive coatings

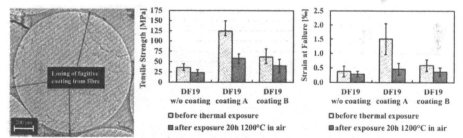

Figure 11. Microstructure of OXIPOL with coating A before thermal exposure (left), tensile strength (middle) and strain at failure (right) of OXIPOL without coating and with different fugitive coatings

Two different LaPO$_4$ coatings were applied to plain weave and DF-11 alumina cloths and OXIPOL plates were then processed as described above. The final fibre coating was analysed via SEM in the final OXIPOL composites (Figure 12). The coating process C leads to an inhomogeneous coating on fibres, it adheres more strongly to the matrix than to the fibre. During pyrolysis, this leads to debonding of the LaPO$_4$ layer from the fibre due to matrix shrinkage. The coating process D, with higher number of coating loops leads to a homogeneous coating with an approximate thickness of 100 nm. A coating thickness gradient can be observed transverse to a cloth thickness. Here also, locally the LaPO$_4$ layer detaches from the fibre, though rarely than with the coating process C.

Three-point bending tests before and after 20 h thermal exposure at 1100°C in air shows that the fugitive coating A and the LaPO$_4$ coating D enhances bending strength and strain to failure compared to uncoated samples (Figure 13 and Figure 14). As expected, the inhomogeneous LaPO$_4$ coating C is not damage tolerant. However, this is not only caused by the coating system C but also due to the reinforcement with plain weave cloths which induce a more brittle behaviour in bending mode. Nevertheless, a slight strength increase is observed compared to uncoated OXIPOL before exposure. The LaPO$_4$ fibre coating D on the contrary, leads to a higher damage tolerance (Figure 14) compared to fugitive coating A. On the other hand the average three-point flexural strength of OXIPOL with coating D is lower (169.5 ± 12.1 MPa) compared to those of OXIPOL with the fugitive coating A (197.9 ± 10.1 MPa).

The CMC with coating A has a higher final fibre volume content (50.5 %) and a lower open porosity (11.7 %) thus a higher strength compared to CMC with fibre coating D where fibre volume content is lower (V$_f$ = 44.8 %) and open porosity is higher (16.6 %). The results show that the use of LaPO$_4$ coating in OXIPOL improves damage tolerance of the material once the homogeneity of the fibre coating is assured. In further developments the matrix densification is a key issue in order to further improve mechanical behaviour and thermal stability.

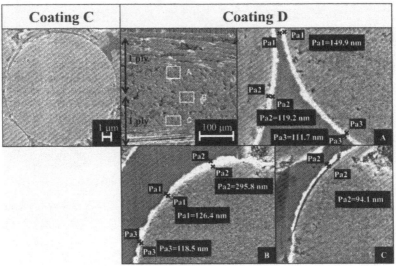

Figure 12. SEM analyses of OXIPOL with LaPO$_4$ fibre coatings C and D before thermal exposure. Micrographs marked with A, B, C represent enlargements from center top viewgraph.

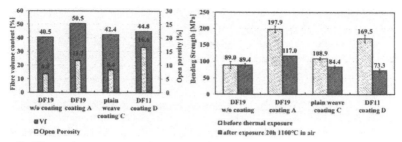

Figure 13. Fibre volume content and open porosity (left), three-point bending strength (right) of OXIPOL without and with different fibre coatings

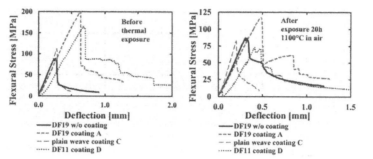

Figure 14. Typical three-point bending stress-displacement curves of OXIPOL with diverse fibre coatings

CONCLUSIONS

Different optimisation parameters of the oxide CMC OXIPOL were investigated like the variation of the fabric density and fibre volume content, the influence of passive SiOC filler particles during pyrolysis, and different fibre coating.

Lighter and thinner alumina 8-harness satin cloths allow a more homogeneous matrix distribution between the cloths and an easier fibre coating. An increase of the fibre volume content leads to an increase of the damage tolerance.

The use of SiOC filler particles shows a degradation of the flexural and interlaminar shear properties, although the interpretation was insufficient due to partial damaging of the rovings. While thin fugitive coating was insufficient for providing the desired improvement, the fugitive coating with a phenolic resin solution of two times 5 mass-% JK60 dissolved in 95 mass-% ethanol (with an intermediate pyrolysis of the coated fabrics) proved to be the best non-oxide based fibre coating discussed here. The oxidation-resistant fibre coating LaPO$_4$ brought high damage tolerance. Improvement of manufacturing process parameters like adjustment of coating thickness, pyrolysis temperature and matrix densification will give further enhancement for gas turbine applications.

ACKNOWLEDGMENTS
This work was partly accomplished in the frame of the HIPOC programme (contract number 03X3528D) within the WING framework. This programme is financial supported by the German Federal Ministry of Education and Research and administrated by the Projektträger Jülich (PTJ). Their support is gratefully acknowledged.

REFERENCES
[1]M. Gerendás, Y. Cadoret, C. Wilhelmi, T. Machry, R. Knoche, T. Behrendt, T. Aumeier, S. Denis, J. Göring, D. Koch, K. Tushtev, Improvement of Oxide/Oxide CMC and Development of Combustor and Turbine Components in the HIPOC Program, *Proc. of the ASME Turbo Expo 2011*, Vancouver, Canada, June 6-10 2011
[2]S. Denis, M. Friess, E. Klatt, B. Heidenreich, Manufacture and Characterization of OXIPOL Based on Different Oxide Fibres, *HT-CMC 7*, 414-419 (2010)
[3]B. Heidenreich, W. Krenkel, M. Frieß, Net-shape Manufacturing of Fabric Reinforced Oxide/Oxide Components via Resin Transfer Moulding and Pyrolysis, *Proc. of the 28th International Cocoa Beach Conference on Advanced Ceramics and Composites*, Florida, USA, January 26-30 2004
[4]H. D. Akkaş, M. L. Öveçoğlu, M. Tanoğlu, Development of Si-O-C Based Ceramic Matrix Composites Produced via Pyrolysis of a Polysiloxane, *Key Eng. Mater.*, 264-268, 961-964 (2004)
[5]G. Stantschev, M. Frieß, R. Kochendörfer, W. Krenkel, Long Fibre Reinforced Ceramics with Active Fillers and a Modified Intra-Matrix Bond Based on the LPI Process, *J. Eur. Ceram. Soc.*, 25, 205-209 (2005)
[6]M. Frieß, B. Heidenreich, W. Krenkel, Long-Fibre Reinforced Oxide Ceramic Composites via the LPI process (translation from German), 14th Symposium "Verbundwerkstoffe und Werkstoffverbunde", July 2-4 2003, Wien, Austria
[7]P. E. D. Morgan, D. B. Marshall, Ceramic Composites of Monazite and Alumina, *J. Am. Ceram. Soc.*, 78 (6), 1553-63 (1995)
[8]G. E. Fair, R. S. Hay, Precipitation Coating of Rare-Earth Orthophosphates on Woven Ceramic Fibers-Effect of Rare-Earth Cation on Coating Morphology and Coated Fiber Strength, *J. Am.Ceram. Soc.*, 91 [8], 2117-2123 (2008)
[9]K. A. Keller, T. Mah, T. A. Parthasarathy, E. E. Boakye, P. Mogilevsky, M. K. Cinibulk, Effectiveness of Monazite Coatings in Oxide/Oxide Composites after Long-Term Exposure at High Temperature, *J. Am.Ceram. Soc.*, 86 [2], 325-32 (2003)
[10]G. E. Fair, R. S. Hay, Precipitation Coating of Monazite on Woven Ceramic Fibers: I. Feasibility, *J. Am.Ceram. Soc.*, 90 [2], 448-455 (2007)
[11]G. E. Fair, R. S. Hay, Precipitation Coating of Monazite on Woven Ceramic Fibers: II. Effect of Processing Conditions on Coating Morphology and Strength Retention of Nextel 610 and 720 Fibers, *J. Am.Ceram. Soc.*, 91 [5], 1508-1516 (2008)

WEAVE AND FIBER VOLUME EFFECTS IN A PIP CMC MATERIAL SYSTEM

Ojard, G.[1], Prevost, E.[1], Santhosh, U.[2], Naik, R.[1], and Jarmon, D. C.[3]

[1] Pratt & Whitney, East Hartford, CT
[2] Structural Analytics, Inc., San Diego, CA
[3] United Technologies Research Center, East Hartford, CT

ABSTRACT

With the increasing interest in ceramic matrix composites for a wide range of applications, fundamental research is needed in the area of multiple weaves and fiber volume. Understanding how the material performs with differing weaves and fiber volume will affect the end insertion application. With this in mind, a series of three panels were fabricated via a polymer infiltration process: 8 harness satin (HS) balanced symmetric layup, 8 HS bias weave, and angle interlock. From these panels, a series of characterization efforts were undertaken with the sample oriented in both the warp and fill direction: with differing fiber volumes. These tests consisted of tensile, interlaminar shear, interlaminar tensile, in-plane thermal expansion and through thickness thermal conductivity. In addition, micro-structural characterization was done. The results from this testing will be presented, trends reviewed and analysis done.

INTRODUCTION

Ceramic Matrix Composites (CMCs) are being considered for an ever wider range of applications where designers and applications can take advantage of their low density and high temperature capability [1,2]. With this increasing interest, the characterization needs to expand beyond the point design approach used for some potential aerospace applications [3-5]. Based on this, a characterization effort was undertaken to look at a variety of weaves and fiber volumes.

For this effort, three panel types were made available for testing: cross ply balanced panel, bias panel with a ratio of 3:1 and an angle interlock panel. Depending on the orientation of the panel during the testing, the fiber volume varied. These series of panels offered an unique opportunity to perform a consistent set of testing in both orientations and compare weaves and fiber volumes. The following is a report on the testing and characterization that was done.

PROCEDURE

Material

With the interest in exploring multiple weaves and fiber volumes, a polymer infiltration pyrolysis system was chosen. This was due to the ability of the process to be easily transferred to the different weaves with no modification to the processing or processing time eliminating that as a variable in any subsequent analysis. The material system used for this effort was the SiC/SiNC system which a non-stochiometric SiC (CG Nicalon™) fiber in a matrix of Si, N and C that is arrived at by multiple iterations via a polymer pyrolysis process. This material has been previously discussed by the authors where the constituent properties were determined [6].

The baseline panel for this effort was a cross ply panel using a 22 ends per inch (epi) 8 HS balanced cloth. The panel was a 6 ply panel with an overall fiber volume set at 40%. The second panel was a Bias weave panel (cross ply layup at 6 plys) where the warp fibers were set at ~3x the fill direction fibers. The overall fiber volume was set at 40% (consistent with the baseline panel). The Angle Interlock panel was an effort to combine a bias weave with a low angle interlock to increase interlaminar properties. The total fiber volume was set at 35%. The manufacturing goals for the fiber volume which shows the nature of the various weaves is found in Table I.

Table I. Manufacturing Goals for Fiber Volume between the Three Panels

Panel Type	Fiber Direction	Fiber Volume (Vf)
		From Manufacture
		(%)
Baseline	Warp	20.1
	Fill	20.1
Bias	Warp	28.8
	Fill	11.1
Angle Interlock	Warp	24.5
	Fill	10.5

Test Matrix

For each weave, the overall panel size was 310 mm x 310 mm. This provided a large enough area that samples could be machined in both the warp and fill direction to not only look at the weave but also fiber volume: warp and fill. For tensile testing, there were 6 samples from each direction with test temperatures of 24°C, 649°C and 982 °C. This resulted in two repeats for each condition. There were two interlaminar tests performed. The interlaminar tensile testing was done at 24°C with 4 repeats per panel. There were 6 samples for the interlaminar shear testing for each direction for testing at 649°C and 982°C. This resulted in three repeats for each condition. For the in-plane thermal expansion, there were 3 repeats for each test direction with the testing done from 24°C to 982 °C. There were only 2 through thickness thermal conductivity tests done for each panel type. In all cases, the testing was done per ASTM standards (as shown in Table II).

Table II. Characterization Matrix for Panel (planned for each panel type)

Test Description	Test Direction	Temp (°C)	Test Environment	Test Method	Reps
Tensile	Warp & Fill	24	air	C1275	2
Tensile	Warp & Fill	649	air	C1359	2
Tensile	Warp & Fill	982	air	C1359	2
Interlaminar Shear	Warp & Fill	649	air	D2344	3
Interlaminar Shear	Warp & Fill	982	air	D2344	3
Interlaminar Tensile	Normal	24	air	D7291	4
In Plane Thermal Expansion	Warp & Fill	24-982	inert	E228	2
In Plane Thermal Expansion	Warp & Fill	24-982	inert	E228	2
Through Thickness Conductivity	Normal	24-815	inert	E1225	2

Characterization:

After both fabrication or testing, characterization efforts were undertaken. Standard microstructural characterization was done samples from the panels looking at the microstructure in the warp and the fill directions. In addition, image analysis was done to determine the fiber volume and percent porosity seen in the images. Scanning electron microscopy was performed on the failed tensile bars to look for differences that would occur due to weave or fiber volume effects.

RESULTS

Tensile Testing
The room temperature tensile results for the three panel types are shown in Figures 1-3. The average data for all the temperatures is shown in Table III. For the baseline panel, the testing does not show any significant difference between the sample directions. This is not the case for the Bias or Angle Interlock panels. These two panel types showed that there was a significant decrease in the strength and the proportional limit due to the decreased fiber volume present in the fill direction (see Table I). The table and curves show that there still was significant strain to failure capability present (as compared to the warp direction for these panels).

Interlaminar Testing
As noted in Table II, there were two interlaminar tests run on the material. The interlaminar shear testing was performed in a 4 point short beam shear test. The test series run did not generate the correct shear failure mode per the ASTM standard and therefore will not be discussed further. There were also a series of 4 interlaminar tensile tests run on each panel. The results of that testing is shown in Table IV. Table IV shows that there were no differences seen between the various panels fabricated.

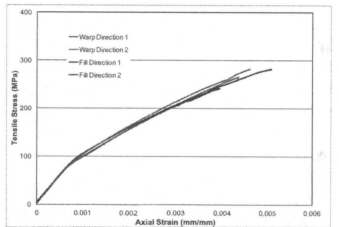

Figure 1. Room Temperature Stress Strain Curved from Baseline 8 HS Cross Ply Panel

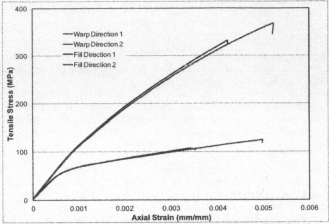

Figure 2. Room Temperature Stress Strain Curved from Bias Panel

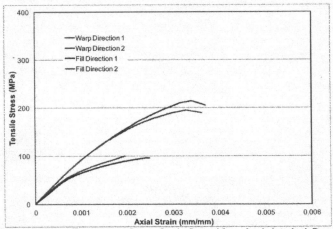

Figure 3. Room Temperature Stress Strain Curved from Angle Interlock Panel

Table III. Average Tensile Data for Three Panel Types

Panel Type	Test Direction	Test Temp (°C)	Proportional Limit (MPa)	UTS (MPa)	Failure Strain (mm/mm)	Modulus (GPa)
Baseline	Warp	24	84.6	253.7	0.0041	114.9
Baseline	Fill	24	92.8	282.3	0.0049	115.9
Baseline	Warp	649	77.3	300.3	0.0052	121.4
Baseline	Fill	649	89.3	298.7	0.0049	125.2
Baseline	Warp	982	82.3	302.3	0.0054	120.1
Baseline	Fill	982	90.6	320.9	0.0063	111.8
Bias	Warp	24	109.0	350.0	0.0047	109.0
Bias	Fill	24	52.1	115.4	0.0042	98.0
Bias	Warp	649	na	na	na	na
Bias	Fill	649	42.8	113.2	0.0047	107.0
Bias	Warp	982	101.0	335.7	0.0047	124.9
Bias	Fill	982	42.2	116.5	0.0055	120.8
Angle Interlock	Warp	24	89.9	204.5	0.0034	95.2
Angle Interlock	Fill	24	53.6	98.0	0.0022	80.4
Angle Interlock	Warp	649	90.4	216.8	0.0039	101.8
Angle Interlock	Fill	649	61.2	108.3	0.0036	66.9
Angle Interlock	Warp	982	88.4	227.8	0.0047	94.9
Angle Interlock	Fill	982	51.9	126.7	0.0054	76.9

na – delamination present in panel reduced the samples present for testing and therefore this temperature was removed from the test matrix

Table IV. Interlaminar Tensile Average for Three Panel Types (with standard deviation)

Panel	Analysis	ILT Strength (MPa)	Failure Location
Baseline	Average	14.1	100% Material
	StDev	1.08	
Bias	Average	12.9	100% Material
	StDev	3.90	
Angle Interlock	Average	14.9	100% Material
	StDev	0.28	

Thermal Testing

A series of in-plane thermal expansion tests were done in both the warp and fill directions. The results of this testing are shown in Figure 4. There is not a clear distinction between the test directions with the known change in fiber volume. The results of the thermal conductivity testing are shown in Figure 5. Here, clear differences are seen with the angle interlock sample showing the greatest conductivity value.

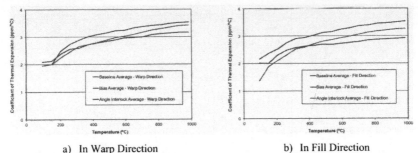

| a) In Warp Direction | b) In Fill Direction |

Figure 4. In-Plane Thermal Expansion Results for the Three Panel Types

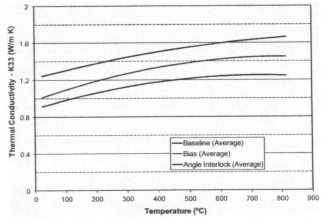

Figure 5. Through Thickness Thermal Conductivity Results

Characterization

As noted previously, micro-structural characterization was performed on the material. From the optical work, image analysis was done to determine fiber volume on the baseline panel and porosity analysis for all the panels. Micro-structural cross sections showing the fill direction for the three panel types are shown in Figure 6. Figure 6 shows that the fill direction has different fiber volumes as expected. From these images, porosity analysis was done and the summary of that work is shown in Table V. The porosity is essentially constant for this effort and will not be discussed further. For this series of work, only the fiber volume for the baseline panel was determined and it was found to be 24.7% which is higher than expected from the manufacturing expectations shown in Table I.

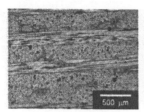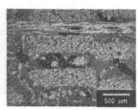

a) Baseline (50x)   b) Bias (50x)   c) Angle Interlock (50x)
Figure 6. Optical Images of Fill Direction for the Three Panel Types

Table V. Porosity Analysis (Image) for the Three Panel Types

Panel Type	Fiber Direction	Porosity (%)
Baseline	Warp	8.4
	Fill	11.6
Bias	Warp	8.2
	Fill	11.1
Angle Interlock	Warp	8.8
	Fill	11.2

After tensile testing, Scanning Electron Microscopy (SEM) was performed on the tensile fracture faces to confirm the presence of fiber pullout. Images from the testing in the fill direction are shown in Figure 7. Fiber pullout is seen consistent with the strain to failures (See Table III.)

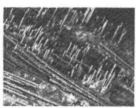

a) Baseline   b) Bias   c) Angle Interlock
Figure 7. SEM Images of Tensile Samples Tested in the Fill Direction
(tested at RT)

Mechanical Characterization

In addition to the optical characterization, the tensile testing was used to determine the fiber volume in the different loading directions. This can be done by looking at the secondary slope of the stress strain curve after the proportional limit. After the proportional limit, all the load is being carried by the fiber and the slope is equivalent to the fiber volume times the modulus of the fiber [7,8]. The results of this analysis are shown in Table VI. There is good agreement with the expectations from manufacturing. The mechanical testing generates higher values in the warp direction and lower values in the fill direction. The testing done for the baseline panel was confirmed using optical image analysis as discussed above.

Table VI. Mechanical Analysis of Fiber Volume for the Three Panel Types

Panel Type	Fiber Direction	Vf - Manufacture	Vf – Tensile Curve
		(%)	(%)
Baseline*	Warp	20.1	24.7
	Fill	20.1	24.7
Bias	Warp	28.8	35.1
	Fill	11.1	6.9
Angle Interlock	Warp	24.5	21.0
	Fill	10.5	8.4

* = optical analysis was done on the baseline panel and the determined fiber Vf was 24.7*

DISCUSSION

Three different panel types were tested in two different directions with some testing done in the through thickness direction of the panels. The breadth of this work allows the data to be compared against fiber volume and weave effects. This is shown in Figures 8-10 where the modulus, proportional limit (0.005% offset) and ultimate tensile strength can be viewed against fiber volume. (For the Elastic Modulus, past work was used for the constituent properties allowing a Rule of Mixtures (ROM) approach to be used [6].) These series of results clearly show that the fiber volume can have a significant impact on the properties. The elastic modulus is consistent with the Rule of Mixtures especially when the porosity in the material was clearly known (See Table V). The change in the fiber volume clearly influences the ultimate tensile strength. With increasing percentage of high strength fibers, there is a clear benefit of the fiber seen. The proportional limit increases with fiber volume and this is most likely due to a decrease in the amount of the weak matrix present.

The interlaminar properties did not clearly differentiate between the panels and samples tested. As noted earlier, the interlaminar shear did not have the correct failure mode so no analysis can be done. The interlaminar tensile testing showed no difference between the panels even with the angle interlock work done (See Table IV). The presence of porosity is not an issue here as the panels were consistent as noted earlier but the effect of porosity as a stress concentrator in these tests needs to be investigated.

While the angle interlock did not improve the mechanical properties in the through thickness direction (as not for the interlaminar tensile), there was a benefit seen for the thermal conductivity (See Figure 5). The presence of fibers in the Z (through thickness) direction allowed improved thermal conductivity over the other panels that relied on the matrix for part of the through thickness thermal conductivity. The thermal expansion testing did not show significant changes but this is most likely due to the expansions between the two phases being the same. This needs to be confirmed with experimental work.

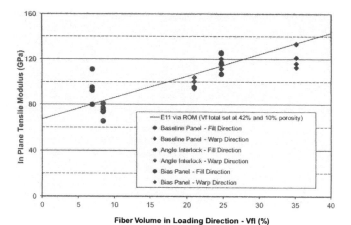

Figure 8. Tensile Modulus versus Fiber Volume
(with ROM Theory Shown)

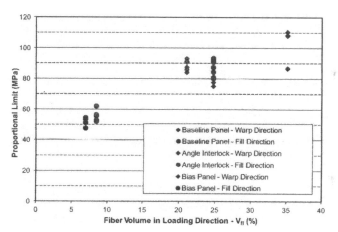

Figure 9. Proportional Limit versus Fiber Volume

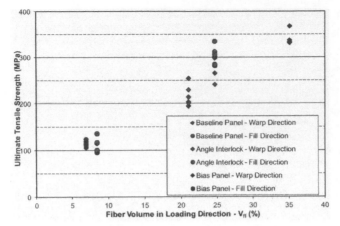

Figure 10. Ultimate Tensile Strength versus Fiber Volume

CONCLUSION

A series of tests were conducted on an 8HS baseline panel, a bias panel and an angle interlock panel. Testing was done in both the warp and fill directions. Fiber volume was determined form the testing and used to look at trends with fiber volume. This testing showed that the mechanical properties were greatly influenced by fiber volume. The interlaminar properties were not affected. There was some effect on thermal properties for the through thickness thermal conductivity for the angle interlock panel due to the presence of Z direction fibers.

This work shows the benefit of looking at a range of weaves and fiber volume as this can impact the design space for future applications.

FUTURE WORK

As part of the effort to understand this data, additional constituent property testing should be done. As noted above, this is clearly the case for the thermal expansion where the data versus temperature is not fully known.

ACKNOWLEDGMENTS

Work performed under the Ceramic Matrix Composite Turbine Blade Demonstration Program, Contract FA8650-05-D-5806 Task 0021, Dr. Paul Jero program manager.

REFERENCES

1. K.K. Chawla (1998), *Composite Materials: Science and Engineering, 2nd Ed.*, Springer, New York
2. Ashby, Michael F. (2005). Materials Selection in Mechanical Design (3rd Edition). (pp: 40. Elsevier, Boston
3. Brewer, D., 1999, "HSR/EPM Combustion Materials Development Program", Materials Science & Engineering, 261(1-2), pp. 284-291.
4. Brewer, D., Ojard, G. and Gibler, M., "Ceramic Matrix Composite Combustor Liner Rig Test:, ASME Turbo Expo 2000, Munich, Germany, May 8-11, 2000, ASME Paper 2000-GT-670

5. Verrilli, M. and Ojard, G., "Evaluation of Post-Exposure Properties of SiC/SiC Combustor Liners testing in the RQL Sector Rig", Ceramic Engineering and Science Proceedings, Volume 23, Issue 3, 2002, p. 551-562.
6. Ojard. G., Rugg, K., Riester, L., Gowayed, Y. and Colby, M., "Constituent Properties Determination and Model Verification for a Ceramic Matrix Composite Systems", Ceramic Engineering and Science Proceedings. Vol. 26, no. 2, pp. 343-350. 2005.
7. Ojard, G., Gowayed, Y., Morscher, G., Santhosh, U., Ahmad, J., Miller, R. and John, R., "Frequency And Hold-Time Effects On Durability Of Melt-Infiltrated SiC/SiC", to be submitted to Ceramic Engineering and Science Proceedings, 2011.
8. Courtney, T.H. (2005), *Mechanical Behavior of Materials*, 2nd Ed., Waveland Press. Long Grove, IL

INNOVATIVE CLAY-CELLULOSIC BIOSOURCED COMPOSITE: FORMULATION AND PROCESSING

Gisèle L. Lecomte-Nana[1,a,b], Olivier Barré[b], Cathérine Nony[b], Gilles Lecomte[a,b], Thierry Terracol[b]

[a] Laboratoire Groupe d'Etude des Matériaux Hétérogènes, Centre Européen de la Céramique, 12 rue Atlantis, 87068 Limoges Cedex, France
[b] Société Bibliontek, AVRUL Dpt Incubateur, 1 avenue Ester Technopole, 87069 Limoges, France

ABSTRACT

The present research work is in line with sustainable and environment conscious development through the use of natural raw materials for the processing of biosourced products. The main scope is to manufacture a new class of products for future application as protection barriers for delicate objects towards fire and flooding. The main requirements that must be fulfilled for such application are: flame resistance, waterproof, good permeability and adequate flexural strength (> 2 MPa).

Kaolinitic-illitic raw clays were selected as starting materials regarding their specific behaviors. Various clay-fiber compositions were prepared and optimized regarding the final required characteristics. Three surfactants, Na_2CO_3, Na_2SiO_3 and HMP (sodium hexametaphosphate) were used to improve the rheological behavior of the system. The couple Na_2CO_3/Na_2SiO_3 for 0.7 mass%/0.05 mass% content within the slurry leads to lower yield stress and to a shear-thinning behavior suitable for the shaping process. Eco-products, consolidated without sintering, can thus be manufactured.

The optimized formulation consists of a clay-fiber mixture with a fiber/clay ratio of 0.14. The key characteristics obtained for this product are: M0 class, resistance towards water seeping of 90 min, heat resistance of $\Delta T \approx 350°C$ and flexural strength of 7 MPa.

1 - INTRODUCTION

Sustainable development is a great challenge for nowadays research in order to preserve the planet's ecosystem. Actually, there is an increasing need to design and manufacture original materials that respond to the needs of the present without inhibiting the ability of future generations to supply their own needs. As a matter of fact, clay materials appear as promising starting components since they result from sustainable resources and can be easily re-used, recycled or disposed on end use without any health or environmental damages. Besides these noticeable characteristics, unfired clay materials exhibit a major drawback in relation with their great affinity for water.[1-4] That is the reason why clays are generally fired or stabilized with hydraulic binders, and besides, improving the mechanical properties of resulting products.[5-7] Moreover natural products, namely natural fibers can be added to clays in order to develop new green composites.

The most used natural fibers are plant-based ones. Such natural fibers are mainly lignocellulosic components from biomass; they are readily available and offer a great potential of valorization through valuable by-products, namely environmental friendly composites. They are therefore well appropriated for economical and sustainable development.

Numerous studies have already been performed on the elaboration of eco-friendly clay-fibers composites.[8-14] Most of these studies deal with sisal, hemp, jute, flax fibers in association with clay minerals such as montmorillonite, sepiolite and kaolinite. The resulting products were devoted to

---

[1] Corresponding Author

building purposes. The key properties herein are thermal stability and fire resistance. The latter was achieved in most cases by using halogen-based retardants.[15]. Another critical point in such composites is the cohesion at clay-matrix/fiber interface. To afford this, the fibers generally undergo a treatment with basic solutions (NaOH, $Ca(OH)_2$) prior to their use, in order to remove the superficial hydrophobic wax layer.[16-18]

The present work aims to develop an innovative green clay-fiber composite with combine improved fireproof and waterproof characteristics. This new class of environmental conscious materials is based on renewable resources and must exhibit suitable physical and chemical properties in relation with future application as protection barriers for delicate objects towards firing and flooding. To our knowledge, such natural plant-based fibers and clay composites are currently under-explored for their potential uses for a wide range of end uses.

2 – EXPERIMENTAL PART

2.1 - Characterization techniques

2.1.1 – chemical analysis

Chemical analysis of raw material has been performed using an Inductively Coupled Plasma Atomic Emission Spectrometer (ICP-AES). Prior to their analyses, the samples are oven-dried for 48 h, then aliquots of 10 mg are dissolved under acidic conditions (solution of 1/3 of HCl and 2/3 of HNO_3 in volume) in a CEM MARS 5 microwave apparatus. The dissolution is performed using a 45 min cycle including a maximum temperature of 180°C under a pressure of 3 MPa The as-obtained solutions are therefore completed to 250 mL and submitted to ICP-AES analyses.

2.1.2 – Rheological behavior

The rheological behaviour of each slurry was characterized using a parallel plate geometry (40 mm diameter steel upper plate and a Peltier lower plate) mounted on a controlled stress rheometer AR1500ex from TA® Instruments. A shear stress (τ) sweep was applied to each suspension varying the shear stress between 0 and 50 Pa at 22°C using a gap of 1 to 1.5 mm. The flow curves were analyzed and fitted to the Herschel - Bulkley model expressed by the following equation:

$$\tau = \tau_u + \left(k \cdot (\dot{\gamma})^n \right) \qquad (1)$$

Where τ and τ_0 are respectively the shear stress and the yield stress, k the consistency, $\dot{\gamma}$ the shear rate, and n the power law exponent.

2.1.3 – Control of released species

Since the composites are assumed to be green materials, they should not emit noxious elements in surrounding environment. Therefore, pH values of the various samples were measured in aqueous solution using a EUTECH Instruments pH-meter. The pH value of a 100 mL deionized-water solution containing 2g of solid sample was followed over a month to verify the stability of the media. Also analyses of evolved gases were performed on the samples in order to ensure the non emission of NOx. These experiments were performed under accelerated thermal degradation of the products. The surrounding atmosphere was analyzed before and after the sample is submitted to the test.

2.1.4 – Scanning electron microscopy

Microstructure observations have been carried out using a scanning electronic microscope (SEM) Stereoscan S260 from Cambridge Instruments. A thin platinum layer was deposited onto the surface of each sample prior to SEM observations. The obtained micrographs allow checking the homogeneity within the sample and the clay-fiber interfaces.

2.1.5 – Flexural strength

Characteristic flexural strengths (σ_{max} see equation 2) have been obtained using three points bending tests with a LLYOD EZ20 device. Samples were 50 mm and 150 mm in width and length respectively, the distance between the lower sills is 100 mm and the loading rate was 0.5 mm/min. For each set of sample, at least ten specimens were tested and the flexural strengths given in this study are always mean values over ten measurements.

$$\sigma_{max} = \frac{3}{2} \times \frac{F_{max} \times L}{b \, h^2} \qquad (2)$$

Where F_{max} represents the maximal load applied to the sample; L denotes the distance between the sills and b and h are respectively the width and thickness of the sample.

2.1.6 – Flame retardancy

Flame resistance has been determined based on the European specification NF EN 60695-11-10/A1. Here the parallepipedic samples (height 50 mm and length 150 mm) are submitted to a Bunsen burner (flame temperature around 700°C) for 1 min. Depending on the behavior of the sample towards this flame as mentioned on Table 1, the adequate label is attributed. The other important point is in relation with the retardancy upon heat propagation. The method use in this study consists in the measurement of the difference in temperature between two opposite surfaces ($\Delta T = 1000°C - T_{outer-surface}(°C)$), one surface being submitted to a temperature of 1000°C and the opposite surface to initial room temperature. A Kimo® measuring set-up is used for these measurements and the thermocouples allow a precision of ± 1°C. Once the temperature is stabilized at 1000°C over the inner surface, the temperature acquisition is performed during 1h over the outer surface. To fulfill the requirements as heat protection barrier, the sample should exhibit at least a difference in temperature $\Delta T \geq 300°C$.

Table 1: summarized classification used to qualify the flame resistance of materials (EN 60695-11-10/A1)

Behavior towards the flame	No flame propagation, no black or red mark	No flame propagation, but small black or red marks	Inflammation, but it stop few seconds after the test is performed		Inflammation and rapid propagation over the sample	
Standard label	M0	M1	M2	M3	M4	M5
Definition	Noncombustible	Nonflammable	Hardly flammable	Moderately flammable	Easily flammable	Very easily flammable

2.1.7 – Durability (Interaction with water)

The samples have been submitted to water permeability tests in order to determine their sensitivity towards water. The experimental system involves a cylindrical column (30 mm in diameter) filled with 300 mm height deionized water, the tested sample (squared section of 100 mm in size) being sealed at the bottom of this cylinder (figure 1(a)). The recorded time until the detection of water on the outer side of the sample is used as a key parameter to estimate the water permeability of samples.

The wettability is estimated through the observation of the liquid-to-solid contact angle on the various samples. In fact, a 50 μL water droplet is put onto the sample and the wetting character is determined using a camera (Figure 1(b)).

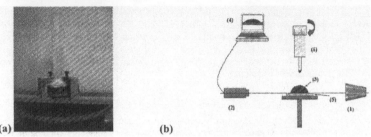

(a) (b)

Figure 1: experimental set up for water permeability (a) and for wettability (b) tests. (1) = light source, (2) = camera, (3) liquid droplet, (4) = acquisition computer, (5) = solid sample, (6) = microsyringe for liquid delivery

2.2 - Materials and processing

The raw clay used in this study for the elaboration of model products are a commercial kaolin from France labeled BIP kaolin and a halloysite, the corresponding chemical and mineralogical compositions are given in table 2. More complex clays labeled A1, A2, A3, A4 and A5 have been used for the optimized formulation. It can be noted from their XRD characterizations (Figure 2) that these clays are predominantly rich in kaolinite, illite and smectites.

The fibers consist of a mixture of short and long cellulosic natural fibers (Figure 3) respectively denoted SCF and LCF. Their global characteristics are presented in table 3. The long cellulosic fibers) are purchase from tropical countries (Ivory Coast, Costa Rica) while the shorter ones are issued from recycling papers and are delivered by Stouls Company.

Table 2: chemical compositions raw kaolin and halloysite (mass %)

	SiO_2	Al_2O_3	Fe_2O_3	CaO	MgO	Na_2O	K_2O	TiO_2	Li_2O	LOI
BIP kaolin	48.1	36.9	0.26	<0.20	0.17	<0.20	1.90	<0.05	0.27	12.21
Halloysite	46.55	39.50								13.96

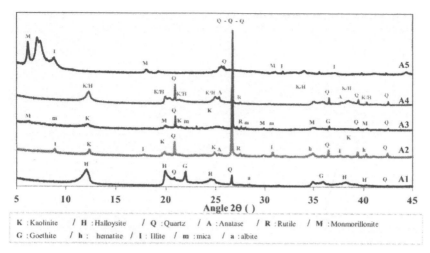

K : Kaolinite / H : Halloysite / Q : Quartz / A : Anatase / R : Rutile / M : Monmorillonite
G : Goethite / h : hematite / I : Illite / m : mica / a : albite

Figure 2: XRD diagrams of raw clays used for optimized products

Figure 3: SEM micrographs of long (a) and short (b) cellulosic fibers

Table 3: composition of cellulosic fibers (mass %)

	Cellulose ($C_6H_5O_5$)	Pentosan ($C_5H_{10}O_5)_n$	Lignin ($C_9H_{10}O_2$)
LCF	64	13	23
SCF	95		< 5

Different formulations have been elaborated for this study, their detailed constitutions and labels are provided in table 4.

The elaboration procedure include a dry mixing of clays and SCF fibers for 10 min at 20 rpm in a Perrier mixer, followed by a wet mixing with appropriate water content (given in table 4) and LCF fibers for 40 min at 30 rpm. The obtained mixture is directly submitted to a cold-rolling process between two felt papers. The as-obtained plastic perform is infiltrate with the corresponding clay slurry regarding clay/(SCF + LCF) ratio. The drying of the resulting green product is performed at room temperature under confined conditions for 48 hours, and then is completed at 40°C for 24 hours.

Table 4: list of various formulations tested in the present study (mass %)

Sample label	Bip kaolin	Halloysite	LCF fibers	SCF fibers	W/S
CF1	80	8	5	7	2
CF2	80	8			2
CF3	78	10	5	7	1.04
CF4	80	8	5	7	1.16
CF5	81	8	4	7	1.6
CF6	71.7	17	4.7	6.6	1.24
CF7 (5.7% of OPC)	73.6	9.4	4.7	6.6	2

3 - RESULTS AND DISCUSSION

3.1 - Flame resistance

In order to have a significant evaluation of our natural thus environmental friendly clay-based composites, we have also characterized the flame resistance of a similar commercial product. This reference product is obtained through sintering processes which involves more energy consumption. The obtained flame resistance results and the corresponding relative evolution is shown are presented in figure 3.1 It appears that all of our natural clay fiber composites exhibit M0 product behavior, which is highly satisfactory in regards to their devoted application as sustainable flame-resistant materials. This general trend can be explained by three main facts:

- In our composites, the raw clays act as flame retardants since the energy provided during firing is first used to eliminate physisorbed water. And furthermore, available energy is used to perform dehydroxylation process, and from literature, the required energy for such reaction ranges from 170 to 400 kJ/mol depending on the nature of phyllosilicates[5-7] (1:1 and 2:1 clay minerals).
- The naturals vegetal fibers, which are very flammable alone, are less sensitive to flame when embedded in a clay matrix due to changes of their surface characteristics
- The presence of connected porosity filled with air, which act as barriers towards flame propagation.

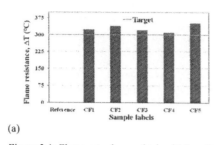
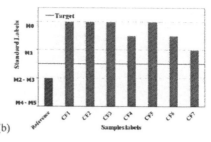

(a)   (b)

Figure 3.1: Flame retardancy obtained (a) and typical flame resistance class (b) for some samples.

SEM micrographs (figures 4(a) and (b)) confirm the occurrence of connected voids into the products with well-coated fibers. Also the observation of the various samples after flame resistance tests shows a slight change of water content. These facts are coherent with the assumption regarding the fundamental role of clays on this peculiar behavior.

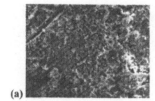
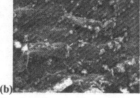

(a)   (b)   (c)

Figure 4: SEM micrographs showing the microstructure of a clay-fibers composite (a) and (b); and the decohesion of fibers after flexural tests (c).

3.2 – Mechanical properties: Three points bending test

The characterization of the respective mechanical properties of all the samples through three points bending tests gives rise to results presented in Figure 5. Actually the target limit flexural strength is 1 MPa and only few samples satisfy this requirement. In order to understand these weak mechanical properties, the samples have been submitted to SEM and FTIR characterizations. Hence, SEM images (figures 4(c)) show that the adhesion at fiber-clay interface is barely effective, thus the observed fibers decohesion. Moreover, the FTIR analysis shows that there is not a significant change of the main chemical functions, thence an evidence of the great amount of non modified fibers in the system. To improve these properties, two main routes may be studied:

- the treatment of fibers prior to their use and/or the optimization of slurries rheology and homogenization

- the use of a more densifying shaping process.

The latter solution has been tested by using, for the best formulation, pressure casting-inspired shaping process. This experiment has been performed at laboratory scale by using cylindrical resin molds under a N_2 pressure of 1 MPa. The bending test performed on the as-obtained products leads to a mean flexural strength of 7 MPa. This result is very satisfactory towards the final application of such composites. Actually, the improvement of mechanical properties here can be explain by the fact that during the shaping process, the constitutive clay particles are mechanically forced to fill-up the available space within the sample leading to a more closely-packed layout within the composite. Thus, the resulting highly cohesive structure obtained in this case, giving rise to a higher flexural strength. Furthermore, a less heterogeneous pore size distribution is obtained with a good interconnectivity that insures the breathable property of the final product.

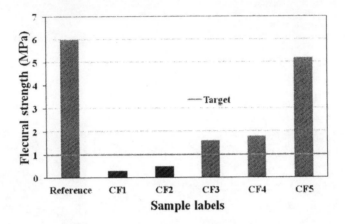

Figure 5: Flexural strength of some characteristic samples (± 5%)

3.3 – Durability - Sensitivity towards water

In order to fulfill the whole requirements for future applications, water absorption tests were performed. The codes indicated a resistance towards water infiltration of at least 90 min, only few of our compositions satisfy this point. Basically, all commercial products found were subject to specific hydrophobic treatment in order to ensure a good resistance towards water seeping, thus the relatively high value obtained (1200 min) for the reference sample. Since no specific treatment has been carried out on our samples, the results are not very surprisingly.

Clays, which are our major constituents, are mostly hydrophilic materials because of the presence of hydroxyl groups which may combine with available water, thus their high affinity towards water. Most of natural fibers behave the same way when their waxy superficial layer is removed, which is probably the case with the recycling paper fibers used here. Despite these preferential trends, water absorption can still be improved in such materials in a sustainable way, by using other specific natural raw materials which exhibit a predominant hydrophobic character. The hydrophobic character of the LCF fibers has been evidenced through wettability test. As can be observed on figure 6, the contact angle between a water droplet and the solid surface of the LCF-material is highly greater than 90°, which is not the case when using SCF fibers. Even though, this last point is not the purpose of the present work, a primary step has been made in this scope by using a great fraction of hydrophobic LCF fibers in some of our samples. The increase of the LCF fibers up to 10 mass % allows to improve the water resistance from 90 min to 200 min. Simultaneously, the flexural strength is decreased from 7 MPa to 2 MPa. Future investigations are necessary to finely tune the fibers proportions regarding water resistance improvement without degrading the other final properties.

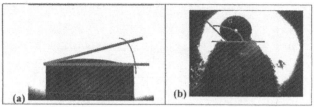

Figure 6: wettability in the case of hydrophilic SCF (a) and hydrophobic LCF (b) fibers.

3.4 – Optimization of the slurry rheology

The rheological behavior of the slurry used in formulation CF5, that exhibit the most satisfactory properties, has been characterized and compared to that of a commonly used "model clay" slurry. The obtained rheograms are shown on figure 7. A significant difference is noted between these curves, namely in relation with the higher yield stress and consistency of slurry CFM3. In order to lowered theses parameters and further decrease the amount of water required for the processing of the slurry, two couple of surfactants are tested. Na_2CO_3/Na_2SiO_3 and Na_2CO_3/HMP (sodium hexametaphosphate) were selected considering their well-known interaction with clay minerals. Sodium carbonate acts mainly on the thixotropy of clay slurries and only a few amount is necessary (generally, from 0.05 to 0.8 mass% with respect to overall solid content. Sodium silicate and HMP have a specific action on the fluidity of clay slurries (lowering the yield stress and the consistency) and greater quantities are commonly used compare to sodium carbonate. The appropriate required amount in this case is determined after rheological measurements using increasing amount of sodium silicate or HMP, keeping constant the amount of silicate carbonate in all cases. Table 5 presents the tested combinations and the related evolution of the rheograms is shown on figure 8. It can be noted that the results which are satisfactory are those of slurries containing of Na_2CO_3/Na_2SiO_3 and of Na_2CO_3/HMP.

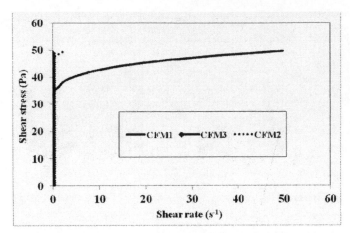

Figure 7: Rheograms obtained with clay-based slurries without fibers (CFM1), with SCF fibers (CFM2) and with both SCF and LCF fibers (CFM3)

Table 5: list of the different slurries characterized through rheological measurements (mass %)

Sample label	CFCX00	CFCX05	CFCX10	CFCX20	CFCX30	CFCX40	CFCX50	CFCX70	CFCX75	CFCX80
X = H for HMP and S for Na$_2$SiO$_3$ (mass %)	0.00	0.05	0.10	0.20	0.30	0.40	0.50	0.707	0.75	0.807

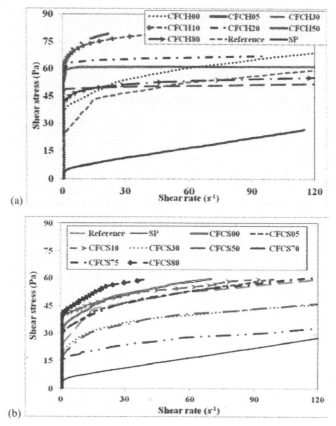

Figure 8: rheograms obtained with CFM1 derived slurries using (a) Na_2CO_3/HMP and (b) Na_2CO_3/Na_2SiO_3 as surfactants.

These slurries have been considered and the amount of water has been reduced from 96 mass% to 70 mass % with respect to the solid. Such modification is very profitable considering further drying step which can induce stress within the product leading to drastic damages. The surfactants (HMP and Na_2SiO_3) amounts are adjusted to obtain the same yield stress and consistency as those mentioned above for the best rheological behaviors. Also the flexural strengths of the final products are checked. As shown on Figure 9, the slurry with 0.7 mass % of sodium silicate exhibit the best balanced with 7 MPa and 650°C ($\Delta T = 350$°C) in thermal resistance. Such result seems to indicate that the predominant clay mineral in our clay mixture is kaolinite-type and therefore, it controls the global rheological behavior. It is also expected that additional interaction may occurs between the cellulosic surface functions and the sodium and silicate ions issued from the dissolution of sodium silicate.

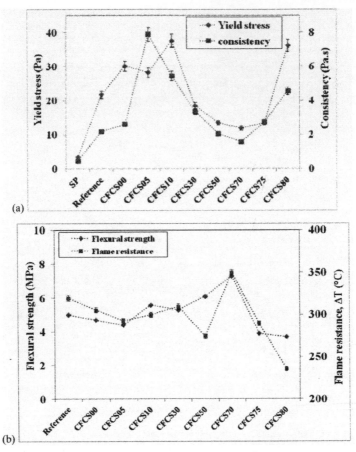

Figure 9: Summarized (a) rheological parameters of slurries and (b) of properties (flexural strength and flame resistance) of composites obtained using the couple Na_2CO_3/Na_2SiO_3 as surfactants.

CONCLUSION

A new clay-cellulosic eco-composite has been processed with the scope of future application as protection barrier for precious objects towards flooding and firing.

The optimized composition consists in a mixture of 88 mass % of kaolinitic-illitic clays, 7 mass% of short cellulosic fibers (from recycling paper industry) and 5 mass % of long cellulosic fibers bers) exhibiting a strong hydrophobic character.

The final product is obtained through a shaping process involving pressure infiltration of a plastic body by a clay slurry containing 0.7 mass % of Na_2SiO_3 and 0.05% of Na_2CO_3.

The as-obtained product after drying exhibit a resistance towards water infiltration of 90 min, a good permeability to air, a maximal flexural strength of 7 MPa and a M0 class behavior considering fireproof. Furthermore a temperature difference of 350°C can be achieved after stabilization at 1000°C on one surface trough a width of 5 mm.

REFERENCES

[1] S.B. Hendricks, Hydration mechanism of the clay mineral montmorillonite saturated with various cations., J. Am. Chem. Soc., 62, 6, 1457-1464 (1940)

[2] S.B. Hendricks, "Lattice structure of clay minerals and some properties of clays." J. Geol., 50, 276-290 (1942)

[3] C.E. Clarke ⁂, J. Aguilar-Carrillo, A.N. Roychoudhury, Quantification of drying induced acidity at the mineral–water interface using ATR-FTIR spectroscopy., Geochimica et Cosmochimica Acta 75, 4846–4856 (2011)

[4] Yukiko O. Aochi∗, Walter J. Farmer, Effects of surface charge and particle morphology on the sorption/desorption behavior of water on clay minerals., Colloids and Surfaces A: Physicochem. Eng. Aspects 374, 22–32 (2011)

[5] D. Prodanovic, Z.B. Zivkovic, S. Radosavljevic, Kinetics of dehydroxylation and mullitisation process of halloysite from Farbani Potok locality, Serbia., Appl Clay Sci 12, :267 - 274 (1997)

[6] E. Mazzucato, G. Artioli, A. Gualtieri, Hogh temperature dehydroxylation of 2M1 muscovite: a kinetic study by in-situ XRPD., Phys Chem Minerals 26:375 - 381 (1999)

[7] G.L. Lecomte, J.P. Bonnet, P. Blanchart, A study of the influence of muscovite on the thermal transformations of kaolinite from room temperature up to 1,100°C. J. Mater. Sci. 42, 8745-8752 (2007)

[8] K. Finlay, M. Gawyrla, D. Schiraldi, Natural fiber reinforced polymer/clay aerogel composites *American Chemical Society, Polymer Preprints, Division of Polymer Chemistry, 49*, 1134-1135 (2008)

[9] X. Huang, A. Netravali, Characterization of flax fiber reinforced soy protein resin based green composites modified with nano-clay particles *Composites Science and Technology, 67*, 2005-2014 (2007)

[10] C. Ludvik, G. Glenn, A. Klamczynski, D. Wood, Cellulose fiber/bentonite clay/biodegradable thermoplastic composites *Journal of Polymers and the Environment, 15*, 251-257 (2007)

[11] M. Alexandre, P. Dubois, "Polymer-layered silicate nanocomposites : preparation, properties and uses of a new class of materials" Materials Science and Engineering, 28, 1-63 (2000)

[12] Z. Lin, S. Renneckar, D. P. Hindman, "Nanocomposites-based lignocellulosique fibers 1. Thermal stability of modified fibers with clay-polyelectrolyte multilayers", Cellulose, 15, 333-346 (2008)

[13] C. Galán-Marín, C. Rivera-Gómez, J. Petric, *Clay-based composite stabilized with natural polymer and fibre,* Construction and Building Materials, 24, 8, 1462-1468, (2010)

[14] N. Guigo, L. Vincent a, A. Mija, H. Naegele, N. Sbirrazzuoli, *Innovative green nanocomposites based on silicate clays/lignin/natural fibres,* Composites Science and Technology 69 1979–1984 (2009)

[15] Z. Zhao, J. Gou, S. Bietto, C. Ibeh, D. Hui, Fire retardancy of clay/carbon nanofiber hybrid sheet in fiber reinforced polymer composites *Composites Science and Technology, , 69,* 2081-2087 (2009)

[16] L. Y. Mwaikambo, et al., Chemical Modification of Hemp, Sisal, Jute, and KapokFibers by Alkalization. Journal of Applied Polymer Science, Vol. 84, 2222–2234 (2002)

[17] L. Y. Mwaikambo, M. P. Ansell, *Hemp fibre reinforced cashew nut shell liquid composites,* Composites Science and Technology 63, 1297–1305 (2003)

[18] M. Le Troedec, C. S. Peyratout, A. Smith, T. Chotard, *Influence of various chemical treatments on the interactions between hemp fibres and a lime matrix, Journal of the European Ceramic Society,* 29(10), 1861-1868 (2009)

PROCESSING AND TESTING RE$_2$Si$_2$O$_7$ MATRIX COMPOSITES

Emmanuel E. Boakye,[1,2] Kristin A. Keller,[1,2] Pavel S. Mogilevsky,[1,2] Triplicane A. Parthasarathy,[1,2] Mark A. Ahrens,[1,3] Randall S. Hay,[1] Michael K. Cinibulk[1]

[1]Materials and Manufacturing Directorate, Air Force Research Laboratory, WPAFB, OH
[2]UES, Inc., Dayton, OH
[3]Wright State University, Fairborn, OH *

ABSTRACT

In prior work, the synthesis of α, β and γ-Re$_2$Si$_2$O$_7$ powders (Re = Y and Ho) at temperatures from 1000° to 1400°C in air was reported, along with the Vickers hardness of dense γ-Y$_2$Si$_2$O$_7$ and γ-Ho$_2$Si$_2$O$_7$ polymorphs. Dense γ-Y$_2$Si$_2$O$_7$ and γ-Ho$_2$Si$_2$O$_7$ pellets were made by a pressureless sintering technique at 1400°C / 8 h. Using the pressureless sintering technique, densification of α- and β-Y$_2$Si$_2$O$_7$ powder compacts at their phase formation temperature (1000° - 1200°C) was not successful and prevented the determination of their hardness. In this work, the field assisted sintering technique (FAST) was used to form dense α-, β- and γ-Y$_2$Si$_2$O$_7$ pellets at a pressure of 20 kN and temperatures of 1050° - 1200°C. Subsequently, their Vickers hardness was measured. SCS-0 fibers were also incorporated into α-, β- and γ-Re$_2$Si$_2$O$_7$ matrixes and densified at 1050°C - 1200°C / 1 h using the FAST approach. Fiber push-out experiments were conducted, and the average sliding stress values were determined.

1.   INTRODUCTION

The application of SiC/SiC ceramic matrix composites (CMCs) is limited by mechanical property degradation in oxidizing environments. Several methods are used to minimize the oxidation of BN or carbon fiber-matrix interphases in SiC/SiC CMCs;[1] however, an ideal solution would be to replace BN or carbon with a material that does not oxidize. The rare-earth disilicates are attractive candidates.[2]

Among the rare-earth disilicates, yttrium disilicate (Y$_2$Si$_2$O$_7$) is the most thoroughly studied. It has five polymorphs and is refractory, melting at 1775°C.[3-8] Its γ-polymorph (γ-Y$_2$Si$_2$O$_7$) has been reported to be a "quasi-ductile" ceramic[5-6] and it has comparable mechanical properties to LaPO$_4$ (monazite), which has been demonstrated to function as a weak fiber-matrix interphase in oxide-oxide CMCs.[6,9-10] Both oxides are soft (Vickers hardness ~6 GPa) and machineable. However, monazite decomposes in the reducing atmospheres typically used for processing SiC-SiC CMCs, and the rare-earth disilicate La$_2$Si$_2$O$_7$ forms as a reaction product of SiO$_2$ and LaPO$_4$ under these reducing conditions.[11-13] An alternate oxide coating, therefore, that is stable in reducing environments is needed. In addition to its softness, γ-Y$_2$Si$_2$O$_7$ is thermochemically compatible with SiC and SiO$_2$,[14] and it has a thermal expansion coefficient (~4x10^{-6}°C^{-1}) that is similar to SiC.[15] This combination of properties motivated the investigation of Re$_2$Si$_2$O$_7$ as a possible fiber-matrix interphase for SiC/SiC CMCs.

The rare-earth disilicates are commonly prepared by a variety of methods, including conventional solid-state reaction of mixed oxides (Re_2O_3 and SiO_2), calcination of rare-earth disilicate sol-gel precursors, and hydrothermal processing.[16-22] In our prior work, the formation of α-, β-, and γ-$Re_2Si_2O_7$ powders (Re = Y and Ho) at temperatures of 1000° - 1400°C in air was reported. Precursors to $Y_2Si_2O_7$ and $Ho_2Si_2O_7$ were made by adding colloidal silica to solutions of yttrium and holmium nitrate. $LiNO_3$ was added to decrease the formation temperature of the γ-polymorphs. Dense γ-$Y_2Si_2O_7$ and γ-$Ho_2Si_2O_7$ pellets were made by pressureless sintering at 1400°C for 8 h and hardness measurements were completed using the Vickers indentation method. Densification of α- and β-Y_2Si_2O powder compacts at their phase formation temperatures of 1000° - 1200°C using pressureless sintering was not successful; therefore, their hardness was not measured. In this work, dense α-, β-, and γ-Y_2Si_2O pellets were made using the field assisted sintering technique (FAST) and hardness measurements were completed using Vickers indentation. Further, SCS-0 fibers were incorporated into α-, β-, or γ-$Re_2Si_2O_7$ matrix and densified at 1050° - 1200°C in 1 h using the FAST approach. Fiber push-out experiments were conducted and the measured sliding stress values are reported and discussed.

2.   EXPERIMENTAL

2.1 Precursor Synthesis and Characterization

Precursors to $Y_2Si_2O_7$ and $Ho_2Si_2O_7$ were made by adding colloidal silica to solutions of yttrium and holmium nitrate, as reported previously.[2] For the $Y_2Si_2O_7$ precursor, 12.7 g of $Y(NO_3) \cdot 6H_2O$ were dissolved in 50 mL of deionized water to make a solution of pH ~1. The sol pH was determined using a pH/ion meter (Corning, Inc., Corning NY). Silica (2.0 – 2.8 g) was added in the form of a sol with a pH of 10. For the $Ho_2Si_2O_7$ precursor, 14.7 g of $Ho(NO_3) \cdot 5H_2O$ were added to 50 mL of deionized water and 2 g of silica was then added. In some cases, 0.12 g of $LiNO_3$ was added to the Y and Ho-derived precursor. The mixtures were dried in an oven for 72 h to form $SiO_2/Y(OH)_3$ and $SiO_2/Ho(OH)_3$ and then heated at 1050°C in air. The subsequent powders were characterized by X-ray diffraction (XRD) with Cu-Kα radiation (Rotaflex, Rigaku Co., Tokyo, Japan).

2.2 Formation and Densification of Pellets

$SiO_2/Y(OH)_3$ and $SiO_2/Ho(OH)_3$ dispersions with and without Li dopant were dried and then heat treated at 1050°C for 1 h. Heat treated powders without Li formed α-$Y_2Si_2O_7$ and α-$Ho_2Si_2O_7$, while heat treated powders doped with Li formed β-$Y_2Si_2O_7$ and β-$Ho_2Si_2O_7$. Powders in the α and β phases were ball milled in isopropanol using alumina milling media. Polyvinyl butyral resin (3 vol%) was added as a binder. After milling, the powder samples were characterized by XRD to confirm the retention of the α and β phases. The milled slurry was separated and dried at 120°C for 18 h. The dried powder was uniaxially pressed into pellets and densified by the FAST approach at a pressure of 20 kN and at temperatures of 1050°, 1100°, and 1200°C. The dense pellets were ~20 mm in diameter and ~3 mm in height, with relative densities greater than 90%. Densities were measured using the Archimedes method.

2.3. Indentation and Characterization

The hardness of each of the sintered pellets was measured by Vickers indentation at loads of 100 g, 500g and 1000 g (Hardness Tester 1600-2007, Buehler, Lake Bluff, IL). Ten measurements were done for each load. The hardness values at loads corresponding to 100g, 500g and 1000g were very similar. To get a better statistics the hardness measured at 100g, 500g, and 1000 g was averaged for each sample and the standard deviation calculated. Indented samples were examined by scanning electron microscopes (SEM, Sirion and Quanta, FEI, Hillsboro, Oregon) and energy dispersive X-ray spectroscopy (EDS) operating at 5 - 20 kV. Foils for transmission electron microscopy (TEM, Philips CM200, FEI, Hillsboro, Oregon) were cut from areas below the Vickers indentations using a focused ion beam (FIB, Dual Beam DB 235 FEI, Hillsboro, Oregon) equipped with a micromanipulator (AutoProbe™ 200, Omniprobe, Dallas, Texas).

2.4 Fiber Push-Out

SCS-0 fibers were sandwiched between two green pellets and densified at 1050 °, 1100°, and 1200°C for 1 h in vacuum using the FAST approach with a force of 20 kN. An approximately 0.4 mm thick cross sectional specimen was prepared for fiber push-out studies; in this case, the fiber axis was oriented perpendicular to the polished surface. A fiber push-out testing apparatus (Process Equipment, Inc., Troy, OH) was used to obtain load displacement curves. The sliding stress was calculated based on a fiber diameter of 140 μm and the minimum applied load, which corresponds to the load when fiber sliding ceases.

3. RESULTS AND DISCUSSION

3.1 Densification of $Y_2Si_2O_7$ and $Ho_2Si_2O_7$

As reported previously, heat treatment of $SiO_2/Y(OH)_3$ and $SiO_2/Ho(OH)_3$ dispersions without Li at 1050°C formed α-$Y_2Si_2O_7$ and α-$Ho_2Si_2O_7$ powders.[2] Alternately, with Li doping, $SiO_2/Y(OH)_3$ and $SiO_2/Ho(OH)_3$ dispersions heat treated at 1050°C formed β-$Y_2Si_2O_7$ and β-$Ho_2Si_2O_7$. The α and β powders formed at 1050° C were uniaxially pressed into pellets and densified by the FAST process at a pressure of 20 kN and at temperatures of 1050°, 1100°, and 1200°C. Table 1 and Fig. 1 summarize the XRD results after densification. X-ray diffraction studies showed that α-$Y_2Si_2O_7$ pellets, densified at 1050°C, retained the α phase, whereas α-$Y_2Si_2O_7$ pellets densified at 1100°C transformed to β- $Y_2Si_2O_7$. In the case of Li-doped β-$Y_2Si_2O_7$ and β-$Ho_2Si_2O_7$ pellets, densification at 1200°C formed γ-$Y_2Si_2O_7$ and γ-$Ho_2Si_2O_7$. (Table 1, Fig. 1). The relative densities of the pellets were determined from the measured and the theoretical densities shown in parentheses. The relative densities were calculated as the percentage of the measured to the theoretical density. The α, β, and γ dense pellets formed by the FAST approach all had relative densities >90%.

Table 1. Formation of dense *Y$_2$Si$_2$O$_7$* and *Ho$_2$Si$_2$O$_7$* pellets using the FAST technique

Starting Powder	FAST Temperature (°C)	Final Phase After FAST	Density Measured (Theoretical)	Relative Density %
α-Y$_2$Si$_2$O$_7$	1050	α	3.8 (4.18)	91
α-Y$_2$Si$_2$O$_7$	1100	β	3.8 (4.02)	94
β-Y$_2$Si$_2$O$_7$ + Li	1200	γ	3.8 (4.04)	94
β-Ho$_2$Si$_2$O$_7$+Li	1200	γ	5.8 (6.30)	92

3.2 Hardness

SEM micrographs of α-, β-and γ-Y$_2$Si$_2$O$_7$ pellets after Vickers indentation are shown in Fig. 2. In addition to the pure rare-earth disilicate, SEM/EDS showed silica as a second phase in all cases. SEM/EDS showed silica-rich dark regions with almost no yttrium in the EDS spectrum and a grey region with very similar yttrium and silica intensities.[2] The experimental hardness values represent the sum of the hardnesses of the rare-earth disilicate and silica phases. The hardness of pure rare-earth disilicate was calculated using the rule of mixtures, assuming that the hardness of pure silica is 8.0 GPa.[23] Results of the experimental measurements and the theoretical determined hardness values for dense α- β- γ-Y$_2$Si$_2$O$_7$ and γ-Ho$_2$Si$_2$O$_7$ pellets are summarized in Table 2. The measured Vickers hardness values were very similar for α- β- γ-Y$_2$Si$_2$O$_7$ and γ-Ho$_2$Si$_2$O$_7$ (5.6-6.9 GPa) and were consistent with our previous reported γ-Y$_2$Si$_2$O$_7$ hardness value of 6.1 GPa for samples densified using a pressureless sintering technique at 1400°C / 8 h.

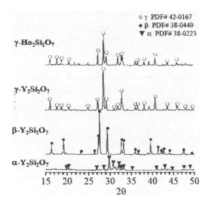

Figure 1. X-ray powder diffraction patterns of densified α, β, γ-Y$_2$Si$_2$O$_7$and γ-Ho$_2$Si$_2$O$_7$ densified at 1050°, 1100°, and 1200°C

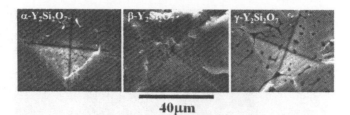

Figure 2. SEM micrographs of α- β- and γ-Y$_2$Si$_2$O$_7$ pellets after Vickers indentation

Table 2. Hardness of α- β- γ-Y$_2$Si$_2$O$_7$ and γ-Ho$_2$Si$_2$O$_7$ densified by FAST technique

Final phase after FAST	Experimental Density	Theoretical Density	Silica volume fraction	Re-silicate volume fraction	Experimental Hardness (GPa)	Calculated Hardness (GPa)
α-Y$_2$Si$_2$O$_7$	3.8	4.18	0.19	0.81	6.7±0.2	6.4±0.2
β-Y$_2$Si$_2$O$_7$	3.8	4.02	0.12	0.88	6.6±0.4	6.4±0.4
γ-Y$_2$Si$_2$O$_7$ + Li	3.8	4.04	0.13	0.87	7.0±0.4	6.9±0.4
γ-Ho$_2$Si$_2$O$_7$ +Li	5.8	6.30	0.12	0.88	5.9±0.2	5.6±0.2

3.3 Deformation Behavior under Indentation

As mentioned previously, the mechanical properties of γ-Y$_2$Si$_2$O$_7$ are comparable to monazite, which is soft and deforms plastically by multiple dislocation glide and twinning systems.[24-26] To study deformation mechanisms that may operate in α-, β-, and γ-Y$_2$Si$_2$O$_7$, TEM foils were machined beneath indented regions of sintered Y$_2$Si$_2$O$_7$ using FIB. A cross-sectional TEM image of an indent made on sintered γ-Y$_2$Si$_2$O$_7$ is shown in Figure 3. It shows extensive fracture and plastic deformation. Preliminary studies of the α- and β-Y$_2$Si$_2$O$_7$ polymorphs also indicated extensive deformation (Fig. 4). More detailed studies of these features are in progress.

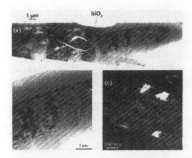

Figure 3. Plastic deformation under an indent made on sintered γ-Y$_2$Si$_2$O$_7$ (cross-sectional TEM image): (a) overview, bright field; (b) multiple slip bands/active dislocation glide, bright field; (c) (11 1̄) staking faults, dark field

Figure 4. Extensive deformation in α-Y$_2$Si$_2$O$_7$

3.4 Fiber Push-Out

Previous results[2] showed that SCS-0 fibers in dense γ-Ho$_2$Si$_2$O$_7$ and γ-Y$_2$Si$_2$O$_7$ matrixes debond and push-out with sliding stresses of 30-60 MPa, within the range reported for C, BN, and LaPO$_4$ coatings. [27-28] Similar to the observation made for LaPO$_4$/alumina composites, SEM showed smearing of the γ-Y$_2$Si$_2$O$_7$ matrix. Furthermore, as with the plastic deformation observed for the LaPO$_4$/alumina interface, TEM observation was consistent with intense deformation at the SCS-0 fiber/matrix interface. These results indicate that the γ-Y$_2$Si$_2$O$_7$ may function as a weak interface for toughening SiC/SiC composites. However, the γ-Y$_2$Si$_2$O$_7$ matrix forms at relatively high temperatures (>1400°C) and may cause fiber damage during coating processing. The α and β polymorphs form at relatively lower temperatures (<1200°C) minimizing the chance of fiber strength degradation. The fact that the α and β polymorphs have similar hardnesses as the γ polymorph implies that the α- and β-Y$_2$Si$_2$O$_7$ may work as well. Pushout measurements were done for SCS-0 fibers in α- and β-Y$_2$Si$_2$O$_7$ matrixes to measure the interface sliding stress. SCS-0 fibers were sandwiched between two green α-Y$_2$Si$_2$O$_7$ pellets and densified at 1050°C and 1100°C for 1 h using the FAST approach. For the SCS-0/α-Y$_2$Si$_2$O$_7$

composite densified at 1050°C, the Y$_2$Si$_2$O$_7$ matrix retained the α phase. However, the α-Y$_2$Si$_2$O$_7$ matrix transformed to β-Y$_2$Si$_2$O$_7$ in composites densified at 1100°C, which is lower than that previously reported.[2]

Fiber push-out results for SCS-0 fibers in the β-Y$_2$Si$_2$O$_7$ matrix are summarized in Figures 5 and 6. The push-out data is classified into 3 groups according to the extent of defocus of the optical micrographs after the fiber push-out. These groups consist of: a) pushed, b) pushed but with fiber damage and c) not pushed. The average sliding stress is plotted and tabulated in Fig. 6. Fibers that pushed with damage (b) showed similar average sliding stresses to fibers that pushed (a). The average sliding stress was 11.7 ± 3.3 and 14.4 ± 5.7 MPa for fibers that pushed and pushed with damage, respectively. SEM investigation of fibers that pushed with damage showed push-out in the front and back-end of the pushed out specimen, confirming fiber sliding after push-out.

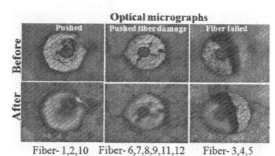

Figure 5. Optical micrographs of pushed SCS-0 fibers in the β-Y$_2$Si$_2$O$_7$ matrix

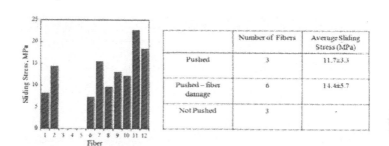

	Number of Fibers	Average Sliding Stress (MPa)
Pushed	3	11.7±3.3
Pushed – fiber damage	6	14.4±5.7
Not Pushed	3	-

Figure 6. Sliding stress of SCS fibers pushed-out in the β-Y$_2$Si$_2$O$_7$ matrix

The data for the tested fibers in the α-Y$_2$Si$_2$O$_7$ matrix was very erratic and was classified into pushed and unpushed sets (Fig. 7). The average sliding stress was 28.0 ± 3.8 MPa. For fibers that pushed, SEM analysis showed fiber sliding in the front end, but the back end showed very little or no sliding (Fig. 8). In the worst case scenario, the matrix was observed to be pulled along with the fiber, which indicates strong fiber-matrix bonding. CTE measurements of the α, β, and γ polymorphs of $Y_2Si_2O_7$ and $Ho_2Si_2O_7$ show a CTE of ~8 for the α phase.[29] The β and γ phases have a CTE of ~4.[29] The composites were formed at 1050°-1200°C. During cooling, the α-$Y_2Si_2O_7$, with a higher CTE, contracts to a greater extent and exerts a radial compressive stress on the fiber, which makes it more difficult to push. The compressive stresses will increase along the fiber length, thus making it progressively more difficult for the fibers to push out.

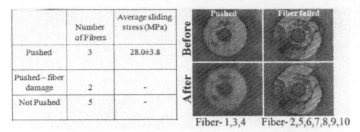

	Number of Fibers	Average sliding stress (MPa)
Pushed	3	28.0±3.8
Pushed – fiber damage	2	-
Not Pushed	5	-

Fiber- 1,3,4   Fiber- 2,5,6,7,8,9,10

Figure 7. Push-out data for SCS-0 fibers in α-$Y_2Si_2O_7$ matrix

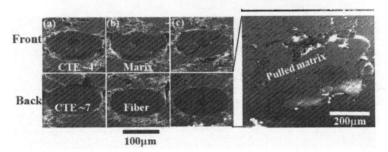

Figure 8. Scanning electron micrographs of pushed SCS-0 fibers in an α-$Y_2Si_2O_7$ matrix

3. SUMMARY AND CONCLUSIONS

Dense α-, β-, and γ-$Y_2Si_2O_7$ pellets were formed using the FAST technique at a pressure of 20 kN and temperatures of 1050° - 1200°C. Vickers hardness values for α-, β-, and γ-$Y_2Si_2O_7$ polymorphs were similar and were in the range (5.6-6.9 GPa). TEM analysis of samples directly beneath the indentations suggested that both extensive dislocation slip and fracture were active. Preliminary results showed that SCS-0 fibers in dense β- and γ-$Y_2Si_2O_7$ matrixes debonded and pushed-out with sliding stresses of 15 - 60 MPa, within the range reported for C, BN, and $LaPO_4$ coatings. Fibers in the α-$Y_2Si_2O_7$ matrix showed poor and erratic pushout behavior due to the radial compressive stresses that developed on the fibers as result of the differences in CTEs. Future and ongoing work involve processing rare-earth disilicate matrix SiC fiber composites.

REFERENCES

1. Kerans, R. J., Hay, R. S., Parthasarathy, T. A. & Cinibulk, M. K. Interface Design for Oxidation Resistant Ceramic Composites. *J. Am. Ceram. Soc.* **85**, 2599-2632 (2002).
2. Boakye, E. E., Mogilevsky, P., Hay, R. S. & Cinibulk, M. K. Rare-Earth Disilicates as Oxidation-Resistant Fiber Coatings for Silicon Carbide Ceramic Matrix Composites. *J Am. Ceram. Soc.* **94**, 1716-1724 (2011).
3. Escudero, A., Alba, M. D. & Becerro, A. I. Polymorphism in the Sc2Si2O7-Y2Si2O7 System. *J. Solid State Chem.* **180**, 1436-1445 (2007).
4. Escudero, A. & Beccerro, A. I. Stability of Low Temperature Polymorphs (y and a) of Lu-doped $Y_2Si_2O_7$. *J. Phys. Chem. Solids* **68**, 1348-1353 (2007).
5. Sun, Z., Zhou, Y. & Li, M. Low Temperature Synthesis and Sintering of g-Y2Si2O7. *J. Mat. Sci.* **21**, 1443-1450 (2006).
6. Sun, Z., Zhou, Y., Wang, J. & Li, M. g-Y2Si2O7, a Machinable Silicate Ceramic: Mechanical Properties and Machinability. *J. Am. Ceram. Soc.* **90**, 2535-2541 (2007).
7. Felsche, J. Polymorphism and Crystal data on the Rare-Earth Disilicates of the Type Re2Si2O7. *J. Less-Common Metals* **21**, 1-14 (1970).
8. Seifert, H. J. *et al.* Yttrium Silicate Coatings on Chemical Vapor Deposition-SiC-Precoated C/C-SiC: Thermodynamic Assessment and High-Temperature Investigation. *J. Am. Ceram. Soc.* **88**, 424-430 (2005).
9. Wang, J. Y., Zhou, Y. C. & Lin, Z. J. Mechanical Properties and Atomistic Deformation Mechanism of g-Y2Si2O7 from First-Principles Investigations. *Acta mat.* **55**, 6019-6026 (2007).
10. Keller, K. A. *et al.* Effectiveness of Monazite Coatings in Oxide/Oxide Composites After Long Term Exposure at High Temperature. *J. Am. Ceram. Soc.* **86**, 325-332 (2003).
11. Boakye, E. E. *et al.* Monazite Coatings on SiC Fibers I: Fiber Strength and Thermal Stability. *J. Am. Ceram. Soc.* **89**, 3475-3480 (2006).
12. Mogilevsky, P., Boakye, E. E., Hay, R. S., Welter, J. & Kerans, R. J. Monazite Coatings on SiC Fiber Tows II: Oxidation Protection. *J. Am. Ceram. Soc.* **89**, 3481-3490 (2006).
13. Cinibulk, M. K., Fair, G. E. & Kerans, R. J. High-Temperature Stability of Lanthanum Orthophosphate (Monazite) on Silicon Carbide at Low Oxygen Partial Pressure. *Am. Ceram. Soc.* **91**, 2290–2297 (2008).
14. Cupid, D. M. & Seifert, H. J. Thermodynamic Calculations and Phase Stabilities in the Y-Si-C-O System. *J. Phase Equil. Diff.* **28**, 90-100 (2007).
15. Sun, Z., Zhou, Y., Wang, J. & Li, M. Thermal Properties and Thermal Shock Resistance of g-$Y_2Si_2O_7$. *J. Am. Ceram. Soc.* **91**, 2623-2629 (2008).
16. Becerro, A. I., Naranjo, M., Perdigon, A. C. & Trillo, J. M. Hydrothermal Chemistry of Silicates: Low Temperature Synthesis of y-Yttrium Disilicate. *J. Am. Ceram. Soc.* **86**, 1592-1594 (2003).
17. Ranganathan, V. & Klein, L. C. Sol-gel Synthesis of Erbium-doped Yttrium Silicate Glass-ceramics. *J. Non-Cryst. Solids* **354**, 3567-3571 (2008).
18. Zhou, P., Yu, X., Yang, S. & Gao, W. Synthesis of $Y_2Si_2O_7$:Eu Nanocrystal and its Optical Properties. *J. Lumin.* **124**, 241-244 (2007).

19   Diaz, M. G.-C. I., Mello-Castanho, S., Moya, J. S. & Rodriguez, M. A. Synthesis of Nanocrystalline Yttrium Dislicate Powder by Sol-Gel Method. *J. Non. Cryst. Sol.* **289**, 151-154 (2001).

20   Diaz, M. Synthesis, Thermal Evolution, and Luminescence Properties of Yttrium Disilicate Host Matrix. *Chem. Mater.* **17**, 1774-1782 (2005).

21   Maier, N., Rixecker, G. & Nickel, K. G. Formation and Stability of Gd, Y, Yb and Lu Disilicates and their Solid Solutions. *J. Solid State Chem.* **179**, 1630-1635 (2006).

22   Parmentier, J. Phase Transformations in Gel-Derived and Mixed-Powder-Derived Yttrium Disilicate, $Y_2Si_2O_7$, by X-Ray Diffraction and ^{29}Si MAS NMR. *J. Solid State Chem.* **149**, 16-20 (2000).

23   Michel, M. D., Serbena, F. C. & Lepienski, C. M. Effect of Temperature on Hardness and Indentation Cracking of Fused Silica. *J. Non-Cryst. Solids* **352**, 3550-3555 (2006).

24   Hay, R. S. Monazite and Scheelite Deformation Mechanisms. *Ceram. Eng. Sci. Proc.* **21**, 203-218 (2000).

25   Hay, R. S. (120) and (12$\underline{2}$) Monazite Deformation Twins. *Acta mater.* **51**, 5255-5262 (2003).

26   Hay, R. S. Climb-Dissociated Dislocations in Monazite. *J. Am. Ceram. Soc.* **87**, 1149-1152 (2004).

27   Rebillat, F. *et al.* Interfacial Bond Strength in SiC/C/SiC Composite Materials, as Studied by Single-Fiber Push-Out Tests. *J. Am. Ceram. Soc.* **81**, 965-978 (1998).

28   Morgan, P. E. D. & Marshall, D. B. Ceramic Composites of Monazite and Alumina. *J. Am. Ceram. Soc.* **78**, 1553-1563 (1995).

29   T. Key, Presley, Boakye, E. E. & Hay, R. S. in *36th International Conference and Exposition on Advanced Ceramics and Composites.*

REACTION BONDED SI/SIB$_6$: EFFECT OF CARBON ADDITIONS ON COMPOSITION AND PROPERTIES.

S. Salamone, M. K. Aghajanian, and O. Spriggs
M Cubed Technologies, Inc.
1 Tralee Industrial Park
Newark, DE 19711

S. E. Horner
PM Soldier Protection & Individual Equipment
Haymarket, VA 20169

ABSTRACT

Silicon Hexaboride (SiB$_6$) is an interesting ceramic compound because of its high hardness and low density. With a density of 2.43 g/cc (lower than B$_4$C) and a hardness equivalent to SiC, composites made from this material should exhibit enhanced properties. Several Si/SiB$_6$ composites were fabricated using a reaction bonding technique. This is a flexible, low temperature (as compared to sintering) process that is well suited for making composites with varying amounts of secondary phases, via in-situ formation. X-ray diffraction has revealed the formation of B$_4$C and SiC phases in composites of reaction bonded SiB$_6$. The ratio of SiB$_6$ to B$_4$C and SiC can be changed with the addition of small quantities of carbon. The compositional changes also affect the physical properties such as density and Young's modulus and the mechanical properties such as hardness. These physical properties are related to the final composition (e.g. remaining SiB$_6$, B$_4$C, SiC and silicon content) of the composite.

INTRODUCTION

The search for materials with low densities and attractive properties has led researchers to boride compounds, namely SiB$_6$ and related composites. Silicon hexaboride has a very low density of 2.43 g/cc and it has been reported that the hardness and electrical conductivity are promising.[1,2] Applications from armor to high temperature thermoelectric and structural products have been investigated.[3-5] However, conventional processing methods such as hot pressing and pressure-less sintering requires complex equipment and high temperatures to achieve relative densities in the range where these properties can be utilized.[2]

Reaction bonding enables one to form composites at relatively low temperatures (<1500°C) and the process also lends itself to infiltrating complex shapes and sizes.[6] Since there is very low (volume) dimensional change, unlike conventional sintering processes, parts can be manufactured to near net shape. This reduces the risk of warping and distortions that can arise through diffusion driven mechanisms.

Reaction bonded composites are produced by starting with a porous preform of ceramic powder and carbon. The preform is transferred to a vacuum furnace and placed in contact with silicon metal. The carbon incorporated into the preform reacts with the infiltrated molten Si to yield SiC. This product, along with the initial particles and other reaction formed phases are bound together in an interconnected ceramic network filled with residual silicon metal. Once cooled, the resultant component is a dense highly loaded ceramic-metallic composite.[7]

The present study produced reaction bonded SiB$_6$ samples with various low levels of added carbon to affect the composition and microstructure of the final composite. The physical, mechanical and microstructural properties were characterized to find a relationship between processing and properties.

EXPERIMENTAL PROCEDURE

Commercially available SiB$_6$ powder was used as the base material for all the experimental samples investigated. Compacts consisting of the starting powder were combined with specific levels of additional carbon. Preforms were consolidated using these SiB$_6$/C mixtures, and then infiltrated in a vacuum furnace with molten Si to yield a Si/ SiB$_6$ ceramic composite.

The phase identification of the initial powder was performed using X-ray diffraction and the average particle size and distribution were measured by a Microtrac S3500 particle size analyzer. The physical properties of the infiltrated composites were measured using several common techniques summarized in Table I. All the microstructures were characterized by examining polished surfaces using a JEOL JSM-6400 Scanning Electron Microscope. The scanning electron microscope (SEM) images were taken in Back-Scattered Mode to differentiate the phases present. Elemental analysis (EDS) was performed in the same SEM after taking the images. The Hardness values (Knoop Indenter) were measured on a Shimadzu-2000 hardness tester.

Samples were sent to an independent laboratory to identify compositional changes occurring over the various thermal conditions. The instrument for detecting the different phases present was a Bruker D4 diffractometer with Cu radiation at 45KV/40mA.

Table I: Summary of Properties and Techniques Used to Quantify the Various Composites.

Property	Technique	Standard
Density	Immersion	ASTM B 311
Elastic Modulus	Ultrasonic Pulse Echo	ASTM D 2845
Elemental Analysis	EDS	
Phase Analysis	Powder XRD	
Knoop Hardness	Indentation	

RESULTS AND DISCUSSION

The powder (measured average particle size of 9.8 μm) was characterized using XRD analysis to determine the phase (or phases) present. The resultant diffraction pattern is shown in Figure 1, superimposed with the stick pattern representing the identified phase. The only phase detected was an off-stoichiometry SiB$_{6-x}$, possessing an orthorhombic structure, which is consistent with the literature.[8]

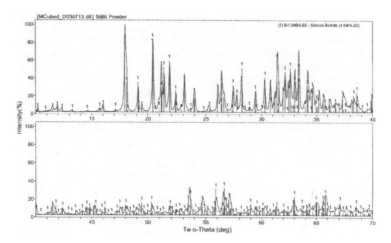

Figure 1: Diffraction pattern of SiB₆ powder along with the stick pattern for SiB₆

Small additions of carbon were added to the silicon hexaboride preforms and subsequently infiltrated with silicon metal. The final densities of the infiltrated composites as a function of initial carbon content are represented in Figure 2. Each point represents one sample. The density increases with increasing carbon content. The magnitude is small, but the trend is consistent. This indicates the formation of a higher density phase.

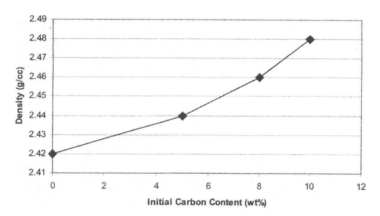

Figure 2: Density of Si/ SiB₆ composites as a function of weight percent of initial carbon

It has been suggested by several authors [4,9] that the silicon boride (SiB₆) phase reacts with free carbon (C) to form silicon carbide (SiC) and boron carbide (B₄C) according to the following reaction:

$$2SiB_6 + 5C = 2SiC + 3B_4C \qquad (1)$$

Powder diffraction was performed on these samples in order to examine the formation of phases. These results are shown in Table II. Rietveld analysis of the phase compositions determined that as the carbon content is increased, the SiB$_6$ phase breaks down and the boron reacts with the excess carbon to form B$_4$C. A small amount of SiC is also formed, but is relatively minor when compared to the boron carbide. An interesting result is that the silicon metal content appears to increase as more carbon is added to the system. There are two plausible explanations for this phenomenon. It is postulated that the carbon is depleted in the reaction with the B$_4$C formation and very little remains to react with the silicon to form silicon carbide. That would leave an excess of silicon, which could account for the increase in free silicon and the rather minor amount of SiC that is formed. The other is simply that as the low density phase reacts to form phases of higher density, there is a volume shrinkage and the infiltrating silicon fills the void.

Table II: Quantitative XRD Analysis of Reaction Bonded SiB$_6$ Composite

	XRD Analysis (wt %)			
	SiB$_6$	SiB$_6$ (~5% C)	SiB$_6$ (~8% C)	SiB$_6$ (~10% C)
Si	33.2 ± 0.3%	34.0 ± 0.5%	35.2 ± 0.6%	37.3 ± 0.9%
SiB$_{6-x}$	59.1 ± 0.7%	41.1 ± 0.7%	25.9 ± 0.6%	18.0 ± 0.5%
SiC (3C)	0.7 ± 0.1%	2.3 ± 0.1%	8.0 ± 0.2%	6.4 ± 0.2%
B$_4$C	7.0 ± 0.5%	22.6 ± 0.6%	30.9 ± 0.7%	38.3 ± 2.4%

The scanning electron microscope (SEM) images were taken in Back-Scattered Mode to differentiate the phases present (compositional differences) - e.g., Si metal (brightest phase), SiB$_6$ (intermediate/gray phase), and B$_4$C (dark phase).

Figure 3 is an image of a Si/SiB$_6$ composite sample with no additional carbon added to the preform. EDAX (elemental analysis) was performed on this sample to aid in the identification of the various phases. This technique, coupled with powder X-Ray diffraction, is an effective way to locate the different phases formed during infiltration. The various regions analyzed are labeled A, B and C on the image. The elemental spectra are shown in Figure 4 for each region along with the calculated weight and atomic ratios of the elements detected.

Region A should be the SiB$_6$ particles. The elemental analysis reveals a large amount of boron present with lower levels of silicon and carbon. The carbon peak most likely corresponds to the SiC present. Region B has high levels of Si present which indicates it consists of the infiltration metal.

The darkest region (C) contains high levels of boron and carbon. This is most likely the B_4C phase. Of note, complete infiltration was achieved leaving samples porosity free.

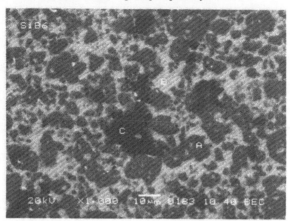

Figure 3: SEM image (B.S.) of the Si/SiB₆ composite without additional carbon.

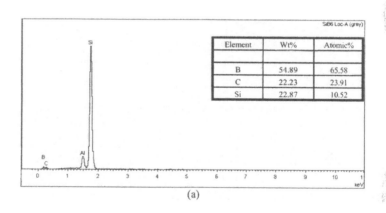

Element	Wt%	Atomic%
B	54.89	65.58
C	22.23	23.91
Si	22.87	10.52

(a)

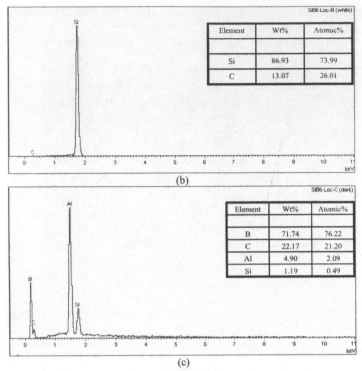

Figure 4: Elemental analysis from regions (a) A, (b) B, and (c) C

Microstructural analysis of the composites reveals spatial differences in the way the silicon phase is distributed throughout the sample. Although the XRD analysis has shown that the silicon content is similar, the microstructure suggests that the silicon metal distribution changes. In the 5% carbon sample the metal appears uniformly dispersed, but as the carbon content is increased and more reactions occur, the metal phase is more finely distributed and there are regions of high concentrations (islands) of the metal phase. Concomitantly, the ceramic phase is more interconnected and the B₄C regions appear smaller and more dispersed. Figure 5 shows this evolution in the microstructure as the carbon content is increased from 5 to 10 weight percent.

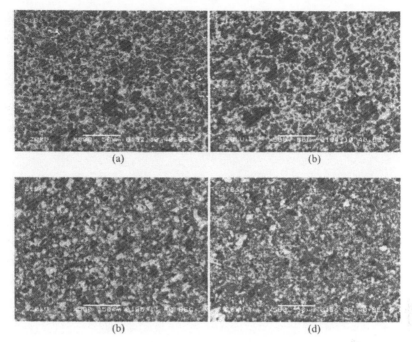

(a)   (b)

(b)   (d)

Figure 5: Comparison of Microstructures for (a) Si/SiB$_6$ (b) Si/SiB$_6$ + 5% C,
(c) Si/SiB$_6$ + 8% C and (d) Si/SiB$_6$ +10% C

      The Knoop Hardness was measured as a function of carbon content at two different indenter loads. The hardness value increases when carbon is added to the system as shown in Figure 6. This trend is followed for both 0.5 and 2 kg loading conditions. The graph shows that these composites exhibit the well documented, indentation size effect (ISE), where the hardness decreases as the indentation load increases.[10] The decrease in hardness at 10 wt% initial carbon is consistent with both loading conditions. This might be due to the increase in final silicon, as shown by the XRD analysis. Comparisons to other materials are made at the ASTM recommended load of 2 kg, for the Knoop indenter.[11] Each point is an average of five hardness measurements.

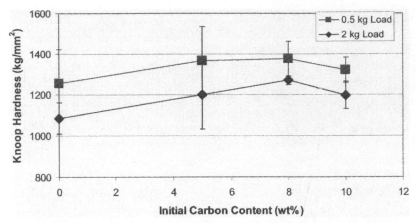

Figure 6: Knoop Hardness as a function of initial carbon content and applied load.

A comparison of the physical and mechanical properties measured is given in Table III. As the density of the reaction bonded SiB$_6$ composite increases (due to addition of carbon) the modulus increases only slightly. The measured hardness value increases by about ten percent with the addition of 10 wt% carbon. The Young's modulus and Knoop Hardness of the RBSiB6 + 10%C is comparable to the values of a standard reaction bonded SiC (RBSiC) with 12 μm particles and roughly the same amount of free silicon metal. However, the density of the RBSiB$_6$ is approximately 18% lower than the RBSiC composite.

Table III: Properties of Reaction Bonded Materials and Constituent Phases

Material	Density (g/cc)	Young's Modulus (GPa)	Knoop Hardness (kg/mm^2) (2 kg Load)
Si	2.34 [12]	113 [12]	849 (200g Load)*
SiC	3.21 [13]	450 [13]	1905
B$_4$C	2.52 [13]	480 [13]	2069
RBBC (~15% Si)	2.56	380	1556
RBSiC (~30% Si)	2.96	335	1184
RBSiB6 (~33% Si)	2.42	290	1085
RBSiB6 + 10%C (~37% Si)	2.48	310	1198

* Si indent load = 200g due to excessive damage @ 2 kg
Note: All Hardness values measured by M Cubed.
RBBC = Reaction Bonded Boron Carbide

SUMMARY

The drive towards low density, high performance materials will continue and the reaction bonded Si/SiB$_6$ composites are an intriguing candidate for further study. RBSiB$_6$ composites were successfully fabricated with various amounts of carbon additions. It was found that as carbon is added to the SiB$_6$ system, B$_4$C is formed and is more finely distributed as the carbon additions increase. The densities of RBSiB$_6$ ceramics are lower than standard reaction bonded SiC and B$_4$C composites yet the hardness values are comparable to certain RBSiC materials.

ACKNOWLEDGEMENT

This work was funded by US Army PM- Soldier Protection & Individual Equipment under contract number W91CRB-11-C-0085.

REFERENCES

[1]J. Matsushita, et al., "Preparation and characterization of silicon hexaboride ceramics", *Processing and Fabrication of Advanced Materials IX; Proceedings of the Symposium*, St. Louis, MO, 17-24, (2001).

[2]D.W. Lee, et al., "Thermoelectric properties of silicon hexaboride prepared by spark plasma sintering method", *J. Ceram. Proc. Res.*, **3**, [3], 182-185 (2002).

[3]M.W. Lindley, et al, "Some New Potential Ceramic-Metal Armor Materials Fabricated by Liquid Metal Infiltration", AMMRC, AD-769 742 (1973).

[4]G.C. Hwang and J. Matsushita, "Preparation of Si infiltrated SiB$_6$-TiB$_2$ composites", *J. Ceram. Proc. Res.*, **11**, [1], 1-5 (2010).

[5]J. Matsushita, et al., "Oxidation behavior of a silicon boride composite", *J. Ceram. Proci. Res.*, **5** [2], 133-35 (2004).

[6]P.G. Karandikar, M.K. Aghajanian and B.N. Morgan, "Complex, Net-Shape Ceramic Composite Components for Structural, Lithography, Mirror and Armor Applications", *Ceram. Eng. Sci. Proc.*, **24** [4], 561-6 (2003).

[7]M.K. Aghajanian, B.N. Morgan, J.R. Singh, J. Mears, R.A. Wolffe, "A New Family of Reaction Bonded Ceramics for Armor Applications", in Ceramic Armor Materials by Design, *Ceramic Transactions*, **134**, J. W. McCauley et al. editors, 527-40 (2002).

[8]M. Vlasse, et al., "The Crystal Structure of SiB$_6$", *J. Solid State Chem.*, **63**, 31-45 (1986).

[9]G.C. Hwang and J. Matsushita, "Fabrication and Properties of SiB$_6$-B$_4$C with Phenolic Resin as a Carbon Source", *J. Mater. Sci. Technol.*, **24**, [1], 102-4 (2008).

[10]J.J. Swab, "Recommendations for Determining the Hardness of Armor Ceramics", *Int. J. Appl. Ceram. Technol.*, **1** [3] 219-25 (2004).

[11]ASTM C1326-96a, "Standard Test Method for Knoop Indentation Hardness of Advanced Ceramics" 2003 Annual Book of ASTM Standards, Vol. 15.01

[12]*Metals Handbook: Desk Addition*, ASM International, Metals Park, OH, (1985).

[13]*Engineered Materials Handbook, Vol. 4, Ceramics and Glasses*, ASM International, Metals Park, OH, (1991).

KINETICS OF PASSIVE OXIDATION OF HI-NICALON-S SiC FIBERS IN WET AIR: RELATIONSHIPS BETWEEN SiO₂ SCALE THICKNESS, CRYSTALLIZATION, AND FIBER STRENGTH

R. S. Hay, G. E. Fair
Air Force Research Laboratory
Materials and Manufacturing Directorate, WPAFB, OH

A. Hart, S. Potticary
U. Cincinnati, Cincinnati, OH

R. Bouffioux
New Mexico Tech. U., Socorro, NM

ABSTRACT

The strengths of Hi-Nicalon[TM]-S SiC fibers were measured after oxidation in wet air between 700° and 1300°C. The oxidation and scale crystallization kinetics were also measured. Thicknesses of amorphous and crystalline scale were measured by TEM. Oxidation initially produces an amorphous scale that starts to crystallize to cristobalite and tridymite in 100 hours at 1000°C or in one hour at 1300°C. Crystallization kinetics for oxidation in wet air were slightly slower than those for dry air. The activation energy of 249 kJ/mol for parabolic oxidation to uncrystallized SiO₂ scale in wet air was indistinguishable from that for dry air oxidation, but the pre-exponential factor was ~2x higher. SiC fiber strength changes with oxidation in dry and wet air were very similar. The fiber strength increased by approximately 10% for SiO₂ scale thickness up to ~100 nm, and decreased for thicker scales. No significant strength degradation was observed for amorphous scales. All fibers with significantly degraded strength had crystallized or partially crystallized scales.

INTRODUCTION

SiC fiber strength is affected by oxidation. Fiber strength defines the maximum attainable CMC strength.[1] Oxidation is affected by fiber impurities, particularly alkali and alkali earths, that increase oxidation rates, reduce scale viscosity, and lower temperatures for scale crystallization.[2-3] Moisture has similar effects.[4-9] Some work suggests that oxidation of SiC fibers reduces their strength.[10-17] However, recent work has shown that thin silica scales (< 100 nm) actually increase SiC fiber strength,[18-21] as might be expected from surface flaw healing and residual compressive stress in the scales. Ambiguous effects of SiC oxidation on strength have also been observed for bulk material.[22-23]

Preliminary data and analysis for the oxidation kinetics, scale crystallization kinetics, and strength of Hi-Nicalon[TM]-S SiC (β-SiC) fiber oxidized in wet air are presented. This paper builds on a thorough description and analysis of Hi-Nicalon[TM]-S SiC fiber oxidation and scale crystallization kinetics in dry air.[20-21] Hi-Nicalon[TM]-S fiber was chosen because it has near-stoichiometric SiC composition (~1 at% oxygen and ~2 at% carbon),[24] and the smoothest surface of currently available SiC fibers.[25] The fiber properties are described in several publications.[11, 24, 26-30]

EXPERIMENTS

Hi-Nicalon[TM]-S fibers have a PVA (polyvinyl alcohol) sizing. To avoid contamination of the fiber surface by sizing impurities, the sizing was removed by two sequential dissolutions in boiling distilled deionized water in a Pyrex glass beaker for one hour.[20-21] The desized fibers were oxidized in flowing wet air by bubbling dry air through distilled water at room temperature (24°C). Water saturation at this temperature yields a water/air molar ratio of 0.03. Fibers were oxidized in an alumina muffle tube furnace with an alumina boat dedicated to these experiments. Both the muffle tube and boat were baked-out at 1540°C for 4 hours in laboratory air prior to use. These bake-outs have been shown to be necessary to prevent contamination by alumina impurities.[31] Fiber oxidation was done at 700 to 1300°C for times up to 100 hours. A total of 20 different heat-treatments were done (Table I). Heat-up and cool-down rates of 10°C/minute were used.

The strengths of the oxidized fibers were measured by tensile testing at least 30 filaments, using published methods.[32] The average and Weibull characteristic value for failure stress were calculated, along with the Weibull modulus. Strengths were calculated using both the original (r_i) and final SiC radius after oxidation (r_f) (Fig. 1). The average fiber diameter measured by optical microscopy and SEM was 12.1 μm.

The uniformity SiO_2 scale uniformity was characterized using reflected light interference fringes observed by optical microscopy. Cross-sectional TEM specimens were prepared from oxidized fibers by published methods.[33-34] TEM sections were ion-milled at 5 kV and examined using a 200 kV Phillips LaB_6-filament TEM and a 300 kV FEI Titan TEM. SiO_2 oxidation product thickness (x) and cracking, fraction of crystallized scale (f), and microstructures of the SiO_2 scale were characterized for a minimum of five filaments, and in many cases more than ten. The crystallized SiO_2 phase (tridymite or cristobalite) was identified from selected area electron diffraction patterns in some scales.

Fig. 1. Diagram showing initial (r_i) and final (r_f) SiC thickness after oxidation and SiO_2 thickness x.

RESULTS AND DISCUSSION

Results for twenty wet oxidation experiments are shown in Table I. All reported strengths are calculated from the final SiC radius present after oxidation.

#	# of Obs.	T (°C)	t (hrs)	x (nm) amorphous	x (nm) crystalline	f	Strength (GPa)	Weib. Char. (GPa)	Weibull Modulus	Comments
1	29	700	100	17.7 ± 5.7	—	0	2.64	2.93	7.71	
2	38	800	1	7.2 ± 1.6	—	0	2.98	3.15	5.89	
3	22	800	10	20.1 ± 11.3	—	0	2.80	3.13	2.96	1 outlier – 0.66 GPa
4	43	800	100	74.1 ± 11.1	—	0	3.05	3.18	4.90	
5	23	900	1	15.9 ± 4.3	—	0	2.77	2.86	5.64	
6	45	900	10	72.0 ± 5.6	—	0	2.73	2.87	6.98	
7	15	900	100	216. ± 8.	—	0	2.77	3.17	4.11	
8	30	1000	1	66.0 ± 5.4	—	0.036	3.14	3.25	6.67	
9	66	1000	10	198. ± 10.	—	0.04	3.15	3.28	6.36	
10	23	1000	30	333. ± 15.	—	0.043	2.82	2.93	7.47	
11	30	1000	100	859. ± 29.	841.. ± 62.	0.15	2.44	2.83	2.70	
12	15	1050	100	—	1470. ± 110.	1	<1	<1	—	6 Tridymite, 1 Cristobalite
13	39	1100	1	167. ± 7.	—	0	2.40	2.77	2.46	
14	22	1100	3	214. ± 17.	—	0	2.82	3.14	5.55	
15	75	1100	10	652. ± 24.	623. ± 53.	0.44	2.14	2.37	3.05	11 Tridymite, 4 Cristobalite
16	5	1100	100	—	1590. ± 79.	1	<1	<1	—	1 Tridymite
17	6	1200	1	274. ± 16.	257. ± 41.	0.2	2.38	2.71	4.06	1 Tridymite
18	5	1200	10	—	790.±90.	1	2.07	2.08	4.68	2 Cristobalite
19	11	1200	100	—	2390. ± 200.	1	<1	<1	—	1 Tridymite
20	13	1300	1	601.±49.	456. ± 17.	0.77	2.55	2.56	4.93	1 Tridymite, 1 Cristobalite

Table I. Hi-Nicalon^{TM}-S wet-air oxidation experiments. f is the fraction of scale that is crystallized.

Fiber Strength

The relationship between oxide scale thickness (x) and the Weibull characteristic strength of the fibers is shown in figure 2. As-received fibers had average strengths of 2.85 GPa and Weibull characteristic strengths of 3.0 GPa; this is plotted at $x = 1$ nm on the logarithmic scale thickness axis. Tensile strengths could not be reliably measured for some fibers with thick, crystallized scales. Extensive experience with filament tensile testing suggests that these fibers have tensile strengths less than 1 GPa, as indicated in Table I and figure 2. Fiber strengths in figure 2 are shown as a function of scale thickness after oxidation at various times at 700 - 1300°C for both dry and wet air oxidation. Dry air is denoted by discs and wet air by circles. Amorphous scales are red, fully crystallized scales are purple, and partially crystallized scales have colors between red and purple, with a hue proportional to the fraction of scale crystallized (f) in Table I. The strengths are plotted for the final SiC fiber radius (r_f) after oxidation, which can be calculated from the original fiber radius (r_i) and the oxide scale thickness (x) from:

$$r_f = [x^2 (\Omega_{SiC}^2/\Omega_{SiO2}^2 - \Omega_{SiC}/\Omega_{SiO2}) + r_i^2]^{1/2} - (\Omega_{SiC}/\Omega_{SiO2})x \qquad [1]$$

where Ω_{SiC} and Ω_{SiO2} are the molar volumes for SiC and SiO$_2$, respectively. The load carried by the SiO$_2$ scale, which has a modulus less than 1/5 that of SiC, is insignificant and beneath measurement error.

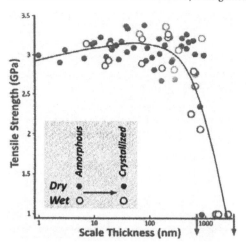

Fig. 2 *Weibull characteristic fiber strength vs. oxide scale thickness.*

Fig. 3 *TEM micrograph of SiO$_2$ crystallization after oxidation in wet air for 10 h at 1100°C.*

Fiber strength increased about 10% for thin oxide scales, with a weakly defined maximum near SiO$_2$ thicknesses of ~50 – 100 nm. As observed for dry air oxidation,[21] there was no relationship between oxidation temperature and fiber strength; scales of similar thickness that formed in a short time at high temperature or a long time at low temperature had similar effects on fiber strength. No significant strength degradation was observed for fibers with amorphous scales. All fibers with Weibull characteristic strengths <2.75 GPa had crystallized or partially crystallized scales (Fig. 2). However, there were some fibers with crystallized or partially crystallized scales that were not significantly degraded in strength. Detailed analysis of Weibull parameters as well as a discussion of possible strengthening and degradation mechanisms will be done in future papers covering dry air, wet air, and active oxidation experiments.

Crystallization

As in dry air oxidation,[20-21] crystallization always nucleated at the scale surface, and growth was comparatively rapid parallel to the surface than through thickness (Fig. 3). EDS measurements for dry air oxidation suggest some carbon incorporated in the amorphous scale was later rejected during crystallization.[20-21] We have not yet checked this for wet-air oxidation.

SiO$_2$ crystallized to cristobalite and various tridymite polymorphs at 1000° - 1300°C (Fig. 4, Table I). Tridymite dominated. This was also observed in dry air oxidation.[20-21] The relative abundance of cristobalite increased at higher temperatures (Table I), as observed in other studies.[35-37] Crystalline silica scales were thinner than corresponding amorphous scales, as observed previously,[38] and the thicknesses of crystalline scales were more variable (Table I). Nucleation was not synchronous, so such variation could be caused by variation in the amount of time spent in amorphous and crystalline states.

Crystallized SiO$_2$ scales often cracked (Fig. 4). Debond cracks between SiO$_2$ scale and SiC were also common. Cracked crystalline scales were previously observed Hi-Nicalon^{TM} and Hi-Nicalon^{TM}-S,[20-21] and were suggested to cause lower fiber strengths.[10, 39] Unlike amorphous SiO$_2$, cristobalite and tridymite have larger coefficients of thermal expansion than SiC, particularly near room temperature.[40-41] Cracks are assumed to form from thermal stress during cooling, and from volume contraction during tridymite phase transformations and the $\alpha \rightarrow \beta$ cristobalite phase transformation.

Thick crystalline scales had wide aperture cracks parallel with the fiber axis that clearly formed during oxidation (Fig. 4). They are inferred to be caused by tensile hoop growth stress that develops during oxidation of cylindrical substrates, which in turn is caused by the 2.2× volume expansion for SiC oxidation.[42-48] When these cracks form the scale is no longer passivating; they are a short-circuit path for O_2 ingress and SiC oxidation rates are increased underneath them.[45] Other cracks, generally with narrower apertures, formed perpendicular to the fiber axis, as observed for dry air oxidation.[20-21]

Kolmogorov-Johnson-Mehl-Avrami (KJMA) analysis was used for silica crystallization kinetics:[49-50]

$$f = 1 - \exp[-Kt^n] \qquad [2]$$
$$K = K_o \exp[-Q/RT] \qquad [3]$$

where f is the fraction crystallized (Table I), K is a rate constant, t is time, n is a the time growth exponent, K_o is a pre-exponential factor and Q an activation energy for growth, and RT has the usual meaning in an Arrhenius expression. A parameter best fit for all t, T, and f yields:

$$Q = 487 \text{ kJ/mol} \qquad [4a]$$
$$K_o = 5 \times 10^{10} \qquad [4b]$$
$$n = 1.6 \qquad [4c]$$

The parameter fit is shown in figure 5, along with the data and fit found previously for dry air oxidation.[20-21] The wet air activation energy Q is slightly lower than that found for dry air (514 kJ/mol), K_o is 40× lower, and the time growth exponents (n) are similar.[20-21] Crystallization kinetics were slightly slower in wet air than in dry air, particularly at high temperatures (Fig. 5). A growth exponent n of ~1.5 is diagnostic of three-dimensional growth from site-saturated nucleation.[50]

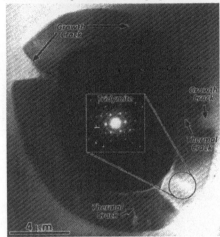

Fig. 4 *TEM micrograph of SiO_2 crystallization after oxidation in wet air for 100 h at 1200°C. Thermal and growth cracks are identified. An inset shows a selected area electron diffraction pattern for tridymite.*

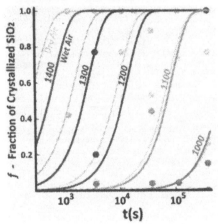

$$t(s)$$

Fig. 5 *Avrami plot of crystallization kinetic data from Table I. Wet air data and parameter fit is in bold; dry air data and parameter fit from earlier work is faded.*

Oxidation Kinetics – Amorphous Scale

No cracks were observed in amorphous scales, even in scales nearly a micron thick. Thick scales were uniform; thin scales had higher relative variability in thickness (Table I). This was also observed for dry air oxidation.[20-21]

Passive oxidation kinetics for SiC in wet air were analyzed using Deal-Grove kinetics for flat plates, with diffusion of molecular O_2 through SiO_2 as the rate limiting step in the parabolic regime.[9, 51] Flat plate geometry was shown to be accurate for scale thicknesses < 1 μm on a 6.5 μm radius fiber.[20-21] A more detailed description of the analysis methods that includes corrections for scale formed during heat-up and cool-down is given elsewhere.[20-21]

The thickness of the SiO_2 scale (x) described by Deal-Grove kinetics for the flat plate geometry is:

$$dx/dt = B/(A + 2x) \qquad [5]$$

$$A = A_0 \exp[-Q_a/RT] \qquad [6]$$
$$B = B_0 \exp[-Q_b/RT] \qquad [7]$$

where A_0 and B_0 are constants and Q_a and Q_b are activation energies. B is the parabolic rate constant and B/A is the linear rate constant. For an initial SiO_2 thickness of x_i, the solution to [5] is:

$$x = \tfrac{1}{2}A\{[1 + (t + \tau)/(A^2/4B)]^{1/2} - 1\} \qquad [8]$$
$$\tau = (x_i^2 + Ax_i)/B \qquad [9]$$

where τ is a time shift that corrects for presence of an initial oxide layer. For long times, [8] becomes the simple expression for parabolic oxidation kinetics:

$$x^2 = B\,t \qquad [10]$$

The best fit for "Deal-Grove" parameters for oxidation kinetics for amorphous SiO_2 scale formation for Hi-Nicalon[TM]-S fiber in wet air are:

$$A_0 = 8.1 \times 10^{-4}\ m \qquad [11a]$$
$$Q_a = 108\ kJ/mol \qquad [11b]$$
$$B_0 = 2.2 \times 10^{-8}\ m^2/s \qquad [11c]$$
$$Q_b\ (parabolic) = 249\ kJ/mol \qquad [11d]$$
$$Q_{b/a}\ (linear) = 141\ kJ/mol \qquad [11e]$$

As found for dry air oxidation, the fit had strong convergence to the Q_b value, but was less sensitive to the other parameters, particularly Q_a and A_0. Data and the parameter fit are plotted in figure 6, along with predicted values for various crystallization fractions (*f*) from [2-4]. Q_a and Q_b values for wet air oxidation were almost identical to those for dry air oxidation, but the pre-exponential factors A_0 and B_0 were ~2× higher. The 2× factor increase with little change in temperature dependence for P_{H2O} of ~ 0.03 i s roughly consistent with other observations of oxidation rate increases in water environments.[52-53] More detailed analysis and discussion will be presented in publications building on this preliminary study.

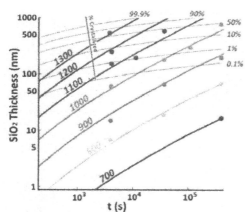

Fig. 6 *Deal-Grove oxidation kinetics parameter fits [5-11] to data in Table I. Data and fit are for amorphous SiO_2 scale growth in wet air for Hi-Nicalon[TM]-S SiC fiber. Avrami crystallization kinetics predictions for various f values [2-4] are superimposed.*

SUMMARY AND CONCLUSIONS

The Deal-Grove oxidation kinetics for amorphous SiO_2 scales, Avrami scale crystallization kinetics, and tensile strength of Hi-Nicalon[TM]-S SiC fiber were measured after oxidation in wet air at 700° to 1300°C for up to 100 hours. The results are generally similar to those previously found for oxidation in dry air. Activation energies for both oxidation and crystallization rates were similar in dry and wet air. However, wet air oxidation had approximately a 2× increase in the pre-exponential factors A_0 and B_0 for linear and parabolic oxidation, respectively, and a decrease in scale crystallization rate at the higher temperatures in comparison with dry air parameters. As observed for dry air oxidation, the fiber strength increased by approximately 10% for SiO_2 scale thickness up to ~100 nm, and decreased for thicker scales. No significant strength degradation was observed for amorphous scales. All fibers with significantly degraded strength had crystallized or partially crystallized scales. Scale crystallization typically began at thicknesses greater than 100 nm. Amorphous scales were uncracked. Crystalline scales were cracked from thermal stress, polymorphic phase transformations, and tensile hoop growth stress. These cracks are inferred to be responsible for lower fiber strength. Future work will try to establish quantitative relationships between fiber strength, oxidation kinetics, crystallization kinetics, and scale residual stress.

REFERENCES
1. Curtin WA, Ahn BK, Takeda N. Modeling Brittle and Tough Stress-Strain Behavior in Unidirectional Ceramic Matrix Composites. Acta mater. 1998;46(10):3409-20.
2. Doremus RH. Viscosity of Silica. J. Appl. Phys. 2002;92(12):7619-29.
3. Pezzotti G, Painter GS. Mechanisms of Dopant-Induced Changes in Intergranular SiO_2 Viscosity in Polycrystalline Silicon Nitride. J. Am. Ceram. Soc. 2002;85(1):91-96.
4. Akashi T, Kasajima M, Kiyono H, Shimada S. SIMS Study of SiC Single Crystal Oxidized in Atmosphere Containg Isotopic Water Vapor. J. Ceram. Soc. Japan 2008;116(9):960-64.
5. Opila EJ, Hann RE. Paralinear Oxidation of CVD SiC in Water Vapor. J. Am. Ceram. Soc. 1997;80(1):197-205.
6. Opila EJ. Oxidation Kinetics of Chemically Vapor-Deposited Silicon Carbide in Wet Oxygen. J. Am. Ceram. Soc. 1994;77(3):730-36.
7. Narushima T, Goto T, Hirai T. High-Temperature Passive Oxidation of Chemically Vapor deposited Silicon Carbide. J. Am. Ceram. Soc. 1989;72(8):1386-90.
8. Maeda M, Nakamura K, Ohkubo T. Oxidation of Silicon Carbide in a Wet Atmosphere. J. Mater. Sci. 1988;23:3933-38.
9. Presser V, Nickel KG. Silica on Silicon Carbide. Crit. Rev. Solid State Mater. Sci. 2008;33:1-99.
10. Takeda M, Urano A, Sakamoto J, Imai Y. Microstructure and Oxidation Behavior of Silicon Carbide Fibers Derived from Polycarbosilane. J. Am. Ceram. Soc. 2000;83(5):1171-76.
11. Shimoo T, Takeuchi H, Okamura K. Oxidation Kinetics and Mechanical Property of Stoichiometric SiC Fibers (Hi-Nicalon-S). J. Ceram. Soc. Japan 2000;108(1264):1096-102.
12. Kim H-E, Moorhead AJ. Strength of Nicalon Silicon Carbide Fibers Exposed to High-Temperature Gaseous Environments. J. Am. Ceram. Soc. 1991;74(3):666-69.
13. Brennan JJ. Interfacial Characterization of a Slurry-Cast Melt-Infiltrated SiC/SiC Ceramic-Matrix Composite. Acta mater. 2000;48(18/19):4619-28.
14. Gauthier W, Pailler F, Lamon J, Pailler R. Oxidation of Silicon Carbide Fibers During Static Fatigue in Air at Intermediate Temperatures. J. Am. Ceram. Soc. 2009;92(9):2067-73.
15. Gogotsi Y, Yoshimura M. Oxidation and Properties Degradation of SiC Fibres Below 850 C. J. Mater. Sci. Lett. 1994;13:680-83.
16. Lara-Curzio E. Stress-Rupture of Nicalon/SiC Continuous Fiber Ceramic Composites in Air at 950 C. J. Am. Ceram. Soc. 1997;80(12):3268-72.
17. Lara-Curzio E. Oxidation Induced Stress-Rupture of Fiber Bundles. J. Eng. Mater. Tech 1998;120:105-09.
18. Hay RS, Fair GE, Urban E, Morrow J, Somerson J, Wilson M. Oxidation Kinetics and Strength Versus Scale Thickness for Hi-Nicalon[TM]-S Fiber. In: Singh Z, Zhou, and Singh, editor. Ceramic Transactions; 2010.
19. Mogilevsky P, Boakye EE, Hay RS, Kerans RJ. Monazite Coatings on SiC Fibers II: Oxidation Protection. J. Am. Ceram. Soc. 2006;89(11):3481-90.
20. Hay RS, Fair GE, Bouffioux R, Urban E, Morrow J, Somerson J, et al. Hi-Nicalon[TM]-S SiC Fiber Oxidation and Scale Crystallization Kinetics. J. Am. Ceram. Soc. 2011;94(11):3983-91.
21. Hay RS, Fair GE, Bouffioux R, Urban E, Morrow J, Hart A, et al. Relationships between Fiber Strength, Passive oxidation and Scale Crystallization Kinetics of Hi-Nicalon[TM]-S SiC Fibers. Ceram. Eng. Sci. Proc. 2011;32(2):39-54.
22. Easler TE, Bradt RC, Tressler RE. Strength Distributions of SiC Ceramics After Oxidation and Oxidation Under Load. J. Am. Ceram. Soc. 1981;64(12):731-34.
23. Badini C, Fino P, Ortona A, Amelio C. High Temperature Oxidation of Multilayered SiC Processed by Tape Casting and Sintering. J. Eur. Ceram. Soc. 2002;22:2017-79.

24. Dong SM, Chollon G, Labrugere C, Lahaye M, Guette A, Bruneel JL, et al. Characterization of Nearly Stoichiometric SiC Fibres. J. Mater. Sci. 2001;36:2371-81.
25. Hinoki T, Snead LL, Lara-Curzio E, Park J, Kohyama A. Effect of Fiber/Matrix Interfacial Properties on Mechanical Properties of Unidirectional Crystalline Silicon Carbide Composites. Ceram. Eng. Sci. Proc. 2002;23(3):511-18.
26. Sauder C, Lamon J. Tensile Creep Behavior of SiC-Based Fibers With a Low Oxygen Content. J. Am. Ceram. Soc. 2007;90(4):1146-56.
27. Bunsell AR, Piant A. A Review of the Development of Three Generations of Small Diameter Silicon Carbide Fibres. J. Mater. Sci. 2006;41:823-39.
28. Ishikawa T. Advances in Inorganic Fibers. Adv. Polym. Sci. 2005;178:109-44.
29. Sha JJ, Nozawa T, Park JS, Katoh Y, Kohyaman A. Effect of Heat-Treatment on the Tensile Strength and Creep Resistance of Advanced SiC Fibers. J. Nucl. Mater. 2004;329-333:592-96.
30. Tanaka T, Shibayama S, Takeda M, Yokoyama A. Recent Progress of Hi-Nicalon Type S Development. Ceram. Eng. Sci. Proc. 2003;24(4):217-23.
31. Opila E. Influence of Alumina Reaction Tube Impurities on the Oxidation of Chemically-Vapor-Deposited Silicon Carbide. J. Am. Ceram. Soc. 1995;78(4):1107-10.
32. Petry MD, Mah T, Kerans RJ. Validity of Using Average Diameter for Determination of Tensile Strength and Weibull Modulus of Ceramic Filaments. J. Am. Ceram. Soc. 1997;80(10):2741-44.
33. Hay RS, Welch JR, Cinibulk MK. TEM Specimen Preparation and Characterization of Ceramic Coatings on Fiber Tows. Thin Solid Films 1997;308-309:389-92.
34. Cinibulk MK, Welch JR, Hay RS. Preparation of Thin Sections of Coated Fibers for Characterization by Transmission Electron Microscopy. J. Am. Ceram. Soc. 1996;79(9):2481-84.
35. Guinel MJF, Norton MG. Oxidation of Silicon Carbide and the Formation of Silica Polymorphs. J. Mater. Res. 2006;21(10):2550-63.
36. Horvath E, Zsiros G, Toth AL, Sajo I, Arato P, Pfeifer J. Microstructural Characterization of the Oxide Scale on Nitride Bonded SiC-Ceramics. Ceramics Int. 2008;34:151-55.
37. Schneider H, Flörke OW. High-temperature transformation of tridymite single crystals to cristobalite*. Zeitschrift für Kristallographie 1986;175(3-4):165-76.
38. Presser V, Loges A, Hemberger Y, Nickel KG. Microstructural Evolution of Silica on Single-Crystal Silicon Carbide. Part I: Devitrification and Oxidation Rates. J. Am. Ceram. Soc. 2009;92(3):724-31.
39. Shimoo T, Hayatsu T, Takeda M, Ichikawa H, Seguchi T, Okamura K. Mechanism of Oxidation of Low-Oxygen SiC Fiber Prepared by Electron Radiation Curing Method. J. Ceram. Soc. Japan 1994;102(7):617-22.
40. Sosman RB. Mechanical and Thermal Properties of the Various Forms of Silica. In: Washburn E, editor. International Critical Tables of Numerical Data, Physics, Chemistry and Technology (Volume 3): McGraw-Hill Book Company, Inc.; 1928. p. 19-25.
41. Mao H, Sundman B, Wang Z, Saxena SK. Volumetric Properties and Phase Relations of Silica - Thermodynamic Assessment. J. Alloys and Compounds 2001;327:253-62.
42. Delph TJ. Intrinsic strain in SiO_2 thin films. J. Appl. Phys. 1998;83(2):786-92.
43. Kao D-B, McVittie JP, Nix WD, Saraswat KC. Two-Dimensional Thermal Oxidation of Silicon - II. Modeling Stress Effects in Wet Oxides. IEEE Trans. Electron. Dev. 1988;35(1):25-37.
44. Rafferty CS, Borucki L, Dutton RW. Plastic Flow During the Thermal Oxidation of Silicon. Appl. Phys. Lett. 1989;54(16):1516-18.
45. Chollon G, Pallier R, Naslain R, Laanani F, Monthioux M, Olry P. Thermal Stability of a PCS-Derived SiC Fibre with a Low Oxygen Content (Hi-Nicalon). J. Mater. Sci. 1997;32:327-47.
46. Hsueh CH, Evans AG. Oxidation Induced Stresses and Some Effects on the Behavior of Oxide Films. J. Appl. Phys. 1983;54(11):6672-86.
47. Brown DK, Hu SM, Morrissey JM. Flaws in Sidewall Oxides Grown on Polysilicon Gate. J. Electrochem. Soc. 1982;129(5):1084-89.
48. Hay RS. Growth Stress in SiO_2 during Oxidation of SiC Fibers. J. Appl. Phys. submitted.

49.   Liu F, Sommer F, Mittemeijer EJ. Analysis of the Kinetics of Phase Transformations; Roles of Nucleation Index and Temperature Dependent Site Saturation, and Recipes for the Extraction of Kinetic Parameters. J. Mater. Sci. 2007;42:573-87.
50.   Liu F, Sommer F, Bos C, Mittemeijer EJ. Analysis of Solid State Phase Transformation Kinetics: Models and Recipes. Int. Mater. Rev. 2007;52(4):193-212.
51.   Deal BE, Grove AS. General Relationships for the Thermal Oxidation of Silicon. J. Appl. Phys. 1965;36(12):3770-78.
52.   Opila EJ. Variation of the Oxidation Rate of Silicon Carbide with Water Vapor Pressure. J. Am. Ceram. Soc. 1999;82(3):625-36.
53.   Yoshimura M, Kase J, Somiya S. Oxidation of SiC Powder by High-Temperature, High-Pressure H_2O. J. Mater. Res. 1986;1(1):100-03.

STUDY ON THE STIFFNESS OF COMELD COMPOSITES JOINTS

Hongjian Zhang[*], Weidong Wen, Haitao Cui

College of Energy & Power Engineering, Nanjing University of Aeronautics and Astronautics, Nanjing, Jiangsu, 210016, China
State Key Laboratory of Intensity and Vibration for Mechanical Structures, Nanjing University of Aeronautics and Astronautics, 210016 Nanjing, Jiangsu, PR China

*Corresponding author: zhanghongjian@nuaa.edu.cn, hongjian_zhang@hotmail.com

ABSTRACT

It is well known that the joint is one of the weakness points. Comeld is a novel technology which can be applied in connecting composites and metals. However, there are no papers or reports about the study on its mechanical properties. As the directions of the fiber and the distributions of the matrix are changed during comeld process, the mechanical properties of comeld composites joints vary from the un-comeld composites. In this paper, the geometry features of the typical microstructure of comeld composites joints are studied. Then a stiffness prediction model is developed and used to predict the stiffness of the comeld composite joints. At last, the sensitiveness of the in-plane stiffness properties to the parameters of the burrs is studied.

1. INTRODUCTION

Composite materials are finding increasing application in industry due to their high strength-to-weight ratio. Typical application areas include: aerospace, marine, motor sport, military and specialist construction. However, in many situations it will not be feasible to construct an entire structure from composites. Thus joints between composites and metallic materials will be required. Historically, the composite to metal joint has presented significant design in challenges in order to achieve high levels of mechanical performance. As a result of this designers have been reluctant to design structures incorporating composite to metal joints, or have adopted highly conservative designs that increase weight and thus negate some of the benefit of using composite materials.

In some instances bolts are used to make joints between composites and metals. Clearly this is highly undesirable as the bolt holes produce stress concentrations into the structure and add weight. Adhesives are also used in metal-composite joints, but adhesively bonded joints do not exhibit any significant elasticity and may fail in a catastrophic fashion with no prior warning. In many cases, in order to address the limitations of both joining techniques, hybrid joints containing both adhesives and mechanical fasteners are used. Such an approach defeats the main objectives of using the composite material i.e. size, weight and cost savings.

Comeld [1-7] is a novel technology which is used to produce joints between fibers reinforced plastic (FRP) composite materials and metals by pre-treating the metallic joint component using Surfi-Sculpt[TM] technology which is recently developed by TWI. Figure 1 is a typical Comeld joints. Some mechanical tests show that Comeld joints can improve the strength and energy to failure.

However, Comeld technology is still at an early stage. The study on its mechanical behaviors is still very rare.

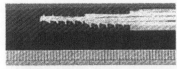

Fig.1 Typeical Comeld joints[1]

The purpose of this paper is to develop the stiffness prediction model based on the investigation the characters of its micro-structure. And the model will be used to predict the stiffness of some typical Comeld joints.

2. DEVELOPMENT OF THE STIFFNESS PREDICTED MODELOF COMELD JOINTS

2.1 Typical micro-structure of Comeld joints

Figure 2 is images of the typical in-plane micro-structure of composites after Comeld. As shown from Fig.1 and Fig.2, the micro-structure of composites after Comeld have the following typical characteristics: (1) the in-plane fibers array around the burrs on metals as waves, which looks like "eyes"; (2)as the directions of fibers are changed, it forms the resin-rich area between waved fibers and burrs. Due to the arrays of fibers and densities of resin in local region are changed obviously during the processing of Comeld technology, the in-plane properties of composites in these regions are different from normal composites. And the in-plane properties are affected by the parameters of the burrs (such as radius R, lateral separation, p, and longitudinal separation, q).

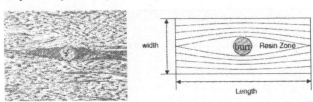

Fig.2 the images of the typical micro-structure of composites after Comeld

2.2 Development of the geometrical model of the Comeld joints

Based on the studies on typical micro-structure of composites after Comeld in section 2.1, it is obviously that the waved fibers array periodically alone their directions when the burrs on metals distribute periodically. And the waved fibers can be simulated with sinusoid or cosine curve. The respective area p×q (lateral separation × longitudinal separation) is selected as the unite cell of the Comeld joints. Due to the symmetry of unite cell, its half part is selected as the representative volume element (RVE) for simply. Figure 3 and Figure 4 are the images of the representative volume element (RVE), in which $\omega = p/2, l = q$.

As show from Fig.4, the representative volume element (RVE) of Comeld joints is constituted by two parts: area Part I (the dashed region) is full with the waved fibers and resin; the resin-rich region and burr are located in area Part II (the unshaded region). The resin and burr are isotropic materials.

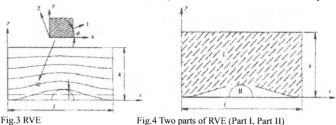

Fig.3 RVE   Fig.4 Two parts of RVE (Part I, Part II)

As the bending degree and the volume content of fibers in Part I is sensitive to location, the following assumptions are proposed before developing the geometrical model of the Comeld joints:

(1) The fibers' arrays alone x-direction (original direction) follow sinusoid or cosine curve with the same-phase. And along y-direction (perpendicular to original direction), the fibers array homogeneously.

(2) The bending degree of fibers varied gradually along the fiber direction. And the bending degree increases with the decrease of the distance to stitch of burr.

(3) The fibers are free during the processing of Comeld technology, which means that the initial strain or damages in fibers can be neglected.

As the waved fibers can be simulated with sinusoid or cosine curve, the in-plane fibers' direction can be described as:

$$\varphi(x,y) = \frac{R}{2}(1-\frac{y}{\omega})\sin(\frac{2\pi x}{l}-\frac{\pi}{2}) \qquad (1)$$

The angle, θ, between the tangent of any point in fibers and x-direction can calculated by derivation of Eq.(1):

$$\tan\theta = \frac{\pi R}{l}\left(1-\frac{y}{\omega}\right)\cos\left(\frac{2\pi x}{l}-\frac{\pi}{2}\right) \qquad (2)$$

In this paper, the parameters, $(\tan\theta)_{max}$, is defined as the fibers' waviness parameter, λ, which is used to indicate the bending of fibers. So, λ can be written as:

$$\lambda = (\tan\theta)_{max} = \frac{\pi R}{l} \qquad (3)$$

And Eq.(2) can be written as:

$$\tan\theta = \lambda\left(1-\frac{y}{\omega}\right)\cos\left(\frac{2\pi x}{l}-\frac{\pi}{2}\right) \qquad (4)$$

So, the angle, θ, can be defined with λ as Eq.(5):

$$\theta = \tan^{-1}\left[\lambda(1-\frac{y}{\omega})\cos(\frac{2\pi x}{l}-\frac{\pi}{2})\right] \qquad (5)$$

As seen from Eq.(5), the angle, θ, varies with location which is agree with actual investigations. And it is the reason why the volume content of fibers differs with location. In order to determine the volume content of fibers of any point in RVE, the infinitesimal volume $dx \times dy$ is built around the point.

As the fibers array alone y-direction homogeneously, the volume content of fibers only depends on x:

$$V_f(x) = \frac{V_f^* \omega}{\omega - \frac{R}{2}\sin(\frac{2\pi x}{l} - \frac{\pi}{2}) - \frac{R}{2}} \qquad (6)$$

where, V_f^* is the volume content of fibers in the composites before Comeld.

2.3 Development of the stiffness prediction model

2.3.1 The elastic properties of a random point in RVE

Based on the equal strain assumption and the volume content of fibers $V_f(x)$ which can be calculated by Eq.(6), the in-plane elastic properties of any point in RVE can be calculated:

$$\left.\begin{array}{l}
E_1(x) = E_{f1}V_f(x) + E_m\left[1 - V_f(x)\right] \\[2mm]
E_2(x) = \dfrac{E_{f2}E_m}{\left(1 - V_f(x)\right)E_{f2} + V_f(x)E_m} \\[2mm]
v_{21}(x) = v_f V_f(x) + v_m\left[1 - V_f(x)\right] \\[2mm]
G_{12}(x) = \dfrac{G_m G_f}{\left(1 - V_f(x)\right)G_f + V_f(x)G_m} \\[2mm]
v_{12}(x) = \dfrac{E_2}{E_1}v_{21}(x)
\end{array}\right\} \qquad (7)$$

where, E_f and E_m are the elastic modules of fibers and resins, respectively.

$E_1(x)$, $E_2(x)$, $v_{21}(x)$, $v_{12}(x)$ and $G_{12}(x)$ are the elastic properties in principle direction of infinitesimal volume around the point.

2.3.2 The equivalent stiffness of RVE in monolayer

By assuming the infinitesimal volume in RVE to be unidirectional composites, the equivalent stiffness of RVE in monolayer is the integral mean in whole RVE volume based on the classical laminate plate theory.

The stress-strain relationship in principal direction of the infinitesimal volume around any point in RVE is:

$$\begin{Bmatrix} \sigma_1 \\ \sigma_2 \\ \tau_{12} \end{Bmatrix} = \begin{bmatrix} Q_{11} & Q_{12} & \\ Q_{12} & Q_{22} & \\ & & Q_{66} \end{bmatrix} \begin{Bmatrix} \varepsilon_1 \\ \varepsilon_2 \\ \gamma_{12} \end{Bmatrix} \tag{8}$$

where, $Q_{ij}(i,j=1,2,6)$ can be calculated as:

$$\begin{aligned} Q_{11} &= \frac{E_1(x)}{1-\nu_{12}(x)\nu_{21}(x)}, \quad Q_{22} = \frac{E_2(x)}{1-\nu_{12}(x)\nu_{21}(x)} \\ Q_{12} &= \frac{\nu_{21}(x)E_2(x)}{1-\nu_{12}(x)\nu_{21}(x)} = \frac{\nu_{12}(x)E_1(x)}{1-\nu_{12}(x)\nu_{21}(x)}, \quad Q_{66} = G_{12} \end{aligned} \tag{9}$$

The coordinate transmitting matrix between the global coordinate system and the local coordinate system is:

$$\mathbf{T} = \begin{bmatrix} \cos^2\theta & \sin^2\theta & 2\sin\theta\cos\theta \\ \sin^2\theta & \cos^2\theta & -2\sin\theta\cos\theta \\ -\sin\theta\cos\theta & \sin\theta\cos\theta & \cos^2\theta - \sin^2\theta \end{bmatrix} \tag{10}$$

Where, θ is the degree of bending of fibers in RVE.

The deflection stress-strain relationship of the infinitesimal volume around any point in RVE can be written as:

$$\begin{bmatrix} \sigma_x \\ \sigma_y \\ \tau_{xy} \end{bmatrix} = \mathbf{T}^{-1} \begin{bmatrix} \sigma_1 \\ \sigma_2 \\ \tau_{12} \end{bmatrix} = \mathbf{T}^{-1}\mathbf{Q} \begin{bmatrix} \varepsilon_1 \\ \varepsilon_2 \\ \gamma_{12} \end{bmatrix} = \mathbf{T}^{-1}\mathbf{Q}(\mathbf{T}^{-1})^T \begin{bmatrix} \varepsilon_x \\ \varepsilon_y \\ \gamma_{xy} \end{bmatrix} \tag{11}$$

For short, $\mathbf{T}^{-1}\mathbf{Q}(\mathbf{T}^{-1})^T$ in Eq.(11) is written as $\overline{\mathbf{Q}}$. Then, the stress-strain relationship in x-y coordinate can be written as:

$$\begin{bmatrix} \sigma_x \\ \sigma_y \\ \tau_{xy} \end{bmatrix} = \overline{\mathbf{Q}} \begin{bmatrix} \varepsilon_x \\ \varepsilon_y \\ \gamma_{xy} \end{bmatrix} = \begin{bmatrix} \overline{Q}_{11} & \overline{Q}_{12} & \overline{Q}_{16} \\ \overline{Q}_{12} & \overline{Q}_{22} & \overline{Q}_{26} \\ \overline{Q}_{16} & \overline{Q}_{26} & \overline{Q}_{66} \end{bmatrix} \begin{bmatrix} \varepsilon_x \\ \varepsilon_y \\ \gamma_{xy} \end{bmatrix} \tag{12}$$

The symmetry matrix $\overline{\mathbf{Q}}$ is the transformation matrix of $\mathbf{Q}$, which is the 2-D stiffness matrix in principal direction. $\overline{Q}_{11}$, $\overline{Q}_{12}$, $\overline{Q}_{22}$ and $\overline{Q}_{66}$ are even functions of θ, $\overline{Q}_{16}$ and $\overline{Q}_{26}$ are odd functions of θ.

Then,

$$C_{ij} = \frac{1}{wl} \int_A \overline{Q}_{ij} dA, (i,j=1,2,6) \tag{13}$$

Where, A is the whole integral domain of RVE. As mentioned before, A includes two parts: Part I and Part II. So, Eq.(13) can be written as:

$$C_{ij} = \frac{1}{wl}\left[\int_0^l \int_0^{\frac{R}{2}\sin(2\pi x/l - \pi/2) + \frac{R}{2}} (Q_m)_{ij}\, dydx + \int_0^l \int_{\frac{R}{2}\sin(2\pi x/l - \pi/2) + \frac{R}{2}}^{w} \overline{Q}_{ij}\, dydx\right] \quad (i, j = 1, 2, 6) \qquad (14)$$

Where, $\overline{Q}_{ij}$ are functions of x and y. $(Q_m)_{ij}$ are constants decided by the elastic properties of resin:

$$\left.\begin{aligned}
(Q_m)_{11} &= \frac{E_m}{1-v_m^2}, \ (Q_m)_{12} = \frac{v_m E_m}{1-v_m^2} \\
(Q_m)_{22} &= \frac{E_m}{1-v_m^2}, \ (Q_m)_{66} = G_m, \ (Q_m)_{16} = (Q_m)_{26} = 0
\end{aligned}\right\} \qquad (15)$$

As $\int_0^l \int_0^{\frac{R}{2}\sin(2\pi x/l - \pi/2) + \frac{R}{2}} dydx = \frac{R}{2}l$, Eq.(14) can be written as:

$$C_{ij} = \frac{R}{2w}(Q_m)_{ij} + \frac{1}{wl}\int_0^l \int_{\frac{R}{2}\sin(2\pi x/l - \pi/2) + \frac{R}{2}}^{w} \overline{Q}_{ij}\, dydx \quad (i, j = 1, 2, 6) \qquad (16)$$

The values of matrix C_{ij} can be calculated by numerical integration of Eq.(16). Flexibility matrix, S_{ij}, can be calculate by finding the inverse matrix of C_{ij}. Then the equivalent stiffness of RVE can be calculated as:

$$\left.\begin{aligned}
E_x &= \frac{1}{S_{11}}, E_y = \frac{1}{S_{22}} \\
v_{xy} &= -\frac{S_{12}}{S_{22}}, v_{yx} = -\frac{S_{12}}{S_{11}}, \ G_{xy} = \frac{1}{S_{66}}
\end{aligned}\right\} \qquad (17)$$

2.3.3 The equivalent stiffness of RVE in laminate

The radius of burr is assumed to be constant in monolayer when calculating the elastic properties of composites laminate. Based on the classical laminate plate theory, the physics equation can be written as:

$$\begin{bmatrix} N \\ M \end{bmatrix} = \begin{bmatrix} A & B \\ B & D \end{bmatrix}\begin{bmatrix} \varepsilon^0 \\ K \end{bmatrix} \qquad (18)$$

Where:

$$\left.\begin{aligned}
A_{ij} &= \sum_{k=1}^n (\overline{Q}_{ij})_k (z_k - z_{k-1}) \\
B_{ij} &= \frac{1}{2}\sum_{k=1}^n (\overline{Q}_{ij})_k (z_k^2 - z_{k-1}^2) \\
D_{ij} &= \frac{1}{3}\sum_{k=1}^n (\overline{Q}_{ij})_k (z_k^3 - z_{k-1}^3)
\end{aligned}\right\} \qquad (19)$$

Then, the elastic properties of composites laminate after Comeld can be calculated:

$$\begin{cases} E_{x'} = A_{11}(1 - v_{x'y'}v_{y'x'})/h, \ E_{y'} = A_{22}(1 - v_{x'y'}v_{y'x'})/h \\ v_{y'x'} = A_{12}/A_{11}, \ v_{x'y'} = A_{12}/A_{22}, \ G_{x'y'} = A_{66}/h \end{cases} \qquad (20)$$

Where, h is the thickness of monolayer.

3. APPLICATIONS AND DISCUSSIONS

3.1 Stiffness prediction of monolayer of Comeld joints

In this section, the stiffness prediction model is used to predict the stiffness of monolayer of Comeld joints. And the sensitiveness of the in-plane stiffness properties to the parameters of the burrs (such as radius R, lateral separation, p, and longitudinal separation, q) of monolayer is studied. The monolayer composite used in this paper is T300/QY8911. And the joint material is TC4. Table 1 lists the stiffness properties for the fibers, resin and joint metal (T300, QY8911, and TC4). The thickness of the monolayer is 0.12mm.

TABLE 1 Stiffness properties

Material	E_1 / GPa	E_2 / GPa	v
T300	221	13.8	0.2
QY8911	3		0.3
TC4	109		0.3

When lateral separation p=5mm, longitudinal separation q=5mm, and radius R=0.25mm, the Comeld-to-Uncomeld ratios of in-plane stiffness properties are listed in Table 2. The parameters with the subscript "C" are the stiffness properties of the composites after Comeld. After Comeld, the in-plane elastic properties are changed obviously: the elastic moduli in 1 direction decrease a little; the elastic moduli in 2 direction increase obviously; the shear moduli and poisson ratios increase greatly. These are caused by the changes of the arrays of fibers and distribution of resin.

TABLE 2 the Comeld-to-Uncomeld ratios of stiffness properties

E_{1C}/E_1	E_{2C}/E_2	G_{12C}/G_{12}	v_{12C}/v_{12}
0.9970	1.1180	1.2207	1.3574

In order to study the sensitiveness of the in-plane stiffness properties to some of the parameters of burrs, the other parameters are designed as constants.

Fig.5 shows the curves of Comeld-to-Uncomeld ratios of the in-plane stiffness properties to radius R over the range of 0.05~0.4mm at the condition of p=q=5 mm. As shown in Fig.5, the elastic modulus ratio in 1 direction is not sensitive to R. But the other three modului ratios increase nonlinear with R obviously. It is because the bending degrees of fibers around the burrs increase with the value of radius R.

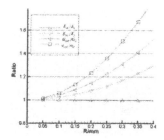

Fig. 5 the curves of the Comeld-to-Uncomeld ratios to R (p=q=5 mm)

Fig.6 shows the curves of Comeld-to-Uncomeld ratios of the in-plane stiffness properties to lateral separation p over the range of 3~10mm at the condition of q=5mm and R=0.25mm. As shown in Fig.6, when p increases from 3mm to 10mm, the elastic modulus ratio in 1 direction remains the same, and the other three modului ratios decrease and have a tendency to arrive asymptotic values. But these three modului ratios are greater than 1, which means they are strengthened by Comeld.

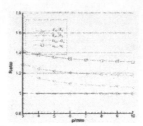

Fig. 6 the curves of the Comeld-to-Uncomeld ratios to p (q=5mm, R=0.25mm)

Fig.7 shows the curves of Comeld-to-Uncomeld ratios of the in-plane stiffness properties to q over the range of 3~10mm at the condition of p=5mm and R=0.25mm. As shown in Fig.7, when q increases from 3mm to 7mm, the elastic modului ratios (except the modulus in 1 direction) decrease sharply. When q increases from 7mm to 10mm, they decrease gently and have a tendency to arrive the asymptotic value which is greater than 1. And the elastic modulus ratio in 1 direction is not sensitive to q. The rules illustrate that the sensitiveness of the in-plane stiffness properties to q is great when q is less than a critical value (such as 7mm).

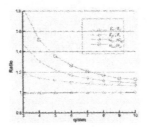

Fig. 7 the curves of the Comeld-to-Uncomeld ratios to q (p=5mm, R=0.25mm)

Based on the analysis of the sensitiveness of monolayer, we can get the following conclusions: (1) when the radius R is constant, the in-plane stiffness properties of monolayer remain the same values when the parameters p and q are greater than critical values; (2) the elastic modului ratios (except the modulus in 1 direction) of laminate increase with R obviously; (3) after Comeld, the elastic modulus in 1 direction is weakened a little, but the other three modului are strengthened greatly.

Based on the simulation results for monolayer, it shows obviously that the in-plane stiffness properties can be increased by optimizing the parameters within their design ranges, such as increasing the value of R and decreasing the values of p and q.

3.2 Stiffness prediction of laminate of Comeld joints

In this section, the stiffness prediction model is used to predict the stiffness of laminate of Comeld joints. And the sensitiveness of the in-plane stiffness properties to the parameters of the burrs of laminate is studied. The material used in this section is same with in section 3.1 (T300/QY8911and TC4). The ply stacking sequence of laminate is $[0/45/0/-45/90/-45/0/45/0]_{2s}$.

Table 3 lists the Comeld-to-Uncomeld ratio of in-plane stiffness properties of the laminate at the condition of lateral separation p=5mm, longitudinal separation q=5mm, and radius R=0.25mm. The parameters with the subscript "C" are the stiffness properties of the composites after Comeld. After Comeld, the in-plane elastic properties of laminate are different from monolayer: the elastic modules in 1 and 2 direction decrease in a small amount; the shear moduli and poisson ratio increase very little. It is because that the stiffness of laminate is the comprehensive average of every layer.

TABLE 3 the Comeld-to-Uncomeld ratio of stiffness properties of laminate

E_{1C}/E_1	E_{2C}/E_2	G_{12C}/G_{12}	$\upsilon_{12C}/\upsilon_{12}$
0.9944	0.9933	1.0161	1.0159

In order to study the sensitiveness of the in-plane stiffness properties of laminate to some of the parameters of burrs, the other parameters are designed as constants.

Fig.8 shows the curves of Comeld-to-Uncomeld ratios of the in-plane stiffness properties of laminate to radius R over the range of 0.05~0.4mm at the condition of p=q=5 mm. As shown in Fig.8, when R increase from 0 to 0.4mm, the elastic modulus ratios in 1 and 2 direction decrease from 1 to 0.98, but the shear moduli and poisson ratios increase from 1 to 1.035. These rules are different from the monolayer. And after Comeld, the elastic modulus in 1 and 2 directions are weakened, but the other two modului are strengthened.

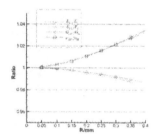

Fig. 8 the curves of the Comeld-to-Uncomeld ratios of laminate to R (p=q=5 mm)

Fig.9 shows the curves of Comeld-to-Uncomeld ratios of the in-plane stiffness properties of laminate to lateral separation p over the range of 3~10mm at the condition of q=5mm and R=0.25mm. As shown in Fig.9, when p increases from 3mm to 6mm, the elastic modulus ratios in 1 and 2 directions increase from 0.98 to 0.995, but the other two modului decrease from 1.04 to 1.01. When p increases from 6mm to 10mm, all of the modului ratios have a tendency to arrive the asymptotic value. And after Comeld, the elastic modului in 1 and 2 directions are weakened, but the other two modului are strengthened.

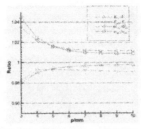

Fig. 9 the curves of the Comeld-to-Uncomeld ratios of laminate to p
(q=5mm, R=0.25mm)

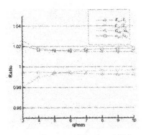

Fig. 10 the curves of the Comeld-to-Uncomeld ratios of laminate to q
(p=5mm, R=0.25mm)

Fig.10 shows the curves of Comeld-to-Uncomeld ratios of the in-plane stiffness properties of laminate to longitudinal separation q over the range of 3~10mm at the condition of q=5mm and R=0.25mm. As shown in Fig.10, when q increases from 3mm to 5mm, the elastic modulus ratio in 2 direction increase from 0.98 to 0.995, but the shear moduli and poisson ratios decrease from 1.02 to 1.015. When q increases from 5mm to 10mm, all of the modului ratios have a tendency to arrive the asymptotic value. The elastic modulus ratio in 1 direction remains the same during the variation of q. And after Comeld, the elastic modulus in 1 and 2 directions are weakened, but the other two modului are strengthened.

Based on the analysis of the sensitiveness of laminate, we can get the following conclusions: (1) when the radius R is constant, the in-plane stiffness properties of laminate remain the same values when the parameters p and q are greater than critical values; (2) the in-plane stiffness properties of laminate vary with R obviously; (3) after Comeld, the elastic modului in 1 and 2 directions are weakened a little, but the other two modului are strengthened a little.

As the stiffness of laminate is the comprehensive average of every layer, the in-plane elastic properties of laminate only change a little when compared with Uncomeld laminate. And the values of the parameters' sensitiveness of laminate are less than the values of monolayer obviously. As the in-plane elastic properties of laminate depend on the ply stacking sequence, the best performance of the joints can be designed by optimizing the parameters (R, p, and q) and the sequence.

4. SUMMARY

In this paper, the typical characteristics of micro-structure of composites after Comeld are studied: (1) the in-plane fibers array around the burrs on metals as waves; (2) the resin-rich area is formed between waved fibers and burrs.

Then, a stiffness prediction model is developed based on the investigation the characters of its micro-structure. And the model is used to predict the stiffness of monolayer and laminate composites after Comeld, respectively.

At last, the sensitiveness of the in-plane stiffness properties to the parameters of the burrs (such as radius R, lateral separation, p, and longitudinal separation, q) of monolayer and laminate composites is studied, respectively. Simulation results show: (1) when the radius R is constant, the in-plane stiffness properties of monolayer and laminate composites remain the same values when the parameters p and q are greater than critical values; (2) the in-plane stiffness properties of monolayer and laminate composites vary with R obviously; (3) as the stiffness of laminate is the comprehensive average of every layer, the Comeld-to-Uncomeld ratios of laminate is less than the ratios of monolayer obviously.

ACKNOWLEDGEMENT

This research has been supported by Nanjing University of Aeronautics and Astronautics Research Funding (NS2010056).

REFERENCES

[1] Dance B.G.I., Buxton A. L.. An introduction to Surfi-Sculpt technology - new opportunities, new challenges. Conf. Proc. 7th Int. Conf. On Beam Technology, Halle, Germany, 17-19th, April 2007:75-84

[2] Smith F., Wylde G. Comeld - An innovation in composite to metal joining. Australasian Welding Journal, 2004, 49:26-27.

[3] Dance B.G.I.. Rapid materials processing and surface sculpting using electron beam and laser processes. TMS 2009 138th Annual Meeting & Exhibition on Proceedings, Volume 3: General Paper Selections, USA: San Francisco, California, 15-19 February 2009:167-174.

[4] Freeman R., Smith F.. Comeld: A novel method of composite to metal joining. Advanced Materials and Processes, 2005, 163(4): 33-40.

[5] Buxton, A.L., Dance, B.G.I.. Surfi-Sculpt - Revolutionary surface processing with an electron beam. Proceedings of the 4th International Surface Engineering Conference, 1-3 August, Minnesota, USA, 2005:107-110.

[6] Mick H.. Micro-sculptures give metal the Velcro touch. New Sci., 2004, 182 (2447) :21-28.

[7] Thomas G., Grant P.S., Coad P., etc. Vacuum Plasma Spraying of Thick W coatings onto surface sculptures for fusion armour applications, Proceedings of the 2006 International Thermal Spray Conference, 15-18 May, Seattle, Washington, USA, 2006:254-230.

HIGH-TEMPERATURE INTERLAMINAR TENSION TEST METHOD DEVELOPMENT FOR CERAMIC MATRIX COMPOSITES

Todd Z. Engel
Hyper-Therm High-Temperature Composites, Inc.
Huntington Beach, CA, United States of America

ABSTRACT

Ceramic Matrix Composite (CMC) materials are an attractive design option for various high-temperature structural applications. However, 2D fabric-laminated CMCs typically exhibit low interlaminar tensile (ILT) strengths, and interply delamination is a concern for some targeted applications. Currently, standard test methods only address the characterization of interlaminar tensile strengths at ambient temperatures, which is problematic given that nearly all CMCs are slated for service in elevated temperature applications. This work addresses the development of a new test technique for the high-temperature measurement of CMC interlaminar tensile properties.

1.0 INTRODUCTION

Hot structures fabricated from ceramic composite (CMC) materials are an attractive design option for the specialized components of future military aerospace vehicles and propulsion systems because of the offered potential for increased operating temperatures, reduction of component weight, and increased survivability. CMC materials exhibit the high-temperature structural capabilities inherent to ceramic materials, but with significantly greater toughness and damage tolerance, and with lessened susceptibility to catastrophic component failure than traditional monolithic ceramic materials – all afforded by the presence of continuous fiber reinforcement. Some potential applications for CMC materials involve the replacement of high-temperature metals in aeroengine components aft of the combustor; examples include convergent and divergent flaps and seals at the engine nozzle, and blades and vanes in the turbine portion of the engine.

Most CMC laminates are fashioned from an assemblage of two-dimensional (2D) woven fabrics. As a result, the thru-thickness direction typically lacks fiber reinforcement and exhibits significantly lower strengths and toughness than the in-plane directions. For this reason, there is concern that thermostructural components fabricated from 2D fabric may have inadequate interlaminar tensile (ILT) strengths for some applications, thereby increasing their vulnerability to interply delamination when subjected to high thru-thickness thermal gradients, acoustic/high cycle fatigue, impact damage, free edge effects, and/or applied normal loads. For these reasons, it is important to fully characterize the ILT strengths of candidate CMC materials systems to ensure robust design when they are being considered for use in high-temperature structural applications.

Currently, the ILT strength of CMCs is evaluated as per ASTM C1468 [1] ("Standard Test Method for Transthickness Tensile Strength of Continuous Fiber-Reinforced Advanced Ceramics at Ambient Temperature"). This test method, commonly referred to as the Flatwise Tension (FWT) test, involves the use of square of circular planform test coupons machined from flat plate stock of representative CMC material. The specimens are typically machined from thin-gage plate material; for this reason it is not possible to directly grip the specimen for the application of the thru-thickness tensile loading required to initiate ILT failure. Instead, loading blocks are adhesively bonded to the opposing faces of the specimen in order to facilitate tensile loading in the thru-thickness direction. The lack of availability of high-strength, high-temperature structural adhesives currently precludes the applicability of this methodology to elevated temperature testing. A robust test method for performing ILT measurements at elevated temperatures is currently lacking and must be addressed to enable the serious consideration of CMC materials for insertion into high-temperature structural applications.

1.1 Proposed High-Temperature Specimen Configuration

This work targets the aforementioned problem by introducing an alternative ILT specimen design that is conducive to testing at elevated temperatures. The body of the proposed specimen is machined from flat CMC plate stock. V-shaped notches are machined into the thickness, and a state of interlaminar tension is induced between the specimen notches with wedge-type fixtures loaded in compression (Figure 1). Because this configuration only requires the application of compressive loading, it is readily adaptable to testing at high temperatures with the use of monolithic ceramic test fixturing. The conceptual specimen geometry, illustrated in Figure 1, shows the four primary geometric parameters – namely the specimen thickness (t), the notch half-angle (φ), the notch spacing (b), and the specimen width (w).

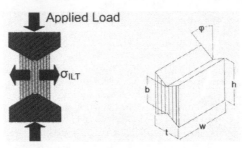

Figure 1. Proposed high-temperature ILT specimen.

The stresses induced in the loaded specimen were determined as a function of specimen geometry using a Strength of Materials analysis approach and assuming: (1) linear-elastic material behavior; (2) small displacement behavior and negligible friction between the specimen and loading fixtures; (3) loading from the fixture is transferred uniformly to the specimen surfaces; (4) effects of the notch tip (i.e. stress concentrations) are ignored. The loaded specimen was decomposed into the individual components of stress in detail in previous work [3]. A formulation for the interlaminar tensile stress of the loaded specimen was derived from this previous analysis, and is provided in Equation 1 as a function of the applied load (P), specimen notch half-angle (φ), specimen width (w), and notch tip spacing (b):

$$\sigma_{ILT} = \frac{P}{wbTA\not{} \varphi} \tag{1}$$

2.0 EXPERIMENTAL PROCEDURES

2.1 Materials

Four (4) distinct CMC material systems were evaluated under the current work; these were a CG Nicalon fiber-reinforced SiC/SiC laminate with a chemical vapor infiltration (CVI)-derived SiC matrix, a T-300 fiber-reinforced C/SiC laminate with a CVI-derived SiC matrix, a Nextel 720 fiber reinforced Oxide/Oxide laminate with an alumina matrix derived from sol-gel processing techniques, and a Hi-Nicalon fiber-reinforced SiC/SiC laminate. A summary of the various material systems is provided in Table I. It should be mentioned that the two SiC/SiC laminates (Materials 1 and 4) are distinctly different material systems; the references to their reinforcing fiber is intended as a descriptor to differentiate the two materials, and there is no intended inference that the fiber itself is a key characteristic responsible for differing mechanical properties.

Table I. CMC Material Systems

Material	Type	Fiber	Matrix	Matrix Processing
1	SiC/SiC	CG Nicalon	SiC	CVI
2	C/SiC	T-300	SiC	CVI
3	Oxide/Oxide	Nextel 720	Alumina	Sol-Gel
4	SiC/SiC	Hi-Nicalon	SiC	CVI

2.2 Flatwise Tension Test Setup

Flatwise tension (FWT) testing was used to establish baseline ILT strength properties for all of the CMC laminates tested; these ILT properties were used as a basis to evaluate the correlation of apparent ILT strengths determined using various configurations of the proposed high-temperature test technique. Flatwise tension testing was conducted on round 19mm-diameter specimens extracted normal to the surface of the laminate. Theses specimens were bonded between steel loading dowels, and subsequently subjected to monotonic tensile loading in an Applied Test Systems (ATS) Series 900 universal test frame under displacement-controlled conditions at a loading rate of 2.5mm per minute. The interlaminar tensile stress (σ_{ILT}) was calculated as a function of the failure load (P) and cross-sectional area of the specimen (A) according to Equation 2:

$$\sigma_{ILT} = \frac{P}{A}$$ (2)

Universal joints were utilized to alleviate undesirable bending stress components during the test. A schematic representation of the test setup is provided in Figure 2.

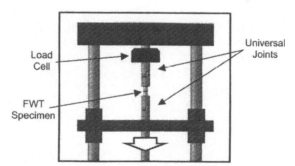

Figure 2. Flatwise tension (FWT) test setup.

2.3 High-Temperature ILT Test Setup

Test fixturing was designed and manufactured to facilitate the wedge loading of the proposed ILT specimen at both ambient and elevated temperatures. A schematic of the fixture assembly, along with an image of the fixture components is provided in Figure 3. The compressive load is transferred into the furnace hot zone during high-temperature testing by way of 12.7mm-diameter alumina rods (Figure 4). An aerosol-based Boron Nitride (BN) coating was applied to the specimen v-notch regions prior to testing. The coating served as a compliant interfacial media to promote more uniform loading at the specimen/loading fixture contact interface, as well as to reduce interfacial friction. The coating was implemented during both room temperature and elevated temperature testing. Testing was performed in an Applied Test Systems (ATS) Series 900 universal test frame under a displacement-controlled loading rate of 2.5mm per minute. This typically resulted in tests lasting 3-5 seconds, whereby failure was indicated by a sharp and significant drop in the measured load. The test rate is an important consideration while testing at elevated temperatures; the loading of the specimen should be rapid enough such that specimen fails in fast-fracture, rather than due to time dependent failure mechanisms (i.e. slow crack growth), such as those demonstrated in interlaminar shear testing by Choi [2].

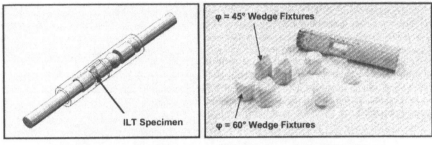

Figure 3. Alumina fixtures for both ambient and elevated temperature ILT testing.

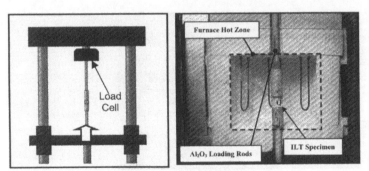

Figure 4. High-temperature ILT test setup.

2.4 Specimen Sizing

Previous development of the proposed test method by the author [3] included a parametric study to evaluate six (6) distinct potential specimen configurations, including two (2) values of the notch half-angle, φ (45° and 60°), and three (3) values of the specimen notch tip spacing to thickness ratio (b/t = ½, 1, and 2). The performance of a particular specimen configuration was evaluated by comparing its resultant apparent ILT strength at room temperature to ILT strengths determined using the ASTM prescribed room temperature FWT test method. This testing was performed upon a 6.4mm-thick SiC/SiC laminate with CG Nicalon fiber reinforcement. The results of this testing demonstrated the best correlation of apparent ILT was achieved with b/t ratios of 1; both notch half-angle configurations (45° and 60°) performed comparably, and no determination could be made as to which was preferred. The results of this preliminary sizing study have been applied to the current work, which focuses on thinner 3.2mm-thick CMC laminates. However, an additional parametric study of notch half-angle (φ) and notch tip spacing to thickness ratio (b/t) was performed to confirm that the previous results and trends would translate from the thicker 6.4mm laminate to the thinner 3.2mm laminate that is the basis of the current work. The additional parametric studies are detailed in §2.6. All specimens tested in the present work had nominal widths (w) of 12.7mm.

2.5 High-Aspect Ratio Tip Notch

During exploratory testing of the proposed ILT specimen with 3.2mm-thick CMC materials, it was qualitatively observed that failures would occasionally occur away from the prescribed fracture plane between the notch tips (Figure 5). This assumed failure plane is the location of maximum ILT stress in the specimen, and serves as the basis for calculating the resultant ILT strength. This is problematic in a sense that it introduces some ambiguity into the results for a few reasons. First, the fracture may be occurring at regions of lower ILT stress than is used for calculating the ILT strength of the material. Secondly, a more significant issue is the close proximity of the fracture to the contact interface between the specimen and loading fixtures; it cannot be known with any certainty the influence of any localized contact stresses on the initiation of failure. Finally, the lack of control over the failure location poses some difficulty with regards to establishing criteria for censoring test results, and determining whether or not a particular test has resulted in a valid failure mode.

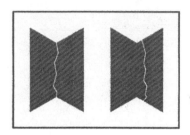

Figure 5. Desirable (left) and undesirable (right) ILT specimen failures.

These concerns were addressed through the introduction of an additional 0.3mm-thick notch feature at the existing notch apex. Two different notch depth-to-width aspect ratios (1:1 and 3:1) for this feature were evaluated. The CVI-based SiC/SiC laminate material with CG Nicalon fiber reinforcement was utilized for this evaluation. Ten (10) specimens with each notch configuration were evaluated under ambient conditions and compared to baseline ILT strengths obtained through FWT testing. These specimens utilized a notch half-angle (φ) of 60°, a b/t ratio of 1, and a width of 12.7mm. The results of the testing are shown in Table II. Both the 1:1 and 3:1 tip notch configurations resulted in comparable average ILT strengths at 14.0 MPa and 14.1 MPa, respectively. These numbers correlate well with the average ILT strength of 14.5 MPa obtained with the FWT testing. The 1:1 configuration demonstrated more significant scatter in the data with a standard deviation and coefficient of variation (COV) of 2.3 MPa and 17.0%; this compares to 1.0 MPa and 6.7% for the 3:1 notch configuration and 0.8 MPa and 5.6% for the FWT specimens. The 3:1 tip notch resulted in specimen failures that were confined to the prescribed fracture plane between the specimen notch tips. However, testing of the 1:1 notch configuration resulted in a number of specimen failures that occurred outside of the prescribed fracture area (Figure 6). For this reason, as well as the reduced scatter in the ILT strength results, the 3:1 tip notch configuration was adopted for the remainder of the testing detailed in this paper. For the post-test images in Figure 6, white typewriter correction fluid was applied to the specimen prior to testing to emphasize the location of fracture.

Table II. Tip Notch Aspect Ratio Test Results

Test Configuration	n	ILT Strength		
		Avg. (MPa)	Std. Dev. (MPa)	COV
FWT	10	14.5	0.8	5.6%
1:1	10	14.0	2.3	17.0%
3:1	10	14.1	1.0	6.7%

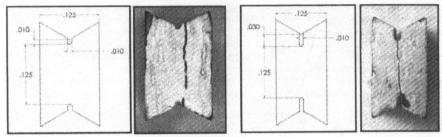

Figure 6. Tip notch aspect ratio configuration of 1:1 (left) and 3:1 (right).

2.6 Parametric Study

Previous work by the author [3] was used to guide initial specimen sizing for the present work. However, this previous testing was performed upon relatively thick 6.4mm CMC laminate. To confirm that the results and trends were still applicable to the thinner 3.2mm laminate that is the basis for the present work, an additional parametric study was performed to assess the effects of the notch half-angle (φ) and the notch tip spacing to thickness ratio (b/t). Parametric studies were performed upon two CMC material systems, namely the C/SiC material (No. 2) and the Hi-Nicalon reinforced SiC/SiC material (No. 4). All specimens tested included the tip notch feature described in §2.5, and were tested in the manner described in §2.3. The parametric study performed with the C/SiC material was a duplication of the previous study, and included two variations of φ (45° and 60°) and three variations of the b/t ratio (½, 1, and 2); this resulted in six distinct specimen geometries. The parametric study performed with the Hi-Nicalon reinforced SiC/SiC material included the same two variations of φ (45° and 60°), but focused on b/t ratios of ½, ¾, and 1. The lower biasing of b/t ratios for this study was motivated by earlier exploratory testing that demonstrated poor correlation of ILT strengths between the proposed technique and FWT technique for materials with relatively high ILT strengths (greater than 17 MPa) and with b/t ratios of 1. In this case, the apparent ILT strength results had been lower than those obtained through FWT tests. Previous analytical modeling [4] had shown that smaller b/t ratios trended towards a more uniform ILT stress distribution and lowered stress concentration effects at the notch tip. In light of the previous results, the reduced b/t ratios were anticipated to be more suitable for the parametric study of the higher strength Hi-Nicalon SiC/SiC laminate.

3.0 EXPERIMENTAL RESULTS AND DISCUSSION

3.1 Parametric Study Results

Six (6) replicates of the C/SiC material were tested in FWT at ambient temperatures, resulting in an average ILT strength of 11.3 MPa, with a standard deviation and COV of 2.1 MPa and 18.1%. Nine (9) C/SiC replicates of the proposed high-temperature ILT specimen were tested at ambient temperatures in both the 45° and 60° notch half-angle configurations and with a b/t ratio of ½. The 45° configuration yielded an average ILT strength of 13.5 MPa, with a standard deviation and COV of 3.6 MPa and 26.6%. The 60° configuration yielded an average ILT strength of 12.5 MPa, with a standard deviation and COV of 3.4 MPa and 27.1%. Ten (10) C/SiC replicates of the proposed high-temperature ILT specimen were tested at ambient temperatures in both the 45° and 60° notch half-angle configurations and with a b/t ratio of 1. The 45° configuration yielded an average ILT strength of 10.6 MPa, with a standard deviation and COV of 2.3 MPa and 22.2%. The 60° configuration yielded an average ILT strength of 11.1 MPa, with a standard deviation and COV of 2.3 MPa and 21.0%. Ten (10) C/SiC replicates of the proposed high-temperature ILT specimen were tested at ambient temperatures in both the 45° and 60° notch half-angle configurations and with a b/t ratio of 2. The 45° configuration yielded an average ILT strength of 7.1 MPa, with a standard deviation and COV of 0.4 MPa and 6.1%. The 60° configuration yielded an average ILT strength of 7.7 MPa, with a standard deviation and COV of 1.2 MPa and 15.9%. These results are listed in Table III, and shown graphically in Figure 7.

Table III. Summary of the C/SiC Parametric Results

Material	b/t	φ	n	ILT Strength		
				Avg. (MPa)	Std. Dev. (MPa)	COV
C/SiC	½	45°	9	13.5	3.6	26.6%
		60°	9	12.5	3.4	27.1%
	1	45°	10	10.6	2.3	22.2%
		60°	10	11.1	2.3	21.0%
	2	45°	10	7.1	0.4	6.1%
		60°	10	7.7	1.2	15.9%
	FWT		6	11.3	2.1	18.1%

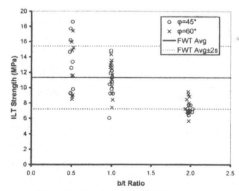

Figure 7. Results of the C/SiC parametric study.

The parametric study of the C/SiC material produced similar trends to those observed in the previous work. The specimens with b/t ratios of 2 exhibited ILT strengths lower than those obtained via FWT testing; this is likely attributable to a significant stress concentration at the notch tips, which resulted in a non-uniform distribution of ILT stress. The specimens with b/t ratios of 1 exhibited the best correlation to the FWT test results in terms of average ILT strength values and their variation. The specimens with b/t ratios of ½ demonstrated the highest average ILT strengths, and greatest degree of variability in the results. The specimen notch half-angle was not a significant factor in the results, as both the 45° and 60° configurations behaved comparably. As a result, the notch spacing to thickness ratio of 1 was determined to be the most suitable specimen configuration for the remainder of the testing performed during the present work, with the aforementioned exception of the relatively high strength Hi-Nicalon SiC/SiC material – a discussion of which is forthcoming.

Five (5) replicates of the Hi-Nicalon reinforced SiC/SiC material were tested in FWT at ambient temperatures, resulting in an average ILT strength of 22.5 MPa, with a standard deviation and COV of 2.6 MPa and 11.6%. Ten (10) Hi-Nicalon reinforced SiC/SiC replicates of the proposed high-temperature ILT specimen were tested at ambient temperatures in both the 45° and 60° notch half-angle configurations and with a b/t ratio of ½. The 45° configuration yielded an average ILT strength of 19.7 MPa, with a standard deviation and COV of 2.9 MPa and 14.8%. The 60° configuration yielded an average ILT strength of 17.5 MPa, with a standard deviation and COV of 4.3 MPa and 24.4%. Ten (10) Hi-Nicalon reinforced SiC/SiC replicates of the proposed high-temperature ILT specimen were tested at ambient temperatures in both the 45° and 60° notch half-angle configurations and with a b/t ratio of ¾. The 45° configuration yielded an average ILT strength of 20.4 MPa, with a standard deviation and COV of 2.1 MPa and 10.5%. The 60° configuration yielded an average ILT strength of 19.5 MPa, with a standard deviation and COV of 2.1 MPa and 10.7%. Ten (10) Hi-Nicalon reinforced SiC/SiC replicates of the proposed high-temperature ILT specimen were tested at ambient temperatures in both the 45° and 60° notch half-angle configurations and with a b/t ratio of 1. The 45° configuration yielded an average ILT strength of 16.4 MPa, with a standard deviation and COV of 2.0 MPa and 12.4%. The 60° configuration yielded an average ILT strength of 17.2 MPa, with a standard deviation and COV of 2.3 MPa and 13.6%. These results are listed in Table IV, and shown graphically in Figure 8.

Table IV. Summary of the Hi-Nicalon SiC/SiC Parametric Results

| Material | b/t | φ | n | ILT Strength | | |
				Avg. (MPa)	Std. Dev. (MPa)	COV
Hi-Nicalon SiC/SiC	½	45°	10	19.7	2.9	14.8%
		60°	10	17.5	4.3	24.4%
	¾	45°	10	20.4	2.1	10.5%
		60°	10	19.5	2.1	10.7%
	1	45°	10	16.4	2.0	12.4%
		60°	10	17.2	2.3	13.6%
	FWT		5	22.5	2.6	11.6

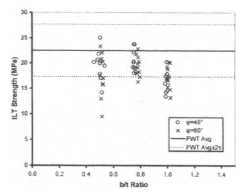

Figure 8. Results of the Hi-Nicalon SiC/SiC parametric study.

In a deviation from the results of the previous parametric study, the specimen configuration with a b/t ratio of 1 did not produce a strong correlation to the results of baseline FWT testing. The transition to a specimen configuration with a b/t ratio of ¾ resulted in shift upwards of the ILT strength distribution. The further transition to a b/t ratio of ½ resulted in increased levels of scatter, especially for the 60° notch configuration (24.4% COV). The increase in experimental scatter may be attributable to the notch spacing parameter (b) being nearly equivalent to the unit cell size of the composite (~1.5mm) for this configuration. The best performing configuration was found to be that with a b/t ratio of ¾, which resulted in a reasonably good correlation with the FWT data in terms of both average ILT strengths and their distribution. Furthermore, this configuration produced very consistent results for both values of the notch half-angle φ (45° and 60°). The results of the testing for the ¾ b/t ratio specimens will be included in the comprehensive listing of room temperature results in §3.2.

3.2 Correlation of Ambient Temperature ILT and FWT Results

The ability of the proposed high-temperature ILT test method to replicate the results of the conventional FWT test method was assessed through the comparative testing at ambient temperatures of four (4) distinct CMC material systems with average ILT strengths ranging from 4.9 to 22.5 MPa. Specimens of the proposed test method featured a tip notch with an aspect ratio of 3:1 (as discussed in §2.5), a b/t ratio of 1 for the C/SiC, CG Nicalon reinforced SiC/SiC, and Oxide/Oxide material systems, a b/t ratio of ¾ for the Hi-Nicalon reinforced SiC/SiC material system, and notch half-angles of both 45° and 60°. The relevant room temperature test results for the C/SiC and Hi-Nicalon reinforced SiC/SiC materials were presented previously in §3.1, and the results for the CG Nicalon reinforced SiC/SiC material was presented previously in §2.5.

Thirteen (13) replicates of the Oxide/Oxide material were tested in FWT, resulting in an average ILT strength of 4.9 MPa, with a standard deviation and COV of 0.9 MPa and 18.7%. Ten (10) replicates of the proposed ILT specimen were tested in the 45° configuration, resulting in an average ILT strength of 0.8 MPa, with a standard deviation and COV of 0.2 MPa and 23.3%. Thirty (30) replicates were tested in the 60° configuration, resulting in an average ILT strength of 5.0 MPa, with a standard deviation and COV of 0.8 MPa and 15.8%.

With the exception of the 45° notch half-angle configuration of the Oxide/Oxide composite, all configurations of the proposed ILT specimen demonstrated good correlation of ILT strength results to those obtained using FWT test techniques. In the case of the C/SiC and Hi-Nicalon reinforced SiC/SiC materials, both notch half-angle configurations tested (45° and 60°) yielded very consistent results. The 45° notch half-angle configuration of the Oxide/Oxide material did not correlate to either the FWT results, or the results of the alternate 60° configuration. The reasons for this are not immediately clear, and are to be a subject of future investigation. A summary of the measured ILT strength data is provided in Table V. Typical FWT fracture surfaces and failed ILT specimens for the four CMC material systems are provided in Figures 9 and 10, respectively.

Table V. Summary of Room Temperature ILT Results

Material	Test Type	n	ILT Strength		
			Avg. (MPa)	Std. Dev. (MPa)	COV
CG Nicalon SiC/SiC	60°	10	14.1	1.0	6.7%
	FWT	10	14.5	0.8	5.6%
C/SiC	45°	10	10.6	2.3	22.2%
	60°	10	11.1	2.3	21.0%
	FWT	6	11.3	2.1	18.1%
Oxide/Oxide	45°	10	0.8	0.2	23.3%
	60°	30	5.0	0.8	15.8%
	FWT	13	4.9	0.9	18.7%
Hi-Nicalon SiC/SiC	45°	10	20.4	2.1	10.5%
	60°	10	19.5	2.1	10.7%
	FWT	5	22.5	2.6	11.6%

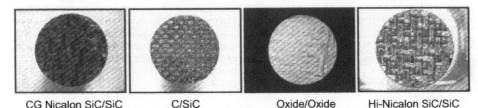

| CG Nicalon SiC/SiC | C/SiC | Oxide/Oxide | Hi-Nicalon SiC/SiC |

Figure 9. Typical FWT fracture surfaces.

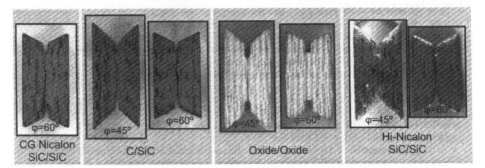

Figure 10. Typical failed high-temperature ILT specimens tested at ambient temperatures.

3.2 High-Temperature ILT Test Results

A limited number of high-temperature ILT tests were performed at 1100°C in air using the C/SiC and Oxide/Oxide materials. Three (3) C/SiC specimens were tested in the 45° configuration, resulting in an average ILT strength of 11.0 MPa, with a standard deviation and COV of 2.1 MPa and 19.1%. Two (2) C/SiC specimens were tested in the 60° configuration, resulting in an average ILT strength of 9.0 MPa, with a standard deviation and COV of 1.8 MPa and 19.9%. Five (5) specimens of the Oxide/Oxide material were tested in the 60° configuration, yielding an average ILT strength of 4.3 MPa, with a standard deviation and COV of 0.9 MPa and 20.7%.

The high-temperature strength results for both materials are within the experimental scatter observed during room temperature testing; therefore, no obvious ILT strength degradation was observed at 1100°C in air for either the C/SiC or Oxide/Oxide materials for the limited data available. Additional high-temperature data will be generated in future work to confirm this. The high-temperature results are summarized in Table VI, and a comprehensive plot of the ILT test data for the four (4) CMC material systems is provided in Figure 11.

Table VI. Summary of High-Temperature ILT Results

Material	Test Type	n	Temperature (°C)	ILT Strength		
				Avg. (MPa)	Std. Dev. (MPa)	COV
C/SiC	45°	10	25	10.6	2.3	22.2%
	45°	3	1100	11.0	2.1	19.1%
	60°	10	25	11.1	2.3	21.0%
	60°	2	1100	9.0	1.8	19.9%
Oxide/Oxide	60°	30	25	5.0	0.8	15.8%
	60°	5	1100	4.3	0.9	20.7%

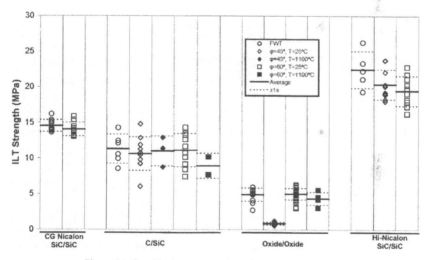

Figure 11. Graphical representation of ILT strength distributions.

4.0 CONCLUSION

A test method was proposed for the interlaminar tension testing of CMC materials at elevated temperatures. The appropriate specimen geometry for the test method was determined empirically and was based upon testing performed in previous [3] and current work. A high aspect ratio tip notch feature was proposed and adopted in the current work as a modification to the specimen geometry that would allow for a more clearly defined "gage section" in the specimen that would produce a more repeatable failure mode. This modification was further motivated by a desire to relocate the fracture surface of the specimen away from the contact interface and loading fixtures, which could introduce additional undesirable components of localized stress. A particularly sensitive feature to the specimen geometry was found to be the notch spacing to thickness (b/t) ratio. This feature was evaluated empirically with a parametric study, and the appropriate specimen design was chosen based upon its demonstrated correlation at ambient temperatures to baseline ILT results obtained through conventional FWT testing. A single b/t ratio of 1 was found to work successfully for all but one of the four materials evaluated; the lone deviation was a relatively high-strength Hi-Nicalon reinforced SiC/SiC laminate for which a b/t ratio of ¾ produced the best results. The C/SiC, Oxide/Oxide, and Hi-Nicalon reinforced SiC/SiC materials were all evaluated using specimens with notch half-angles (φ) of both 45° and 60°, while the CG Nicalon material was evaluated only at the 60° notch half-angle. All specimen configurations of the four materials demonstrated good correlation to the FWT test results, with the exception of the 45° configuration of the Oxide/Oxide material. The ILT strengths of this specimen variant were significantly lower than those of both the FWT testing, and the 60° specimen configuration. The reason for this result is uncertain at this time.

Elevated temperature testing was performed at 1100°C in air on both the C/SiC and Oxide/Oxide materials. Both materials yielded ILT strengths that were comparable to those observed during room temperature testing; therefore, these materials don't appear to suffer any significant degradation of ILT strength up to 1100°C in air. Additional replicates will be tested in future work to confirm this result, and the ILT properties of these materials will be characterized at various additional temperatures.

The levels of scatter observed in the ILT test data were consistent for both the FWT testing and the proposed high-temperature ILT technique in all four CMC material systems evaluated. It was initially anticipated that the proposed technique would produce some level of increased scatter in the ILT strengths results. Typically, ceramic materials exhibit statistical strength distributions that relate to the size of their stressed area or volume. Provided that the ILT failure mode is a brittle fracture, and that the stressed area of the FWT specimen is seven times that of the high-temperature ILT specimens tested in this work, some discrepancy in the distribution of strengths between the two methods was anticipated. However, this result was not observed. Despite the difference in specimen sizes, the level of scatter was comparable for both techniques for all four of the CMC material systems that were evaluated.

ACKNOWLEDGEMENTS

We would like to acknowledge that this work has been funded by NAVAIR under contract number N68335-08-C-0491, and monitored by Dr. Sung Choi.

REFERENCES
[1]ASTM Standard C 1468-06: Standard Test Method for Transthickness Tensile Strength of Continuous Fiber-Reinforced Advanced Ceramics at Ambient Temperature, *Annual Book of ASTM Standards*, Vol. 15.01, ASTM International, West Conshohocken, PA, 2009.
[2]Choi, S.R., "Life Limiting Behavior of Ceramic Matrix Composites Under Interlaminar Shear at Elevated Temperatures," *Processing and Properties of Advanced Ceramics and Composites III*, The American Ceramic Society, Westerville, OH; Ceramic Transactions, **225** (2011).

[3]Engel, T., W. Steffier, and T. Magaldi, "High-Temperature Interlaminar Tension Test Method Development for Ceramic Matrix Composites," *Processing and Properties of Advanced Ceramics and Composites III*, The American Ceramic Society, Westerville, OH; Ceramic Transactions, **225** (2011).
[4]Engel, T. (2010, October). *High-Temperature Interlaminar Tension Test Method Development for Ceramic Matrix Composites*. Presented at Materials Science and Technology 2010 Conference and Exhibition, Houston, TX.

TENSILE FRACTURE MECHANISM OF SILICON IMPREGNATED C/C COMPOSITE

Akio Ohtani, Ken Goto

Institute of Space and Astronautical Science, Japan Aerospace Exploration Agency

3-1-1 Yoshinodai, Sagamihara, Kanagawa, Japan

ABSTRACT

Tensile fracture behavior of a silicon infiltrated cross-ply carbon-carbon (C/C-Si) laminate was examined at room temperature. Unlike C/C composites, C/C-Si has a non-linear stress-strain response with one third of the strength of a C/C composite before infiltration of Si. At 90% of the ultimate strength, the modulus decreased by 10% against initial value. Cross-sectional observation by optical microscopy and observation of the fracture surface by SEM were carried out to investigate the fracture mechanism. It was found that cracks in C/C-Si initiated from the tip of the process cracks generated after Si infiltration in 90° layer and propagated into the 0° layer cutting the 0° fibers. From the other process cracks, almost no new cracks were generated. From the in-situ observation during the loading-unloading test, the crack propagation rate from the tip of the process cracks increased with an increase in applied load. The length of new cracks correlated with the modulus degradation and the unrecoverable strain increase. This fact means that new cracks growing just from the process cracks caused by the Si infiltration process led to the low tensile strength and modulus degradation of the C/C-Si.

INTRODUCTION

Carbon/carbon composite (C/C) has been used in structural applications such as space vehicles, nuclear reactors, aircraft brakes, and racing car brakes, because of its light weight, high strength and toughness at very high temperatures. However, its interlaminar shear strength and anti-oxidation properties are very poor. C/C composites also have poor gas-barrier properties because of the network of porosity generated by the shrinkage of matrix resin into the carbon during the pyrolysis process. In order to address some of these shortcomings, silicon infiltrated C/C composites (C/C-Si) have been developed. One prevalent manufacturing process of C/C-Si is the liquid silicon infiltration of a C/C by molten silicon. The molten silicon partially reacts with carbon matrix to form silicon carbide. The infiltration is mostly applied by only capillary forces. The process temperatures are at least above the melting point of the silicon (1415°C). By infiltration of the silicon, C/C-Si was expected to improve the above mentioned problems of C/C as well as wear resistance. Actually, many studies for C/C-Si have been investigated from the standpoint of friction properties.[1-3] Therefore, C/C-Si is one of the candidate materials for the usage in high temperature applications. However, C/C-Si has poor mechanical properties. Especially the strength of C/C-Si is much lower than the base C/C composite. In previous research, a few reports for the mechanical properties of C/C-Si had been reported. The mechanical properties under the tensile loading, and fatigue loading were investigated with cross-ply laminates.[4-6] It was clarified that while Young's modulus of C/C-Si was almost the same as C/C composite, the tensile strength of C/C-Si was much lower than C/C. Fracture surfaces in macro and micro scopic views were quantitatively examined in comparison with C/C. In another study, fracture mechanisms of C/C-Si were discussed compared with C/C from room temperature to 1600°C.[7] However, the mechanism of the degradation for C/C-Si was not identified.

Some of the possible reasons for lower tensile strength of C/C-Si: degradation of carbon fiber strength by siliconization, the stress concentration at the tip of initial defects, and an increase in the bond strength between fiber and matrix. But the fracture mechanism and the reason of the degradation of C/C-Si have not been investigated in the previous study even at room temperature.

In this study, tensile tests and observation of fracture behavior of cross-ply C/C-Si composites under tensile loading was carried out in order to clarify the dominant mechanism for the lower tensile strength of C/C-Si. Furthermore, to confirm the degradation of the reinforcing fiber, bending test just for a 0 degree layer was performed.

EXPERIMENTAL PROCEDURE

Materials

In this study, a silicon infiltrated C/C composite was used. The C/C before Si infiltration was supplied by Across Corp., Japan, and was fabricated using the preformed yarn method,[8] where the yarns of PAN-based carbon fiber (TORAYCA®M40, TORAY, Japan) bundled into 6000 filaments were used for reinforcement. The C/C was symmetrically laminated in a cross-ply stacking sequence [0°/90°]8s, and densified by the carbon matrix to a fiber volume fraction of 50% after heat treatment at ≈ 2273°K. After the fabrication of C/C, Si was infiltrated into the porous part of C/C at 1600 °C to fabricate the C/C-Si.

Mechanical testing and observation

The specimens for the static tensile test were prepared from 300×300mm plates (thickness=3.3mm) to double dumbbell shape using a conventional milling machine. One of the fiber axes in the specimens was set parallel to the loading direction. Hereafter, the plies, fibers that run parallel or normal to the loading direction, is denoted to 0° plies or 90° plies, respectively.

In order to investigate the mechanical properties, static tensile tests for the C/C-Si at room temperature were carried out using a screw-driven universal testing machine (AG-5000A, Shimadzu Corp., Japan) at a constant crosshead speed of 0.1mm/min in air at 298°K. The strain during the tests was measured using two strain gauges attached parallel to the loading direction on both sides and center of specimens. By the same procedure, a cyclic loading test was also performed to investigate the degradation of Young's modulus. The applied cyclic load was increased by 10% of the maximum load obtained from the static tensile test in each cycle, and Young's modulus in each hysteresis curve was estimated. Cross-sectional observation was performed by an optical microscope before and after the tensile test. During the cyclic loading test, microstructure observation was also conducted in order to elucidate the fracture mechanism of C/C-Si. After the tests, fracture surfaces of the specimens were also observed by a scanning electron microscope (SEM).

RESULTS AND DISCUSSION

Static and cyclic tensile test

Stress-strain curve of C/C-Si is shown in Fig.1. The curve of C/C was also shown in the figure by dotted line for comparison. This C/C was the same as the base of C/C-Si before high temperature infiltration of the silicon. While the curve of C/C is linear, nonlinear behavior can be seen in C/C-Si from the beginning of the curve. The modulus and the strength of each specimen are shown in Fig.2. The modulus of C/C-Si was obtained from the strain 0% to 0.01% of the curve. The modulus in each specimen was similar, however, the strength of C/C-Si was less than 1/3 of the C/C.

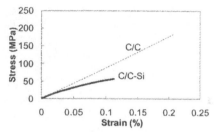

Figure 1. Stress-strain curves of C/C and C/C-Si.

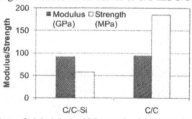

Figure 2. Modulus and Strength of C/C and C/C-Si.

The stress-strain curve during the cyclic loading test was shown in Fig.3. The loading-unloading cycles were repeated 10 times up to the ultimate fracture with increasing applied stress. As shown in the figure, the unrecoverable strain at the unloaded point in each curves increased with an increase in peak load. The relationship between Young's modulus obtained from the slope at the top of each unloading curve and that of C/C was shown in Fig.4 as a function of the applied stress normalized by the strength. From the figure, the modulus was constant from the beginning to 30% of the applied stress, and then, decreases with increase in the applied stress. After 90% of fracture strength applied, the Young's modulus dropped to 90% of the initial value. On the other hand, C/C had constant modulus up to the ultimate fracture. These results indicated the fracture mechanism of C/C-Si was different from that of C/C. Some damage to C/C-Si is apparent beginning at less than 30% of the ultimate strength, and accumulating up to the final fracture.

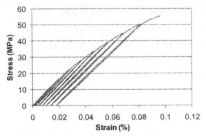

Figure 3. Stress-strain curves of cyclic tensile test for C/C-Si.

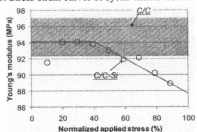

Figure 4. Modulus degradation by applied stress.

Observation for side end of specimen

In order to investigate crack initiation and propagation during loading, the side surface of the specimen was observed in detail. Figure 5 and 6 show the polished side surface before loading and a schematic drawing of 3 types of cracks in C/C-Si. The 0° and 90° plies, and many defects defined as "process cracks" can be seen in these figures. One type of crack, designated "type 1", seems to occur during the pyrolysis process of C/C composite before the infiltration of silicon. These cracks are relatively large, and all of them were filled with Si and SiC. Second was "type 2", which was a relatively thin crack with similar size and orientation as the type 1 cracks. The other process cracks, "type 3", are small cracks in the SiC rich region running perpendicular to the fiber axis. Crack types 2 and 3 are generated due to the mismatch in thermal expansion properties of carbon and the infiltrated

Si and SiC during infiltration process of silicon and also cooling stage. Almost all process cracks were aligned not in longitudinal but in thickness and width direction.

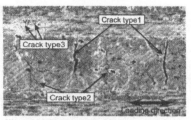

Figure 5. Cross-sectional photo of C/C-Si.

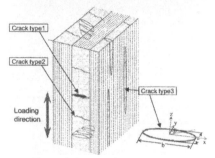

Figure 6. 3 types of cracks in C/C-Si.

Here, one of a pair of magnified cross-sectional photos of a specimen before and after loading was shown in Fig.7(a) and (b). From these images, a new crack generated during the tensile test could be seen. The new crack generated in 0° ply from the tip of a process crack and propagated into the 0° ply propagating through the reinforcing fibers. From the detailed observation shown in Fig.7, many new cracks were observed. From these observations, the crack propagation mechanism appears to be as follows: all new cracks initiated from the tips of process cracks and propagated through the 0° fibers. The number of new cracks generated from the tips of the three types of process cracks, determined by the comparison of the images before and after loading, were counted are shown in Table 1. This table shows process crack size, the number of new cracks generated from the observed process cracks (the number of these cracks were indicated in case arc), and probability of crack occurrence which was calculated from the number of new cracks divided by the number of observed process cracks. Crack length of type 1 and 2 correspond to the thickness of one ply. The shape of crack type 3 was approximated by an ellipse as shown in Fig.7, with the long ellipse dimension corresponding to the ply thickness. Short ellipse dimension is the mean value obtained from the crack size observations. For example, the 2nd line in the table means, 33 cracks initiated from type 2 cracks out of 839 type 2 cracks observed. In the case of type 3 cracks, the propagation not only in thickness direction but in width direction was also investigated by observing the 0° ply. From these results, almost all new cracks were initiated by type 2 cracks.

Next, cack propagation in a cyclic tensile test was observed at each stress level. Figure 8 shows the same part of the specimen where new crack initiated from Crack type 2 and propagated under each cyclic loading. In each load cycle, the length of the every observable new crack was measured.

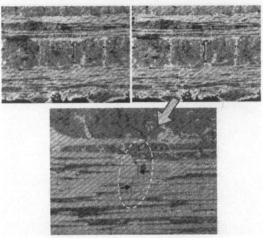

Figure 7. New crack generated from the tip of Crack type 2.

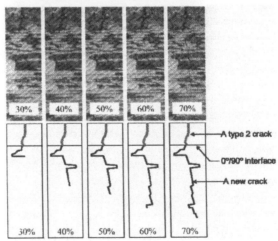

Figure 8. Crack propagation from Crack type 2 under cyclic loading.

Table 1. Number of new cracks.

	Crack size (μm)	No. of new cracks	Probability of crack occurrence (%)
Crack type 1	225	0 (/983)	0.0
Crack type 2	225	33 (/839)	3.9
Crack type 3	a 27 b 225	2(/1124)	0.18

Figure 9 shows the modulus and the unrecoverable strain as a function of the mean crack length for each cyclic load level. From these relationships, the modulus decreased and the unrecoverable strain increased with an increase in the mean crack length. From this result, new cracks generated from Crack type 2 into 0° ply were one of the dominant factors to decrease the modulus of the C/C-Si.

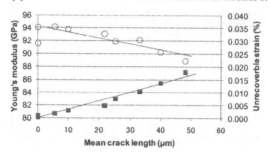

Figure 9. Relationship between Young's modulus and mean crack length.

The reason why almost all new cracks were initiated from crack type2 is explained as follows. Comparing crack type 1 with crack type 2, the tip of crack type 2 is sharper than type 1. So the stress at the tip of crack type2 should be larger than that of crack type 1. In the case of crack type 2 and 3, stress intensity factor K for each type of process crack were calculated by following equations respectively.

$$K = \sigma\sqrt{\pi a} \qquad (1)$$

$$K = \frac{F(\theta)}{\Phi}\sigma\sqrt{\pi b} \qquad (2)$$

$$\left\{ \Phi = \int_0^{\pi/2}\sqrt{1-\left(\frac{c^2-b^2}{b^2}\right)\sin^2\varphi}\,d\varphi, \qquad F(\theta) = \left[\left(\frac{r_0}{b}\right)^2\cos^2\theta + \left(\frac{r_0}{c}\right)^2\left(\frac{b}{c}\right)^2\sin^2\theta\right]^{1/4} \right\}$$

Where, θ; angle between certain point at the edge of elliptical crack and y axis, r_0; distance between certain point at the edge of elliptical crack and center of ellipse, φ; eccentric angle, β; π/2-θ.

From eq(1) for crack type 2, K/σ=18.8, and from eq(2) for crack type 3, K/σ≤13.1. From these results, stress intensity factor of crack type2 was larger than that of crack type3. So consequently, the stress at the tip of crack type2 should be larger than that of crack type 1 and type 3. That's why the number of new cracks generated from crack type 2 was probably largest.

SEM observation for fracture surface

Fracture surfaces of the C/C and the C/C-Si were observed by SEM. The typical fracture surface of C/C-Si was shown in Fig.10 (a) and (b). Fiber bundle pull out and delamination between 0° and 90° ply could be seen in C/C. Here, it is well known in C/C that delaminations occur and propagate in 0°/90° ply interface under tensile loading because interlaminar strength of C/C is very weak. Finally fiber fracture of 0° plies occurs. The typical fracture surface of C/C-Si was shown in Fig.11(a) and (b). One was the relatively flat fracture surface as shown in Fig.11(a), and fiber bundle pull out was observed in the other Fig.11(b). In Fig.11(a), there were no fiber bundle pull-out region and no delamination between 0° and 90° layer. The fracture surface of 0° and 90° layer was almost the same level. In Fig.11 (b), the fracture surface around the 0°/90° interface was relatively flat and the delamination was not observed between 0° and 90° layer. Fiber bundle pull-out occurred not from the interface but the inside of 0° layer. This fracture aspect indicated the crack generated from a tip of Crack type 2 in 90° layer propagated into 0° layer as mentioned in section 3.2. On the other hand, the fracture surface of SiC rich region between fiber bundle pullout region was observed in Fig.11 (b). The fracture surface of these SiC rich regions should be corresponding to Crack type 3. Cut fibers were observed just adjacent to these surfaces. Therefore, from this fracture aspect, Crack type 3 would not propagated far in the thickness direction and width direction as mentioned in the results of Table1

From these results, the fracture mechanism of C/C-Si could be described as shown in Fig.12. In the case of C/C-Si, Crack type 2 in 90° ply propagated not along the interface between 0°and 90° plies but into 0° ply to cut a part of 0° fibers because interfacial adhesive properties between 0° and 90° plies were improved by the infiltration of Si. In addition to this degradation, the mechanical properties of a part of 0° fiber near interface between 0° and 90° plies might be degraded by the siliconization. Consequently, the crack propagation into 0° ply from the tip of Crack type 2 was the one of the main reason for the degradation of the mechanical properties as shown in Fig.4.

(a) Fiber bundle pull out   (b) Delamination at 0°/90° interface

Figure 10. Typical fracture surfaces of C/C.

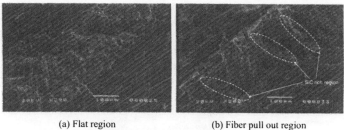

(a) Flat region   (b) Fiber pull out region

Figure 11. Typical fracture surfaces of C/C-Si.

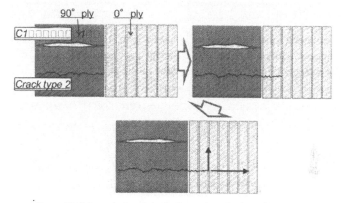

Figure 12. Schematic drawings for fracture mechanism of C/C-Si.

Bending test of a layer

In order to investigate the mechanical properties of 0° ply, 3-point bending tests for a ply extracted from cross-ply laminates of C/C-Si were carried out. Uni-directional specimens with 0.3mm thickness were extracted from the laminates by machining process with a rotary surface grinder. Span length was 10mm and radius at the tip of the indenter was 1mm.

Bending properties of C/C-Si and C/C were shown in Fig.13. Bending modulus and strength of C/C-Si were 16% and 53% lower than that of C/C specimen. Fracture aspects of these specimens were tensile fracture at the tensile side for both the C/C and C/C-Si specimens and no compressive failure was observed. These results indicate that the tensile properties of 0° ply for C/C-Si was lower, especially the strength was less than half compared to that of C/C specimen.

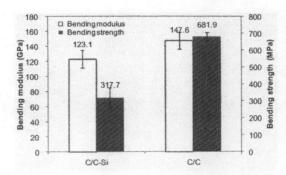

Fig.13 Bending modulus and strength of UD specimen

From these results, a possible explanation for the reduced mechanical properties of C/C-Si 0°/90° laminates is discussed as below. In the case of the 0°/90° specimen, the new cracks initiated from the process cracks propagated into 0° ply enough to decrease the modulus as shown in Fig.4. Then, the load bearing area in 0° ply was decreased due to the new crack propagation and also the initial strength of 0° ply was almost half to C/C specimen. Therefore, the strength of 0°/90° specimen was reduced to less than 1/3 compared to C/C specimen as shown in Fig.2.

CONCLUSIONS

1. The modulus degradation of C/C-Si was mainly caused by the propagation of the cracks in 0° plies from the process cracks formed after the Si infiltration process.

2. The strength degradation of 0°/90° C/C-Si was mainly due to the crack propagation into 0° plies and the strength degradation of 0° ply by siliconization of carbon fiber.

3. Fracture mechanism of C/C-Si is process cracks after infiltration in 90° ply propagated into 0° ply and to fracture a portion 0° fibers.

REFERENCES

[1] W. Krenkel, F. Berndt, "C/C–SiC composites for space applications and advanced friction systems", Materials Science and Engineering: A, Volume 412, Issues 1-2, 5, Pages 177-181 (2005).

[2] W. Krenkel, B. Heidenreich, R. Renz, "C/C-SiC composites for advanced friction systems. Advanced Engineering Materials", 4(8): 427–436(2002).

[3] Hussam Abu El-Hija, Walter Krenkel and Stefan Hugel, "Development of C/C–SiC Brake Pads for High-Performance Elevators", Int. J. Appl. Ceram. Technol., 2 [2] 105– 113 (2005).

[4] Wen-Xue Wang, Yoshihiro Takao, Terutake Matsubara, "Tensile strength and fracture toughness of C/C and metal infiltrated composites Si–C/C and Cu–C/C", Composites Part A: Applied Science and

Manufacturing, Volume 39, Issue 2, February 2008, Pages 231-242

[5]Masahiro Moriyama, Yoshihiro Takao, Wen-Xue Wang, Terutake Matsubara, "Strength of metal impregnated C/C under tensile cyclic loading", Composites Science and Technology, Volume 66, Issue 15, 1 December 2006, Pages 3070-3082

[6]Masahiro Moriyama, Yoshihiro Takao, Wen-Xue Wang, Terutake Matsubara, "Fatigue characteristics of metal impregnated C/C composites with slots for load transfer", International Journal of Fatigue, Volume 32, Issue 1, January 2010, Pages 208-217

[7]K. Fujimoto, T. Shioya and K. Satoh "Mechanical Properties of Silicon Impregnated C/C Composite Material at Elevated Temperature", Advanced Composite Materials, Vol. 11, No. 4, 2003, pp. 393-403.

[8]T.Chang, T.Nakagawa, and A.Okura, "Studies on a New Manufacturing Process of Carbon Fiber Reinforced Carbon Matrix(C/C) Composites", Rep.Inst.Ind.Sci., Univ. Tokyo, 35 [8] (1991).

EFFECTS OF TARGET SUPPORTS ON FOREIGN OBJECT DAMAGE IN AN OXIDE/OXIDE CMC

D. Calvin. Faucett, Jennifer Wright, Matthew Ayre, Sung R. Choi[†]
Naval Air Systems Command, Patuxent River, MD 20670

ABSTRACT

Foreign object damage (FOD) aspects of an oxide/oxide ceramic matrix composite (CMC) were determined using spherical steel ball projectiles in impact velocity ranging from 100 to 340 m/s. The oxide/oxide CMC test targets were ballistically impacted at a normal incidence angle while sustained in cantilever support. Surface and subsurface impact damages were in the forms of craters, fiber breakage, delamination, cone cracks, and backside protrusion, typically observed in many CMCs in their ballistic impact responses. Qualitative impact damage was also assessed through post-impact residual strength measurements. Compared to two other types of target supports (full and partial supports), the cantilever support configuration resulted in the greatest impact damage of all the three. The impact damage in cantilever support is believed to have been involved both by quasi-static impression and by stress wave interactions, while the impact damage in partial and full supports was involved mainly by quasi-static dynamic impression.

INTRODUCTION

The brittleness of ceramic materials, either monolithic ceramics, ceramic matrix composites (CMCs), ceramic environmental barrier coatings (EBCs), or ceramic thermal barrier coatings (TBCs), has raised concerns on structural damage when subjected to impact by foreign objects. This has prompted the propulsion community to take into account foreign object damage (FOD) as an important design parameter when those materials are intended to be used for aeroengine hot-section components. A considerable amount of work on impact damage of brittle monolithic materials has been carried out during the past decades experimentally or analytically [1-14], including gas-turbine grade toughened silicon nitrides [15-17].

Ceramic matrix composites have been used in some aeroengines and are targeted as enabling materials for higher-temperature propulsion components of advanced civilian or military gas-turbine engines. A span of FOD work has been conducted to determine FOD behavior of gas-turbine grade CMCs such as state-of-the-art melt-infiltrated (MI) SiC/SiC [18], N720™/aluminosilicate (N720/AS) oxide/oxide [19], N720/alumina (N720/A) oxide/oxide [20], and 3D woven SiC/SiC [21]. Hertzian indentation responses were also determined to simulate a quasi-static impact phenomenon [22]. Unlike their monolithic counterparts, all the CMCs investigated have not exhibited catastrophic failure at impact velocities up to 400 m/s. However, the degree of damage, when assessed via residual strength, was still substantial, particularly at higher impact velocities $\geq$340 m/s by hardened steel ball projectiles with a diameter of 1.59 mm.

[†] Corresponding author. Email address: sung.choi1@navy.mil

The mode of target supports has shown significant effects on the degree of impact damage. The partial support always resulted in greater impact damage than its full support counterpart, regardless of types of CMCs [18-20]. Also, the uniaxial target support yielded significant impact damage [23-24]. A new target support, cantilever support, was introduced in a previous study for an MI prepreg SiC/SiC CMC [25]. The result showed that among the three different target supports, the cantilever support yielded the greatest impact damage, the partial support did the intermediate damage, and the full support yielded the least damage [25].

The current paper is to extend the previous work [25] to determine the effects of target supports on impact damage using an N720/alumina oxide/oxide CMC. A cantilever target support was used and their results were compared with those determined previously in both full and partial supports [20]. The impact damage was estimated in conjunction with residual strength and impact morphologies. The prediction of impact force in three different target supports was made and assessed using the available impact damage size data.

EXPERIMENTAL PROCEDURES
Materials
The oxide/oxide material used in this work has been described elsewhere [20] and is repeated here briefly. The composite was a commercial, 2-D woven, N720™ fiber-reinforced alumina matrix CMC, fabricate by ATK/COIC (San Diego, CA; Vintage 2011). N720™ oxide fibers, produced in tows by 3M Corp. (Minneapolis, MN), were woven into 2-D 8 harness-satin fabrics. The fabrics were cut into a proper size, slurry-infiltrated with the matrix, and 12 ply-stacked followed by consolidation and sintering. The fiber volume fraction of the final composite panels was about 0.45. No interface fiber coating was employed. Porosity was about 25%. Basic physical and mechanical properties of the oxide/oxide CMC are shown in Table 1. Target specimens measuring 12 mm in width, 50 mm in length, and about 2.7 mm in as-furnished thickness were machined from the composite panels.

Table 1. Basic physical and mechanical properties of target oxide/oxide CMC and projectile materials at ambient temperature [20,23,24]

Material		Fiber/matrix	Bulk density (g/cm^3)	Fiber volume fraction	Elastic modulus[1] E (GPa)	Tensile strength[2] (MPa)	ILT Strength[3] (MPa)	ILS strength[4] (MPa)
Target	Oxide/oxide	Nextel™ 720 /alumina (2D woven)	2.74	0.45	81	140	3.8 (1.6)[#]	8.6(0.6)
Projectile	Hardened chrome steel (SAE52100)	-	7.78	-	200*	>2000*	-	-

Notes: 1. By the impulse excitation technique, ASTM C 1259; 2. By dog-boned tensile specimens, ASTM C1275; 3. By ASTM C1468; 4. By ASTM C1292. [#]The numbers in parentheses indicate ±1.0 standard deviation. * Data from the literature. ILT: interlaminar tension; ILS: interlaminar shear

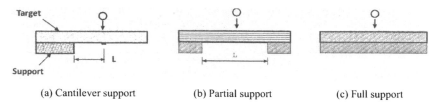

(a) Cantilever support   (b) Partial support   (c) Full support

Figure 1. Three different types of target supports used in an N720/alumina oxide/oxide CMC: (a) Cantilever support used in this work; (b) Partial support and (c) full support both used in a previous study [20]. The span either in (a) or (b) was L=20mm.

<u>Foreign Object Damage Testing</u>

A ballistic impact gun, as described elsewhere [15,16,18], was used to conduct FOD testing. Hardened (HRC≥60) chrome steel-ball projectiles with a diameter of 1.59 mm were placed into a gun barrel. Helium gas and relief valves were utilized to pressurize and regulate a reservoir to a specific level, depending on prescribed impact velocity. Upon reaching a specific level of pressure, a solenoid valve was instantaneously opened accelerating a steel-ball projectile through the gun barrel to impact onto the 12mm-wide side of a CMC target. The target specimens were held in cantilever support with L=20 mm, as illustrated in Fig. 1. The figure also shows two other supports (full and partial), used previously [20] for comparison. Three different impact velocities of 150, 200, and 340 m/s were employed. For a given impact velocity, three to four target specimens were utilized. Some additional target specimens were used for impact only having their impact morphologies as well as cross-section features characterized with respect to different impact velocities. The use of the three target supports employed in this and previous work was to simulate the configurations of gas-turbine airfoil components such as blades, vanes, and shrouds, etc. ·

<u>Post-Impact Residual Strength Testing</u>

Post-impact residual strength testing was conducted to determine residual strengths of impacted target specimens from which the severity of impact damage was better assessed. Strength testing was done in four-point flexure with 20mm-inner and 40-mm outer spans using an MTS servohydraulic test frame (Model 312) at a crosshead speed of 0.25 mm/min.

RESULTS AND DISCUSSION

<u>Projectiles</u>

The steel ball projectiles exhibited no visible damage or plastic deformation upon impact even at the highest impact velocity of 340 m/s. This was attributed to their higher hardness (and higher elastic modulus), as compared to the open and soft structure as well as lower elastic modulus (=81GPa) of the target oxide/oxide material. This was consistent with the previous observations in either partial or full support [19,20]. It should be noted that the steel projectiles were significantly flattened or fragmented particularly at impact velocities ≥350 m/s when impacted on harder target materials such as silicon nitrides [15-17], MI SiC/SiC [18], or MI prepreg SiC/SiC [25].

Impact Morphology of CMC Targets

Front Impact Damage:
Impact damage as observed from the front impact sites of the CMC targets included features such as crater formation, impressions, densification or compaction, fiber/matrix and fiber-tow breakage, etc. A typical example of an impact site generated at 340 m/s in cantilever support is shown in Fig. 2, where a 3-D topological image depicts how the damage was formed accompanying an irregular crater. The image was taken by a digital microscope via Extended Focus 3D Synthesis (Model KH-7700, HiRox, Japan). Figure 3 presents frontal impact damage with respect to impact velocity and includes for comparison the impact damage obtained previously in partial and full supports [20]. For a given impact velocity, the dependency of impact damage on the type of supports was hardly distinguishable from the impact sites. The ballistic impact did generate surface and subsurface damage including craters with fiber/matrix breakages and some material removal with their severity being dependent on impact velocity. Figure 4 depicts a summary of frontal impact damage size (diameter, d) against impact velocity in cantilever support from this study and partial and full supports from the previous study [20]. The impact size increases with increasing impact velocity but is almost independent of the type of target supports.

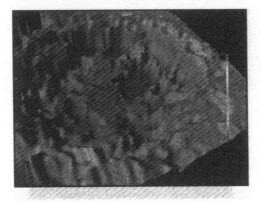

Figure 2. A typical example of a 3-D topological image taken from an impact site generated at 340 m/s in cantilever support in an N720/alumina oxide/oxide CMC impacted by 1.59-mm hardened chrome steel ball projectiles. The impact direction v as from top to bottom.

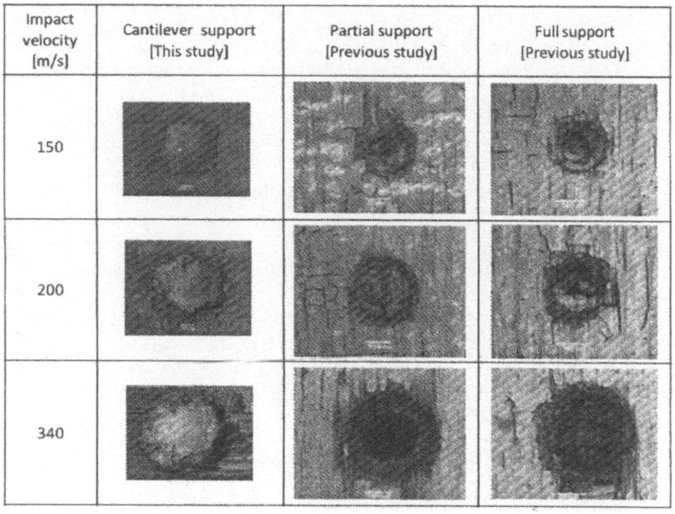

Impact velocity [m/s]	Cantilever support [This study]	Partial support [Previous study]	Full support [Previous study]
150			
200			
340			

Figure 3. Frontal impact damages with respect to impact velocity for three different types of target supports in an N720/alumina oxide/oxide CMC impacted by 1.59-mm hardened chrome steel ball projectiles. Cantilever support in this work; Partial and full support from the previous study [20].

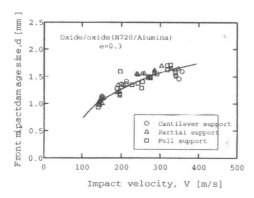

Figure 4. Frontal impact damage size (d) as a function of impact velocity (V) in an N720/alumina oxide/oxide CMC impacted by 1.59-mm hardened chrome steel ball projectiles. Cantilever support data from this work; Partial and full support data from the previous study [20]. The line was arbitrarily drawn.

Backside Impact Damage:
The backside impact damage started even at the lowest velocity of $V=150$ m/s, and was developed more with increasing impact velocity. In particular, the backside damage at 340 m/s was significant with fibers (tows) protrusion, delamination, matrix spalling, and scabbing, as shown in Fig. 5. Similar observations were also made in partial support [20], as seen in the figure. The full support, by contrast, exhibited no visible back damage since the target backside was in direct contact with a rigid steel support. At $V=340$ m/s in cantilever support, the backside damage was several times greater than the frontal impact counterpart.

It is important to note that the degree of the backside damage was a little greater in cantilever support than in partial support. It was initially speculated that the backside damage would be much more significant in partial support than in cantilever support since there was the presence of additional tensile stresses upon impact in the backside of the targets in partial support while there was supposed to be no backside tensile stresses at all in cantilever support because of „zero" moment of arm at impact sites. However, the outcome was quite opposite in reality, indicating that the impact in cantilever support would have been related to violent ballistic, dynamic effects rather than quasi-static dynamic impression events.

Cross-Sectional Views
Figure 6 shows a summary of the cross-sections of target specimens in association with different impact velocities and support types. The cross-sections provided many important features such as the shape and size of craters, damage and densification beneath impact sites, the formation of cone-cracks, interlaminar delaminations, the mode of penetration, backside damage and scabbing, etc. It can be also observed how the overall damage proceeded with respect to impact velocity and support type. A complete penetration by the projectiles would be expected to take place at >340 m/s in both cantilever and partial supports. Despite significant damage in either cantilever or partial support, the target specimens still survived, without any catastrophic failure upon impact, as observed in many CMCs [18-20]. The formation of cone cracks or cone-shaped damage is noted and has been commonly observed in ballistic impact by spherical projectiles in either CMCs, monolithic ceramics, or glasses [15-21].

Post-Impact Residual Strength
All the impacted target specimens in cantilever support failed form their impact sites in post-impact strength testing. The results of post-impact residual strength testing for impacted target specimens are summarized in Fig. 7, where post-impact strength was plotted as a function of impact velocity. Included in the figure was the residual strength determined in both partial and full supports [20]. The as-received flexure strength (σ_f) of the oxide/oxide CMC was also included. Despite some inherent data scatter, the overall trend is obvious such that post-impact strength decreased with increasing impact velocity. The greatest strength degradation occurred in cantilever support, the intermediate degradation took place in partial support, and the least strength degradation occurred in full support.

The results of post-impact residual strength were consistent with the morphological observations with regard to the degree of impact damage. The more impact damage yields the greater strength degradation and *vice versa*. Therefore, a combined effort to obtain both impact morphology and residual strength could provide enhanced, detailed qualitative and quantitative information and insights regarding the nature and degree of related impact damage.

Impact velocity [m/s]	Cantilever support [This study]	Partial support [Previous study]	Full support [Previous study]
150	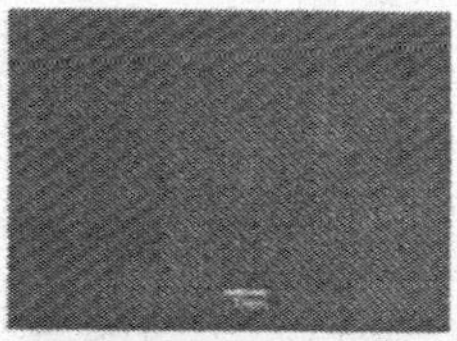		N/V
200			N/V
340	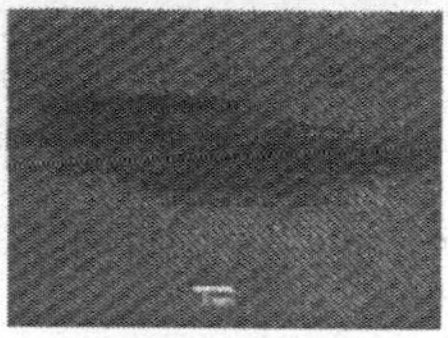		N/V

Figure 5. Backside impact damages with respect to impact velocity for three different types of target supports in an N720/alumina oxide/oxide CMC impacted by 1.59-mm hardened chrome steel ball projectiles. The cantilever support data were determined from this study while the partial and full support data from the previous work [20]. No visible damage was observed in full support. "N/V": „no visible".

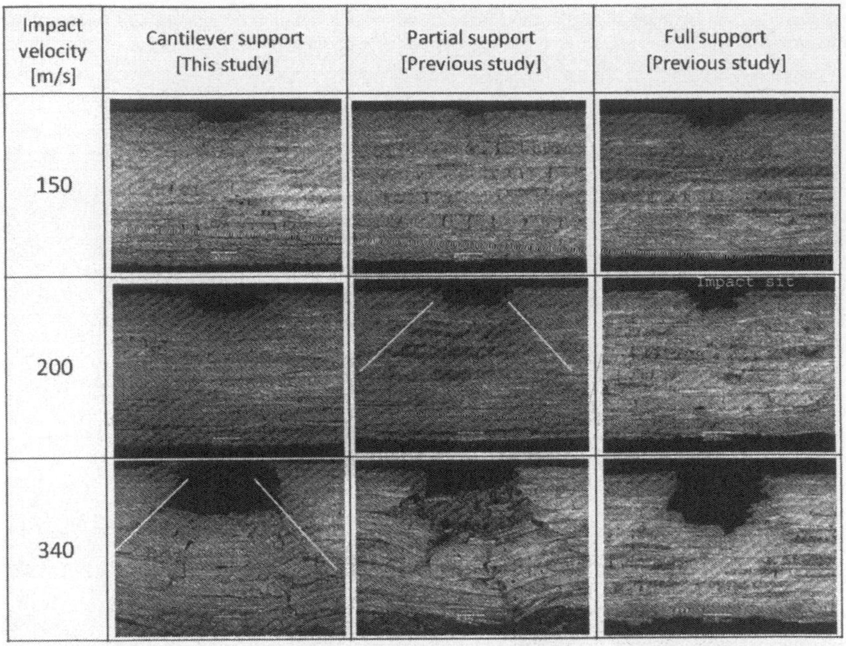

Impact velocity [m/s]	Cantilever support [This study]	Partial support [Previous study]	Full support [Previous study]
150			
200			
340			

Figure 6. Cross-sectional views of impact damage with respect to impact velocity in three different types of target supports in an N720/alumina oxide/oxide CMC impacted by 1.59-mm hardened chrome steel ball projectiles. The cantilever support data were determined from this study while the partial and full support data from the previous work [20]. The representative cone crack or cone-shape damage configurations are presented as solid lines.

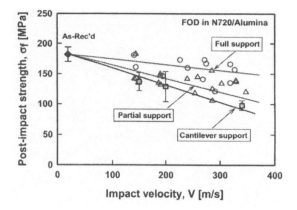

Figure 7. Post-impact strength as a function of impact velocity for three different types of target supports determined in an N720/alumina oxide/oxide CMC impacted by 1.59-mm hardened chrome steel ball projectiles. The data on cantilever support were from this study, while the data on both partial and full supports were from the previous study [20]. The "As-rec" indicates ,as-received" flexure strength.

Prediction of Impact Force

One may wonder as to how much impact force was involved in impact under different target supports. By assuming that the impact event was quasi-static, a first-order approximation of impact force was made using the energy balance principle in conjunction with the contact yield pressure analysis that were used in monolithic ceramics and CMCs [17,26]. The energy balance in impact may be written as follows:

$$U_k = U_{el} + U_{pl} + U_L + U_{k.bc} + U_{k.tp} \qquad (1)$$

where U_k is the kinetic energy of a projectile, U_{el} is the elastic deformation energy of projectile and target, U_{pl} is the plastic deformation energy of projectile and target, U_L is the energy loss associated with friction and cracking such as spallation and delamination, $U_{k.bc}$ is the bouncing-back kinetic energy of a projectile, and $U_{k.tp}$ is the kinetic energy of target in contact with projectile right after impact but before the projectile"s bouncing-back. Assume that the impacting kinetic energy is mainly consumed to deform the target both plastically and elastically with little energy loss, since the projectile is much more rigid than the soft targets. The related energies may be expressed as follows:

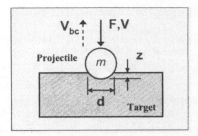

Figure 8. A schematic showing a spherical projectile impacting onto a target specimen under an idealized quasi-static impact condition.

$$U_k = mV^2/2$$
$$U_{pl} = \int F dz$$
$$U_{k,bc} = mV_{bc}^2/2 \qquad (2)$$
$$U_{el} = \alpha F^2 L^3/(2EI)$$

where m is the mass of a projectile, V is the impact velocity, F is the time-independent impact force, V_{bc} is the bouncing-back velocity of a projectile, L is the span length of a target support defined in Fig. 1, E is the elastic modulus of the target, I is the 2nd moment of inertia of the target, and α is a constant depending on the type of target supports ($\alpha =1/3$, 1/48, 0, respectively in cantilever, partial, and full supports). The U_{el} is simply an elastic energy associated with target deflection due to F. The impact depth z is defined in Fig. 8 and is geometrically related to the impact impression diameter d and the projectile diameter D assuming no violent impact to occur[‡]

$$d = (4Dz - 4z^2)^{1/2} \qquad (3)$$

A relationship between impact velocity and bouncing-back velocity of a projectile is defined as

$$e = -V_{bc}/V \qquad (4)$$

where e is a coefficient of restitution. Considering the impact event as a quasi-static indentation, the impact force may be expressed as

---

[‡] It was observed from a previous study [26] on the oxide/oxide material that the impact damage was greater and more violent than the static indentation counterpart. The geometrical relationship of Eq. (3) used in impact was thus idealized for analysis purpose.

$$F = (\pi d^2/4)\, p_y = [\pi(Dz - z^2)]\, p_y \qquad (5)$$

where p_y is the contact yield pressure which can be estimated through static indentation onto the target material by the ball projectiles [xx]. A linear relationship between contact impression area ($A = \pi d^2/4$) and indent force (P) indicates the average „contact yield pressure" ($=P/A$) remains constant, so that [26]

$$p_y = \frac{dP}{dA} \approx \frac{\Delta P}{\Delta A} \qquad (6)$$

The value of p_y for the oxide/oxide material was estimated to be $p_y \approx 1240$ MPa from the previous study [26]. Using Eqs. (1) to (5), one can obtain the following equation:

$$m(1 - e^2)V^2/2 = \pi\, p_y (Dz^2/2 - z^3/3) + \alpha \pi^2 p_y^2 (Dz - z^2)^2 L^3 / (2EI) \qquad (7)$$

This quartic equation of z in Eq. (6) can be solved numerically or analytically as a function of impact velocity (V) for the given values of m, e, p_y, D, α, L, E, and I. Once z is solved, the impact force F can be calculated using Eq. (5).

The resulting prediction of impact force depicts in Fig. 9 with a value of $e=0.3$. As seen in the result, F increases monotonically or linearly with increasing V. For a given V, the impact force is highest in full support, intermediate in partial support, and lowest in cantilever support. However, the difference between the partial and full supports was not substantial. At V=300 m/s, the impact force amounts up to 2.2, 1.9, 0.9 kN, respectively, in full, partial, and cantilever supports. It should be noted that the difference in F between $e=0.0$ and 0.5 was negligible with an only about 4% difference, indicative of insensitivity of e on F.

Since pertinent impact-force measurements (e.g., strain gaging) were not available due to target"s geometrical limitation as well as the violent nature of impact, the prediction of impact force made was not verified directly. Instead, the data on front impact damage size (d) shown in Fig. 4 were used for an indirect verification. This result is presented in Fig. 10, where the predicted d (solid lines) is compared with the experimental data for the three different target supports. The prediction in both partial and full supports is in good agreement with the experimental data. This indicates that despite many simplifying assumptions and uncertainties associated with the violent and random nature of impact events, the prediction of impact force was still reasonable at least as a first-order approximation. On the contrary, the prediction in cantilever support deviates significantly from the experimental data. This would give a challenging issue in view of the fact that the greatest damage occurred in cantilever (as seen in Fig. 7) while the least impact force occurred in cantilever support (Fig. 9 or 10). There might be some dynamic effects unique in cantilever support such as stress wave interactions, visco-elastic/plastic aspects, or strain-rate sensitivity, which would make the quasi-static approach quite inapplicable. This issue (and others, too) would require a further study including appropriate force-measuring instrumentations, dynamics modeling, more statistically reliable data generation, and so on.

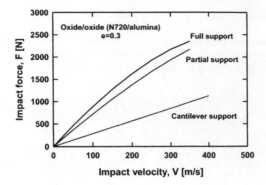

Figure 9. Predicted impact force as a function of impact velocity in three different target supports with *e*=0.3 for an N720/alumina oxide/oxide CMC impacted by 1.59-mm hardened chrome steel ball projectiles.

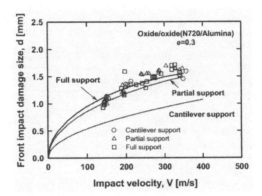

Figure 10. Comparison in front impact size (diameter) between the prediction (solid lines) and the experimental data obtained in three different target supports for an N720/alumina oxide/oxide CMC impacted by 1.59-mm hardened chrome steel ball projectiles. *e*=0.3.

Consideration Factors in FOD in CMCs

Exploration of FOD behavior in CMC materials or CMC airfoil components is an enormous task involving a variety of complicated variables. It is also associated with a high degree of complexity due to materials" various architectures and dynamic effects. Multidisciplinary approaches in materials, design, service/operation environments, mechanics, modeling, and fabrication/processing are needed to systematically assess FOD behavior of CMCs/components. A schematic showing numerous affecting factors is illustrated in Fig. 11. Included in the figure are the factors such as target, target support, protective coatings, FOD condition, projectile/particle, operational environments, and dynamic effects.

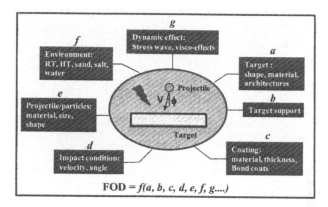

Figure 11. A schematic showing various affecting factors in foreign object damage (FOD) in CMCs/airfoils.

CONCLUSIONS

The overall impact damage of the N720/alumina oxide/oxide CMC was dependent not only on impact velocity but also on the type of target supports. The oxide/oxide CMC shows significant impact damage occurring particularly at 340 m/s in either full, partial or cantilever support. Of the three different target supports, the cantilever support resulted in the greatest impact damage, the partial support yielded the intermediate damage, and the full support gave rise to the least impact damage. The results of post-impact residual strength were all consistent with the impact morphological observations as well. The prediction of impact force was in reasonable agreement with the experimental data both in partial and in full supports. However, the prediction in cantilever support deviated substantially from the experimental data, which was contradictory to our basic understanding of mechanics. A further study is needed to address this issue particularly with an emphasis on impact dynamics (stress wave interactions).

Acknowledgements
The authors acknowledge the support by the Office of Naval Research and Dr. David Shifler.

REFERENCES
1. S. M. Wiederhorn, B. R. Lawn, "Strength Degradation of Glass Resulting from Impact with Spheres," *J. Am. Ceram. Soc.*, **60**[9-10] 451-458 (1977).
2. S. M. Wiederhorn, B. R. Lawn, "Strength Degradation of Glass Impact with Sharp Particles: I, Annealed Surfaces," *J. Am. Ceram. Soc.*, **62**[1-2] 66-70 (1979).
3. J. E. Ritter, S. R. Choi, K. Jakus, P. J. Whalen, R. G. Rateick, "Effect of Microstructure on the Erosion and Impact Damage of Sintered Silicon Nitride," *J. Mater. Sci.*, **26** 5543-5546 (1991).
4. Y. Akimune, Y. Katano, K. Matoba, "Spherical-Impact Damage and Strength Degradation in Silicon Nitrides for Automobile Turbocharger Rotors," *J. Am. Ceram. Soc.*, **72**[8] 1422-1428 (1989).
5. C. G. Knight, M. V. Swain, M. M. Chaudhri, "Impact of Small Steel Spheres on Glass Surfaces," *J. Mater. Sci.*, **12**, 1573-1586 (1977).
6. A. M. Rajendran, J. L. Kroupa, "Impact Design Model for Ceramic Materials," *J. Appl. Phys.*, **66**[8] 3560-3565 (1989).
7. L. N. Taylor, E. P. Chen, J. S. Kuszmaul, "Microcrack-Induced Damage Accumulation in Brittle Rock under Dynamic Loading," *Comp. Meth. Appl. Mech. Eng.*, **55**, 301-320 (1986).
8. R. Mouginot, D. Maugis, "Fracture Indentation beneath Flat and Spherical Punches," *J. Mater. Sci.*, **20**, 4354-4376 (1985).
9. A. G. Evans, T. R. Wilshaw, "Dynamic Solid Particle Damage in Brittle Materials: An Appraisal," *J. Mater. Sci.*, **12**, 97-116 (1977).
10. B. M. Liaw, A. S. Kobayashi, A. G. Emery, "Theoretical Model of Impact Damage in Structural Ceramics," *J. Am. Ceram. Soc.*, **67**, 544-548 (1984).
11. M. van Roode, et al., "Ceramic Gas Turbine Materials Impact Evaluation," *ASME Paper No.* GT2002-30505 (2002).
12. D. W. Richerson, K. M. Johansen, "Ceramic Gas Turbine Engine Demonstration Program," Final Report, DARPA/Navy Contract N00024-76-C-5352, Garrett Report 21-4410 (1982).
13. G. L. Boyd, D. M. Kreiner, "AGT101/ATTAP Ceramic Technology Development," *Proceeding of the Twenty-Fifth Automotive Technology Development Contractors' Coordination Meeting*, p.101 (1987).
14. M. van Roode, W. D. Brentnall, K. O. Smith, B. Edwards, J. McClain, J. R. Price, "Ceramic Stationary Gas Turbine Development – Fourth Annual Summary," ASME Paper No. 97-GT-317 (1997).
15. (a) S. R. Choi, J. M. Pereira, L. A. Janosik, R. T. Bhatt, "Foreign Object Damage of Two Gas-Turbine Grade Silicon Nitrides at Ambient Temperature," *Ceram. Eng. Sci. Proc.*, **23**[3] 193-202 (2002); (b) S. R. Choi, et al., "Foreign Object Damage in Flexure Bars of Two Gas-Turbine Grade Silicon Nitrides," *Mater. Sci. Eng.* **A 379** 411-419 (2004).
16. S. R. Choi, J. M. Pereira, L. A. Janosik, R. T. Bhatt, "Foreign Object Damage of Two Gas-Turbine Grade Silicon Nitrides in a Thin Disk Configuration," *ASME Paper No.* GT2003-

38544 (2003); (b) S. R. Choi, et al., "Foreign Object Damage in Disks of Gas-Turbine-Grade Silicon Nitrides by Steel Ball Projectiles at Ambient Temperature," *J. Mater. Sci.*, **39** 6173-6182 (2004).

17. S. R. Choi, "Foreign Object Damage Behavior in a Silicon Nitride Ceramic by Spherical Projectiles of Steels and Brass," *Mat. Sci. Eng.*, **A497** 160-167 (2008).

18. S. R. Choi, "Foreign Object Damage Phenomenon by Steel Ball Projectiles in a SiC/SiC Ceramic Matrix Composite at Ambient and Elevated Temperatures," *J. Am. Ceram. Soc.*, **91**[9] 2963-2968 (2008).

19. (a) S. R. Choi, D. J. Alexander, R. W. Kowalik, "Foreign Object Damage in an Oxide/Oxide Composite at Ambient Temperature," *J. Eng. Gas Turbines & Power*, Transactions of the ASME, Vol. **131**, 021301 (2009). (b) S. R. Choi, D. J. Alexander, D. C. Faucett, "Comparison in Foreign Object Damage between SiC/SiC and Oxide/Oxide Ceramic Matrix Composites," *Ceram. Eng. Sci. Proc.*, **30**[2] 177-188 (2009).

20. S. R. Choi, D. C. Faucett, D. J. Alexander, "Foreign Object Damage in An N720/Alumina Oxide/Oxide Ceramic Matrix Composite," *Ceram. Eng. Sci. Proc.*, **31**[2] 221-232 (2010).

21. K. Ogi, et al., "Experimental Characterization of High-Speed Impact Damage Behavior in A Three-Dimensionally Woven SiC/SiC Composite," *Composites Part A*, **41**[4] 489-498 (2010).

22. V. Herb, G. Couegnat, E. Martin, "Damage Assessment of Thin SiC/SiC Composite Plates Subjected to Quasi-Static Indentation Loading," Composites Part A, **41**[11] 1677-1685 (2010).

23. D. C. Faucett, S. R. Choi, "Foreign Object Damage in An N720/Alumina Oxide/Oxide Ceramic Matrix Composite under Tensile Loading," *Ceramic Transactions* **225**, pp. 99-107, Eds. N.P. Bansal, J.P. Singh, J. Lamon, S.R. Choi (2011).

24. D. C. Faucett, S. R. Choi, "Effects of Preloading on Foreign Object Damage in An N720/Alumina Ceramic Matrix Composite," *Ceramic Eng. Sci. Proc.*, **32**[2] 89-100 (2011).

25. D. C. Faucett, J. Wright, M. Ayre, S. R. Choi, "Effects of the Mode of Target Supports on Foreign Object Damage in An MI SiC/SiC Ceramic Matrix Composite," presented at MS&T Conference, October 16-20, 2011, Columbus OH. To be published in *Ceramic Transactions* (2012).

26. D. C. Faucett, D. J. Alexander, S. R. Choi, "Static Contact Damage in An N720/Alumina Oxide/Oxide Ceramic Matrix Composites with Reference To Foreign Object Damage," *Ceram. Eng. Sci. Proc.*, **31**[2] 233-244 (2010);

EXPERIMENTAL AND NUMERICAL STUDY ON APPLICATION OF A CMC NOZZLE FOR HIGH TEMPERATURE GAS TURBINE

Kozo Nita[1], Yoji Okita[1], Chiyuki Nakamata[1]
Advanced Technology Department, Research & Engineering Division, Aero-Engine & Space Operations, IHI Corporation
229, Tonogaya, Mizuho-machi, Nishitama-gun, Tokyo, 190-1297, JAPAN

ABSTRACT

Ceramic Matrix Composites (CMC) is the promising material because its allowable temperature is by 200degC higher than those of conventional Ni-base super alloys. A lot of attempts to apply this ceramic for a hot section of advanced gas turbine engines have been done. However in order to gain full performance benefits from this advanced material, many things to be confirmed are remained.

In this study, a thermal shock test using a nozzle cascade rig and numerical thermal stress analyses have been conducted under an engine emergency shutdown (thermal shock) condition, which is the most severe condition of any gas turbine engine operations.

In the thermal shock test, gas temperature dropped immediately to idle condition within a few second by fuel cut-off and the temperature fields of the CMC nozzle surface were measured by using infrared (IR) cameras. Important dimensionless numbers, such as Reynolds number of main gas flow, are approximately equivalent to those of typical gas turbine engines.

The numerical thermal stress analyses are conducted at the actual thermal distribution to calculate the locations where a high thermal stress is generated.

Experimental and numerical results are described and discussed in detail in this paper.

INTRODUCTION

Turbine inlet gas temperature (TIT) has been becoming higher and higher to improve an efficiency of the gas turbines. Although TIT is getting higher, less cooling air consumption is expected to achieve higher gas turbine efficiency. Therefore, high temperature gas turbines have been developed with progress in advanced cooling and material technologies. Ceramic Matrix Composite (CMC), such as SiC/SiC, can be applied at the higher temperature and has lower density compared with conventional nickel alloys. So, it is an appropriate candidate for the next generation aero-engine. Thus, recently, the capability of CMC used for such as turbine shroud has been demonstrated in the world. F.Zhong and G.L.Brown [1] tested the cooling effectiveness of multi-hole cooling on a CMC plate. Burst test [2] or thermal cycle test [3,4] with CMC vanes have been conducted.

IHI Corporation has been tested CMC's capability [5-9]. And it is participating in the engineering research and development programs for the realization of a CMC, which is a contracted research with the Ministry of Economy, Trade and Industry (METI) started in 2009. In 2010, CMC nozzles were designed and manufactured on trial.

A thermal shock condition where the engine is suddenly decelerated from the maximum power to the lowest idle condition is the most stringent because of the very large stress in the thermally. And this condition could occur in an actual engine mainly because of emergency reasons. So, the thermal shock test is very important steps considering a practical use. In this study, the CMC nozzle in this test is supposed under no cooling condition, the static pressure is lower than actual and Reynolds number of main gas flow is also lower. However in advance of thermal shock test, a confirmatory experiment and numerical study were performed and this thermal shock test can be simulated that under an actual condition.

This paper reports the results of high temperature test for the CMC turbine nozzle performed in 2010-2011.

Design of CMC turbine nozzle

To study an applicability of CMC to the turbine parts, a nozzle part is selected. Nozzles are served at the higher temperature than other parts, that is to say, the largest stress is occurred under "thermal shock" condition. So, considering stresses occurred on the CMC nozzles is very important.

The basic structure of the CMC nozzle, that shape is the typical gas turbine nozzle, is described in Fig.1, This nozzle is made of SiC/SiC and long fibre reinforced. As determined that SS (Suction side) is plus % Chord and PS (Pressure side) is minus, there are impinge holes on the insert corresponded to from -45%Chord to 70%Chord of the turbine nozzle.

In the following, leading edge is described as LE and trailing edge is described as TE.

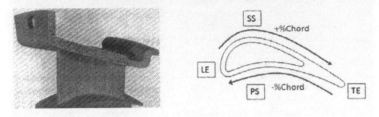

Fig.1 Structure of CMC nozzle

High temperature cascade test rig

To evaluate the design of CMC turbine nozzle under the gas turbine-representative operating conditions, a high-temperature cascade test rig was newly developed.

Fig.2 shows the layout of the test facility. The facility for thermal shock test can raise the gas flow temperature up to 1700degC as high as that of actual gas turbine engines and is the proven facility.

Fig.3 depicts cross section of the rig and Fig.4 shows the test rig setup with the test facility for the test. In this test, air supplied from compressor is heated in the combustor. The front duct of the test section is cooled to protect from heat and measured the metal wall temperature by thermocouples. Furthermore, three IR cameras set on the sight tubes (observation windows) are used for measuring the thermal fields of the nozzle pressure surface and suction surface of the CMC nozzle.

As shown in Fig.5, in this test section of the rig, the cascade has five nozzles and two side walls. One CMC turbine nozzle is placed in the center of the cascade and the others were dummy metal nozzles (SUS).

Three total temperature and pressure probes are installed in a circumferential direction at the inlet of the cascade section. Each probe has also three thermal and pressure censors in a radical direction to measure the temperature distribution of the main flow gas (nine temperature and pressure censors at all). Fig.6 shows the viewpoints of the three IR cameras. In this paper, the symbols (①~ ③) are correspond to these viewpoints. Blackbody paint whose emissivity is 0.94 was applied to the surfaces of the turbine nozzle to improve the emissivity and avoid reflections from circumstance. Additionally, four thermocouples are sprayed on the CMC nozzle surfaces to calibrate the temperature fields captured by IR cameras. One of them is attached on the pressure side of the trailing edge where IR camera can't capture due to a dummy nozzle at the front side. And there are each two markings to figure out the surface length on each IR camera view. On the endwall of the CMC nozzle, there are four holes to let into the thermocouples. And there are many markings on the surfaces of the CMC nozzle to estimate the point easily through IR camera view. The stays of IR cameras have a traverse function to focus on when the temperature of nozzle surfaces are raising. The frame rate of the IR camera is 1[fps], accuracy of measurement is ±10degC and the number of the pixels is 320 × 240. The accuracy of K-thermocouples is ±2.5degC.

Each cooling plenum is instrumented with air thermocouples and pressure probes. The temperature measured by these thermocouples are monitored and saved by PC every 0.5 sec.

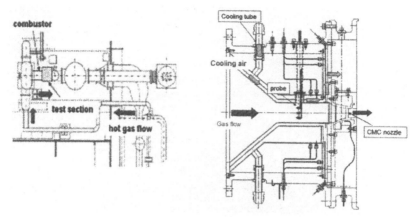

Fig.2   Layout of the test facility   Fig.3 Schematic of the cross section

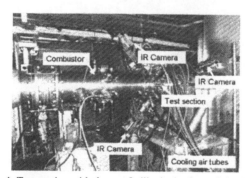

Fig.4   Test section with the test facility in the thermal shock test

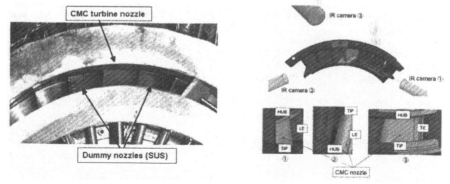

Fig.5 Cascade of the turbine nozzles   Fig.6 View of each IR camera

Each cooling plenum, including the insert for CMC nozzle, is instrumented with air thermocouples and pressure probes. The temperature measured by these thermocouples are monitored and saved by PC every 0.5 sec.

Flow in the cascade

Before the thermal shock test, velocity distribution in the cascade was calculated by CFD as shown in Fig.7. It is calculated using k-ω realizable model with Fluent®[10]. It indicates the periodicity of flow field in the cascade. From Mach number on the nozzle surface, static pressure field both at the inlet of the cascade and at the outlet, flow rate and velocity vector, a flow field condition around the CMC nozzle can be equivalent to that in actual engines. Besides, function checks of the test rig were conducted using metal nozzle instead of the CMC nozzle and the counting system of the rig was improved to raise the precision because of the usage of higher-responsible pressure and thermal probes.

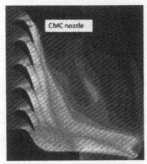

Fig.7 Velocity distribution (CFD)

Thermal shock test condition

The CMC turbine nozzle is tested under a thermal shock condition. To protect from heat, the test rig is cooled except for the CMC nozzle.

After the CMC nozzle becomes in a steady condition, suddenly fuel is cut-off. The air temperature of main gas flow drops to idle condition. Consequently transient gas inlet temperature and transient temperature fields on the CMC nozzle surface are obtained.

Numerical analysis

Using the thermal boundary conditions obtained in the experimental result, unsteady heat transfer analysis by 3D CFD is carried out. A thermal analysis is conducted by Patran Thermal®[11] is used and a stress analysis is conducted by ANSYS®[12].

As with an experimental test, inner band of the nozzles is freely supported. And in this CFD and FEM analyses, isotropic properties of CMC are used.

Maximum thermal stress is assumed to be generated when the difference between maximum and minimum surface temperatures is the largest. So at first, a temperature of the nozzle in the transient condition is analyzed and as a next step the locations where a high thermal stress is generated and its stress value is calculated.

RESULTS AND DISCUSSION (Thermal shock test)

The temperature field on the nozzle surface captured by IR cameras under a thermal shock condition is shown in Fig.8. It is calibrated using the temperatures measured by the K-thermocouples on both SS and PS.

The temperature of SS falls down faster than that of PS. This is because Mach number and heat transfer coefficient on SS are higher than on PS, SS is susceptible to the effect of temperature drop of main gas flow.

After the thermal shock test, the CMC nozzle is decomposed from the cascade as shown in Fig.9 and visual inspection is conducted. Additionally, the examination of the interior of the CMC nozzle by X-ray computed tomography (CT) scanner is also conducted to find internal cracks and no cracks were occurred after the thermal shock test.

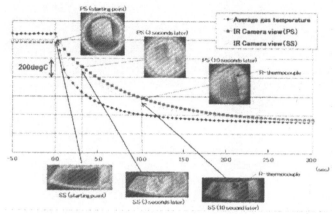

Fig. 8 Transient temperature field of the nozzle surfaces

Fig.9 CMC nozzle after the test

RESULTS AND DISCUSSION (Numerical analysis)

Transient heat transfer analysis is carried out using the results of 3D CFD.

Thermal evaluation points of the CMC nozzle are shown in Fig.10 and Fig.11 shows the temperature difference curve for 20 seconds after the fuel cut at each point. The temperature difference between main gas side and secondary air side of the airfoil (difference between the both sides of the CMC layer) is also drawn. The great temperature difference between the two sides causes large thermal stresses.

The time point that the difference becomes greatest is 2 seconds after fuel cut at the SS and LE of the nozzle, 3 seconds at the SS of the nozzle. While at the band front, the time point is lower: 3 seconds after at inner band rear and 4 seconds after at outer band rear. And at the band front, the time is lowest: 6 seconds after at inner band front and 9 seconds after at outer band front.

These lags are caused by the difference of the heat transfer coefficient at the gas sides. For example, at the band front, the heat transfer coefficient is smaller and the smallest at the outer band front.

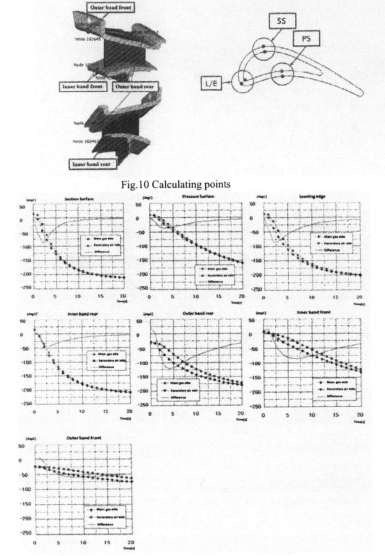

Fig.10 Calculating points

Fig.11 Temperature difference curves in transient field

Fig12(A)~(C) show each thermal field at 0,3,8 seconds after the fuel cut. As shown in Fig.12 (B), the temperature of PS is lower than that of SS. And shown in Fig12(C), from the view point of the span direction, the temperature is lowest at mid span of the airfoil. Above mentioned, the difference of the heat transfer coefficient causes the temperature distribution of CMC nozzle.

As shown an encircled point in Fig.12(C), there is the thermally lowest point near the middle point of SS. The heat transfer coefficient is almost same but the thickness grows more than double behind there because it is intergradation point from hollow section to solid section. So, thermal capacity is varied there and the temperature is also varied. On the other hand, TE is the thinnest and a thermal capacity is also the smallest. So, the temperature drops faster than other points.

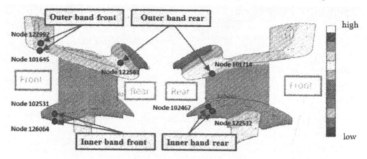

Fig.12 (A) Temperature field at fuel cut off

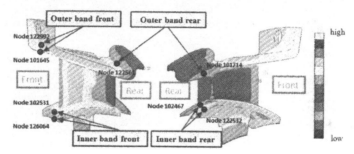

Fig.12 (B) Temperature field at 3 seconds after fuel cut off

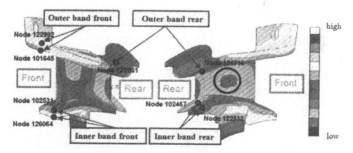

Fig.12 (C) Temperature field at 8 seconds after fuel cut off

Using this temperature distribution as boundary conditions, a stress analysis was conducted. Fig.13 (A) (B) shows thermal stresses with aerodynamic loads in the hoop direction, span direction and through-thickness direction at 2 seconds after the fuel cut. And Fig.14 (A) (B) shows at 6 seconds after.

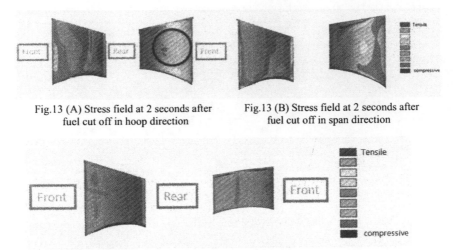

Fig.13 (A) Stress field at 2 seconds after fuel cut off in hoop direction

Fig.13 (B) Stress field at 2 seconds after fuel cut off in span direction

Fig.13 (C) Stress field at 2 seconds after fuel cut off in through-thickness direction

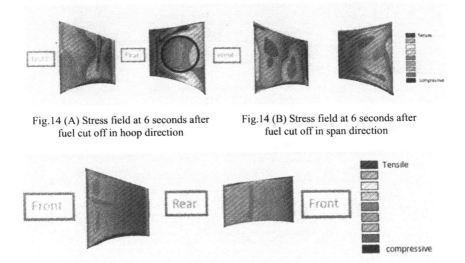

Fig.14 (A) Stress field at 6 seconds after fuel cut off in hoop direction

Fig.14 (B) Stress field at 6 seconds after fuel cut off in span direction

Fig.14 (C) Stress field at 6 seconds after fuel cut off in through-thickness direction

Among 2 cases, the biggest stress occurs at tip side of TE. As shown in Fig.12 (B) and(C), the temperature of the airfoil is much lower than that of the band under transient condition. Additionally, as mentioned with Fig.11, the temperature drops the slowest at tip side so thermal expansions are different in a span direction. Thus a high stress occurs especially at the tip side of the TE.

In the Fig.13 (A), tensile hoop stress appears on the airfoil of the SS on the semicircle around the encircled point which is the thermally lowest point shown in Fig.12(C). On the PS, at a symmetrical point with the SS, tensile hoop stress occurs along with but only in span direction. 4 seconds later, the stress grows larger as shown in Fig.14 (A). This is caused by the difference of the thickness and temperatures of the airfoil.

On the other hand, as shown in Fig.13 (B), tensile stress in the span direction occurs at boundaries with bands (both at hub side and tip side) and at tip side is larger. In the Fig.14 (B), compressive stress appears on the airfoil of the SS on the semicircle around the thermally lowest point. And at 6 seconds after fuel cut off, a large compressive tensile occurs at the entire TE surface.

Fig.13(C) and Fig.14(C) show the stress in through-thickness direction. Compared to other directions, a much smaller stress occurs on the airfoil.

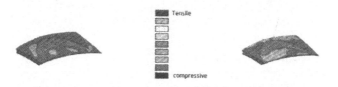

Fig.15 (A) Stress field of the band at 2 seconds   Fig.15 (B) Stress field of the band at 2 seconds
after fuel cut off in hoop direction   after fuel cut off in span direction

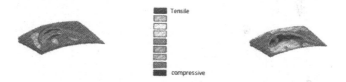

Fig.16 (A) Stress field of the band at 6 seconds   Fig.16 (B) Stress field of the band at 6 seconds
after fuel cut off in hoop direction   after fuel cut off in span direction

As shown in Fig.15 (A) (B) and Fig.16 (A) (B), in the span direction, compressive stress occurs at the secondary air side. The difference of the temperatures on both sides causes a deflection (retroflexion) of the band and a large displacement occurs at TE mainly. Thus, a large stress occurs there.

CONCLUSIONS

This paper dealt with the experimental test and numerical analysis with the CMC nozzle in transient hot gas flow. We have the following findings:

In the experimental result
  1. A CMC nozzle has been tested under a thermal shock condition; transient temperature fields on the nozzle surface have been obtained. After the test, no cracks can be seen on the CMC nozzle.

In the numerical analysis
  2. The temperature of the airfoil is different from that of the band, so thermal expansions are

different in a span direction. Thus a high stress occurs especially at the tip side of the TE.

3. The difference of thickness occur the difference of temperature of the airfoil, and it causes high thermal stress.

4. At the band, the temperature of main gas side drops faster than that of secondary air side, so a deflection occurs there. Thus a high stress occurs at the tip side and the hub side (especially TE and LE)

ACKNOWLEDGMENTS

The authors would like to express their thanks to the Ministry of Economy, Trade and Industry (METI).

REFERENCES

[1] F. Zhong and G.L.Brown, "Experimental study of multi-hoe cooling for integrally-woven,"Inernational Journal of Heat and Mass Transfer 52 (2009) pp. 971-985.

[2] D.N.Brewer et al., "Ceramic matrix composite vane subelement burst testing," GT2006-90833, ASME Turbo Expo 2006, May 2006.

[3] V.Vedula et al., "Ceramic matrix composite turbine vanes for gas turbine engines," GT2005-68229, ASME Turbo Expo 2005, June 2005.

[4] M.Verrilli and A.Calomino,"Ceramic matrix composite vane subelement testing in a gas turbine environment,"GT2004-53970, ASME Turbo Expo 2004, June 2004.

[5] T.Nakamura, "Development of a CMC Thrust Chamber," 23rd Annual Cocoa Beach, Conference and Exposition on Advanced Ceramics and Composites Volume 20 Issue 4 (1999) pp. 39 −46

[6] H.Murata, et al., IHI Engineering Review, Vol.46, Number3,2006,101-108 (Japanese)

[7] T.Tamura and S.Yamawaki, "Research of Reducing Thermal Stress Generated in MGC Turbine Nozzles," The 30th GTSJ Gas Turbine Symposium Proceedings Oct. 2002 pp.263-268

[8] S.Fujimoto, "Research of Reducing Thermal Stress Generated in MGC Turbine Nozzles," Proceedings of Asian Joint Conference on Propulsion and Power, March 2004 pp.603-608

[9] S.Fujimoto and Y.Okita, "Experimental and numerical research on application of MGC material for high temperature turbine nozzles,"GT2005-68648, ASME Turbo Expo 2005, June 2005

[10] Fluent is a registered trademark of ANSYS, Inc.

[11] Patran Thermal is a registered trademark of MSC Software Corporation

[12] ANSYS is a registered trademark of ANSYS, Inc.

Author Index

Author Index

Printed and bound by CPI Group (UK) Ltd, Croydon, CR0 4YY

28/10/2024

14581334-0001